Frontiers in Earth Sciences

Series Editors

J. P. Brun, Clermont-Ferrand, France

Onno Oncken, Potsdam, Brandenburg, Germany

Helmut Weissert, Zürich, Switzerland

Wolf-Christian Dullo, Paleoceanography, Helmholtzzentrum für Ozeanforschrung | G, Kiel, Germany

More information about this series at http://www.springer.com/series/7066

Gilles Ramstein • Amaëlle Landais •
Nathaelle Bouttes • Pierre Sepulchre •
Aline Govin
Editors

Paleoclimatology

Volume 2

Editors
Gilles Ramstein
LSCE/IPSL, CEA-CNRS-UVSQ
Université Paris-Saclay
Gif-sur-Yvette, Ariège, France

Nathaelle Bouttes
LSCE/IPSL, CEA-CNRS-UVSQ
Université Paris-Saclay
Gif-sur-Yvette, Ariège, France

Aline Govin
LSCE/IPSL, CEA-CNRS-UVSQ
Université Paris-Saclay
Gif-sur-Yvette, Ariège, France

Amaëlle Landais
LSCE/IPSL, CEA-CNRS-UVSQ
Université Paris-Saclay
Gif-sur-Yvette, Ariège, France

Pierre Sepulchre
LSCE/IPSL, CEA-CNRS-UVSQ
Université Paris-Saclay
Gif-sur-Yvette, Ariège, France

ISSN 1863-4621   ISSN 1863-463X   (electronic)
Frontiers in Earth Sciences
ISBN 978-3-030-24981-6   ISBN 978-3-030-24982-3   (eBook)
https://doi.org/10.1007/978-3-030-24982-3

Translated from the original French by Mary Minnock
© Springer Nature Switzerland AG 2021

This work is subject to copyright. All rights are reserved by the Publisher, whether the whole or part of the material is concerned, specifically the rights of translation, reprinting, reuse of illustrations, recitation, broadcasting, reproduction on microfilms or in any other physical way, and transmission or information storage and retrieval, electronic adaptation, computer software, or by similar or dissimilar methodology now known or hereafter developed.

The use of general descriptive names, registered names, trademarks, service marks, etc. in this publication does not imply, even in the absence of a specific statement, that such names are exempt from the relevant protective laws and regulations and therefore free for general use.

The publisher, the authors and the editors are safe to assume that the advice and information in this book are believed to be true and accurate at the date of publication. Neither the publisher nor the authors or the editors give a warranty, expressed or implied, with respect to the material contained herein or for any errors or omissions that may have been made. The publisher remains neutral with regard to jurisdictional claims in published maps and institutional affiliations.

This Springer imprint is published by the registered company Springer Nature Switzerland AG
The registered company address is: Gewerbestrasse 11, 6330 Cham, Switzerland

Foreword

A brief history of paleoclimates

Climate is undeniably a topical issue of utmost importance. It has been the focus of attention for several decades now, during which the study of ancient climates (paleoclimatology) has progressed and gained a solid reputation. Currently, this work has become fundamental to our understanding of how the climate system functions and to validate the models used to establish future projections. Thanks to the study of past climates, a database documenting a much greater diversity of climate changes than during recent centuries, has been created. This diversity makes it possible to test climate models in situations that are vastly different from those we have known over the last 150 years and, in some cases, for climates that are closer to those that await us in the future if we apply the conclusions of the Intergovernmental Panel for Climate Change.

The Earth's climate changes have always changed over time and will continue to change in the future. While we are all aware of the weather phenomena that condition our daily lives, few of us are aware of what climate really is. Climatology is the science that explores the great variability in meteorological conditions over time and space, throughout history. This word comes from the Greek word *klima* meaning inclination, in this case referring to that of the rays of the Sun. Since the dawn of our civilization, therefore, we have linked the variations in climate and in the energy received from the Sun in a relationship of cause and effect. For a long time, the term 'climate' was reserved to describe the characteristics of air temperature and precipitation particular to different parts of the globe. This description was based on meteorological measurements and their averages conducted over a few decades. It is only recently understood that climate also varies over much longer time scales and therefore concerns more than just the atmosphere. At present, specialists studying climate and its variations analyze all of the fluid and solid envelopes of the Earth. Along with the atmosphere, we associate the hydrosphere and the cryosphere which, together, represent the systems where water exists in solid (snow, glaciers, and ice sheets) and liquid form (rivers, lakes, and seas), the continents where plate tectonics and volcanic activity occur, and finally at the surface, the whole living world (biosphere) that influences nature, the properties of soil cover, and the biogeochemical cycles.

Climatology has evolved from being a descriptive discipline to become a multidisciplinary science involving five complex systems and their various interactions. It is therefore not surprising that the resulting climate studies vary on scales ranging from the season to millions of years. Although it is only in the last few decades that this science has exploded, the first discovery and study of climate change beyond the annual and decadal scales date back to the eighteenth century. It was at this time that the presence of erratic boulders in the mountainous landscape became associated for the first time with the massive extension of glaciers. In 1744, the Grenoble geographer Pierre Martel (1706–1767) reported that the inhabitants of the Chamonix Valley in the Alps attributed the dispersion of *roches moutonnées* to the glaciers themselves, which would have extended much further in the past. This was a revolutionary idea, because until then, most scientists still referred to the myth of the Biblical Flood to explain landscape structures. This was the case of Horace Benedicte de Chaussure (1740–1799) from Geneva, the French paleontologist, Georges Cuvier (1769–1832) and the Scottish geologist,

Charles Lyell (1797–1875), who continued to assume that these boulders were carried by the strength of strongly flowing waters. However, the location and nature of these boulders and other moraines led some scientists to admit that ice transport would provide a better explanation for the various observations. The Scottish naturalist, James Hutton (1726–1797), was the first to subscribe to this idea. Others followed his lead and detected the imprint of climatic changes in the fluctuations of the extent of the glaciers. These pioneers were the Swiss engineer, Ignace Venetz (1788–1859); the German forestry engineer, Albrecht Reinhart Benhardi (1797–1849); the Swiss geologist, Jean de Charpentier (1786–1855); and the German botanist, Karl Friedrich Schimper (1803–1867), who introduced the notion of ice ages. But it was the Danish-Norwegian geologist, Jens Esmark (1763–1839), who, in pursuing his analysis of glacier transport, proposed in 1824, for the first time, the notion that climate changes could be the cause and that these could have been instigated by variations of Earth's orbit.

It was the work of these pioneers that led the Swiss geologist, Louis Agassiz (1801–1873) to make the address to the Swiss Society of Natural Sciences of Neufchatel in 1837 entitled 'Upon glaciers, moraines and erratic blocks'. It was also at the beginning of the nineteenth century that the Frenchman Joseph Adhémar (1797–1862), not content with studying the polar ice caps, attempted to explain in his book, *Révolutions de la Mer, Déluges Périodiques* (1842), the pattern of ice ages stemming from the precession of the equinoxes. The astronomical theory of the paleoclimates was born and would be continued, thanks to the development of celestial mechanics, by the Frenchmen, Jean le Rond d'Alembert (1717–1783), Jean-Baptiste Joseph Delambre (1749–1822), Pierre-Simon Laplace (1749–1827), Louis Benjamin Francoeur (1773–1849), and Urban Le Verrier (1811–1877). In parallel, other advances were made with the first calculations of the long-term variations in the energy received from the Sun, variations due to the astronomical characteristics of the eccentricity of the Earth's orbit, the precession of the equinoxes, and the obliquity of the ecliptic. This was demonstrated by the work of John Frederick William Herschel (1792–1871), L.W. Meech (1821–1912), and Chr. Wiener (1826–1896), supported by the work of the mathematicians André-Marie Legendre (1751–1833) and Simon-Denis Poisson (1781–1840).

This sets the stage for James Croll (1821–1890) to develop a theory of ice ages based on the combined effect of the three astronomical parameters, a theory according to which winter in the northern hemisphere played a determining role. This theory was much appreciated by the naturalist, Charles Robert Darwin (1809–1882), and was taken up by the Scottish geologist brothers, Archibald (1835–1924) and James (1839–1914) Geikie, who introduced the notion of the interglacial. It is also the basis for the classification of alpine glaciations by Albrecht Penck (1858–1945) and Edward Brückner (1862–1927) and American glaciations by Thomas Chowder Chamberlin (1843–1928). However, geologists became increasingly dissatisfied with Croll's theory and many critics of it emerged. Many refuted the astronomical theory and preferred explanations that related to the Earth alone. The Scottish geologist, Charles Lyell (1797–1875), claimed that the geographical distribution of land and seas explained the alternation of hot and cold climates, while others turned to variations in the concentration of certain gases in the atmosphere. Hence, the French physicist, Joseph Fourier (1786–1830), expounded on the first notion of the theory of the greenhouse effect. He was followed by the Irish chemist, John Tyndall (1820–1893), to whom we owe the first experiments on the absorption of infrared radiation and the hypothesis of the fundamental role played by water vapor in the greenhouse effect. Later, the Italian, Luigi de Marchi (1857–1937) and the Swedish chemist, Svante Arrhenius (1859–1927) proposed, along with other scientists of their time, that the ice ages were caused by decreases in atmospheric carbon dioxide concentration. In 1895, Arrhenius suggested, in an article published by the Stockholm Physics Society, that a 40% reduction or increase in CO_2 concentration in the atmosphere could lead to feedback processes that would explain glacial advances or retreats.

A revival of the astronomical theory became, however, possible with advances in the calculation of astronomical elements by the American astronomer John Nelson Stockwell (1822–1920) and the Serbian astronomer Vojislava Protich Miskovitch (1892–1976) and of solar irradiation (1904) by the German mathematician, Ludwig Pilgrim (1879–1935). It was Joseph John Murphy (1827–1894), however, who, as early as 1869, proposed that cool summers of the northern hemisphere had instigated the ice ages. This original idea was taken up in 1921 by the German paleoclimatologist Rudolf Spitaler (1859–1946), but was popularized by the Serbian geophysicist engineer, Milutin Milankovich (1879–1958), mainly through his books *Mathematical Theory of Thermal Phenomena Produced by Solar Radiation* (1920) and *Kanon der Erdbestrahlung und seine Anwendung auf des Eizeitenproblem* (1941). Milankovitch was a contemporary of the German geophysicist Alfred Wegener (1880–1930) with whom he became acquainted through the Russsian-born climatologist Wladimir Köppen (1846–1940), Wegener's father-in-law (Thiede, 2017). The modern era of astronomical theory was born, even if there remained much criticism related to the lack of reliable paleoclimatic data and of a reliable timescale, both by geologists and meteorologists. It was not until the 1950s and 1960s that new techniques made it possible to date, measure, and interpret the climate records contained in marine sediments, in ice and on land. In 1955, the American, Cesare Emiliani (1922–1995), proposed a stratigraphy, which still applies today, based on the succession of minima and maxima of the oxygen-18 / oxygen-16 isotopic ratio measured in the foraminiferal shells found in sediments taken from the deep ocean. The interpretation of this isotopic ratio in terms of salinity was made by Jean-Claude Duplessy (1970), and in terms of temperature and volume of ice (1973) by Nicholas Shackleton (1937–2006) and Niels Opdyke (1933–2019). Mathematical tools made it possible to establish transfer functions to quantitatively interpret information collected in the oceans in 1974 by the American paleoceanographers John Imbrie (1925–2016) and Nilva Kipp (1925–1989),) and in tree rings (Harold Fritts, 1968). Efforts by the CLIMAP group (1976) resulted in the first seasonal climate chart of the Last Glacial Maximum and the pivotal article by James Hays, John Imbrie, and Nicholas Shackleton (1976). The arrival of big computers allowed the first climate simulations to be conducted using general circulation models (Fred Nelson Alyea, 1972), and further astronomical calculations led to the establishment of a high-precision time scale reference, as well as the determination of the daily and seasonal irradiation essential for climate modeling (André Berger, 1973 and Berger and Loutre, 1991).These calculations of the astronomical parameters were based on the 1974 and 1988 developments of the orbital elements by the French astronomers Pierre Bretagnon (1942–2002) and Jacques Laskar, respectively. These are valid over a few million years. The Laskar solution was extended over a few tens of millions of years by Laskar et al. (2011) and over the whole Mesozoic with the American paleobiologist Paul Olsen and colleagues (2019).

This evolution and the recent advances in paleoclimatology show the difficulties involved in tackling the study of the climate system. Overcoming these difficulties requires high-quality books to improve understanding and to update the range of disciplines involved. It is with this perspective in mind that this book was written. Written originally in French, it unquestionably fills a gap in the field of graduate and postgraduate third-level education that goes far beyond its description. It provides an overview of the state of knowledge on a number of key topics by outlining the information necessary to understand and appreciate the complexity of the disciplines discussed, making it a reference book on the subject. The first of the two volumes is devoted to the methods used to reconstruct ancient climates, the second to the behavior of the climate system in the past. Many of the thirty-one chapters are written by researchers from the *Laboratoire des Sciences du Climat et de l'Environnement* and associated research laboratories each focusing on his or her area of expertise, which ensures a reliable document founded on solid experience.

Understanding the evolving climate of the Earth and its many variations is not just an academic challenge. It is also fundamental in order to better understand the future climate and its possible impacts on the society of tomorrow. Jean-Claude Duplessy and Gilles Ramstein have achieved this huge feat by bringing together fifty or so of the most highly reputed researchers in the field.

The book they have written is a whole, providing both the necessary bases on the reconstruction techniques of ancient climates, their chronological framework, and the functioning of the climate system in the past based on observations and models. This book will allow all those who want to know more, to explore this science, which, although difficult, is hugely exciting. It will also give them the essential information to establish an objective idea of the climate and its past and future variations.

You may find most of the references and pioneering studies mentioned in this preface in BERGER A. 2012. A brief history of the astronomical theories of paleoclimates. In: "Climate change at the eve of the second decade of the century. Inferences from paleoclimates and regional aspects". Proceedings of Milankovitch 130th Anniversary Symposium, A. Berger, F. Mesinger, D. Sijacki (eds). 107–129. Springer-Verlag/Wien.DOI 10.1007/978-3-7091-0973-1.

André Berger
Emeritus Professor at the Catholic University of Louvain
Louvain la Neuve, Belgium

Introduction

For a long time, geology books devoted only a few lines to the history of past climates of our planet, mostly to establish the deposition framework for the sediments that geologists found on the continents, the only area of enquiry available to them. Scientists soon realized that the copious coal deposits of England, Belgium, Northern France, Germany, and Poland resulted from the fossilization of abundant vegetation facilitated by a warm and humid equatorial climate that reigned over Western Europe, some 350 million years ago (an illustrated insert in Chap. 2 volume I provides a diagram of continental drift since 540 Ma). Fifty million years later, the sediments of these same regions, red sandstone, poor in fossils and associated with evaporites testify to the replacement of forests by desert areas, dotted with occasional highly saline lakes, similar to what we currently find in Saharan Africa. Humidity gave way to intense aridity and we had no idea why. It was not until the discovery of plate tectonics that we realized that Europe had slowly drifted toward the tropics. This transformation of the face of the Earth due to tectonics is illustrated through 16 maps in Chap. 2 volume I.

The discovery of glaciations was a revelation for the geologists of the nineteenth century. A major polemic broke out at the Swiss Society of Natural Sciences in Neuchâtel when, in 1837, its president Louis Agassiz presented his explanation, incredible at the time, for the presence of gigantic boulders that dot the Jura mountains. He daringly claimed that these erratic boulders were not the remnants of the Biblical Flood, but rather enormous rocks transported over long distances by gigantic glaciers which used to cover the high latitudes of our hemisphere.

The controversy died down quickly, when European and American geologists discovered traces of glaciers all over the Northern Hemisphere, just as Agassiz imagined. In Europe, as in North America, mapping of the terminal moraines left behind by glaciers when they melted showed proof of the presence of gigantic ice caps in a past that seemed distant. especially since there was no idea how to date them.

As the idea of the Biblical Flood fell out of favor, a new theory, based on astronomical phenomena, soon appeared. Scientists like Joseph Adhémar and James Croll realized that there were small, quasi-periodic variations over time in the movement of the Earth around the Sun and suggested that associated mechanisms could periodically cause glacial advances and retreats. Finally, it was Milutin Milankovitch, a professor in Belgrade, who would lay the foundations for a complete mathematical theory of glaciations, the legitimacy of which was proven when paleoceanographers found the frequencies of orbital parameters reflected in the isotopic analysis of marine cores. We now know that the last one of these glacial periods culminated only 20,000 years ago and was preceded by many others.

The great contribution by Milankovitch was to plant a new idea within the scientific community: Ancient climates are not only of immense curiosity to geologists; they obey the same physical laws as those governing the current climate.

This intellectual revolution has had far-reaching consequences and has profoundly altered the approach to the study of ancient climates making paleoclimatology a science with many links to geology, geochemistry, oceanography, glaciology as well as the approach to the physical and dynamic dimensions of the climate. The first part of this book describes the physical, chemical, and biological phenomena that govern the functioning of the climate system and shows how it is possible to reconstruct the variations in the past at all timescales.

This is the work of paleoclimatologists. As soon as the means became available to them in the second half of the twentieth century, they undertook to track down all traces of climate change so as to establish a planetary vision. This led them to develop new methods of sampling continental sediments, marine sediments in the context of major oceanographic campaigns, and ice cores by carrying out large-scale drilling campaigns of mountain glaciers and the ice sheets of Greenland and of Antarctica. The level of resources that needs to be mobilized is such that the drilling campaigns of polar ice and of marine sediments from all the world's oceans could only be carried out in an international cooperative framework which makes it possible to coordinate the efforts of the various teams.

This scientific investment has produced an abundant harvest of samples containing records of past climates. On the continents, lake sediments; peat bogs; concretions in caves; and fossil tree rings have provided many indicators of environmental conditions, especially of the behavior of vegetation and the atmosphere. In the ocean, samples have been taken from all of the large basins and cores are able to trace the history of the last tens of millions of years. Finally, the large drillings in the ice sheets have provided information not only on polar temperatures, but also on the composition of the atmosphere (dust and the concentrations of greenhouse gases, such as carbon dioxide and methane).

Unfortunately, nature has no paleothermometer or paleopluviometer, and therefore, there is no direct indicator of the changes in temperature or precipitation: Everything has had to be built from scratch, not only to reconstruct the climates, but also to date them. Extracting a reconstruction of the evolution of the climate from these samples has necessitated considerable developments using the most innovative methods from the fields of geochemistry, biology, and physics. Firstly, it was essential to establish a timeframe to know which period was covered by each sample. Many methods were developed, and they are the subject of the second part of this book. Radioactive decay, which is governed by strict physical laws, plays a vital role. It has made it possible to obtain timescales converted into calendar years, and it has provided clarification on stratigraphic geology. Other more stratigraphic approaches have been implemented: identification of characteristic events that need to be dated elsewhere; counting of annual layers; or modeling of ice flow. It has thus been possible to establish a chronological framework, and paleoclimatologists are now trying to make it common to all data via an on-going effort to make multiple correlations between the various recordings. Few climatologists rely on one indicator. The confidence that they have in reconstructing a climate change at a given time is obtained by intersecting reconstructions from independent indicators but also by confronting them with results from models. Methods of reconstructing the evolution of the different components of the climate system from geological indicators then had to be developed. These are extremely varied, and their description constitutes the main and third part of volume I. Many use the latest developments in paleomagnetism, geochemistry, and statistical methods to empirically link the distribution of fossil plants and animals with environmental parameters, primarily air and water temperature. Reconstructions achieved in this way have now reached a level of reliability such that, for certain periods, not only qualitative variations (in terms of hot/cold, dry/wet) can be obtained, but even quantitative ones with the associated uncertainties also quantified. This is the level of climate reconstruction necessary to allow comparison with climate models.

The use of climate models also gained momentum during the second half of the twentieth century. First established to simulate atmospheric circulation, they have progressed by integrating more and more efficiently the physics, processes, and parameterization of the radiative budget and the hydrological cycle, in particular, by incorporating satellite data. However, the atmosphere only represents the rapid component of the climate system.

The late 1990s dramatically demonstrated the need to link atmospheric models to global patterns of the ocean and vegetation to reconstruct climate change. Indeed, teams from the GISS in the USA and from Météo-France bolstered by their atmospheric models that had succeeded in reconstructing the current climate, independently tried to use the disruption of the radiative budget calculated by Milankovitch to simulate the last entry into glaciation 115,000 years ago. In both cases, it was a total fiasco. The changes induced by the variation of the orbital parameters in these models were far too small to generate perennial snow. The components and feedbacks related to the ocean and terrestrial vegetation needed to be included. Developing a model that couples all three of these components is what modelers have been striving to achieve over the last 20 years, and these are the models that now contribute to the international IPCC effort.

Today, the so-called Earth system models that incorporate aspects from the atmosphere–ocean–terrestrial and marine biosphere, chemistry, and ice caps are used to explore the climate of the future and the climates of the past. Spatially, they are increasingly precise, they involve a very large number of processes and are run on the largest computers in the world. But, the flip side of this complexity is that they can only explore a limited number of trajectories because of the considerable computing time they require. Also, from the beginning, climate modelers armed themselves with a whole range of models. From behemoths like the 'general circulation models' to conceptual models, with models of intermediate complexity in-between. From this toolbox, depending on the questions raised by the paleoclimatic data, they choose the most appropriate tool or they develop it if it does not exist. With the simplest models, they can explore the possible parameter variations and, by comparing them with the data, try to establish the most plausible scenario. All of these modeling strategies are described in detail in volume II, which constitutes the last and fourth part of this book.

This investigative approach at each step of the research work, dating, reconstruction, modeling, and the back and forth between these stages allows us to develop and refine the scenarios to understand the evolution of the past climates of the Earth. We are certain that this approach also allows us, by improving our understanding of the phenomena that govern the climate of our planet and through continuous improvement of the models, to better predict future climate change. This comparison between models and data, which makes it possible to validate numerical simulations of the more or less distant past, is an essential step toward the development of climate projections for the centuries to come, which will, in any case, involve an unprecedented transition.

Acknowledgement

This book would never have been possible without the very efficient help of Guigone Camus and Sarah Amram. We are grateful to Mary Minnock for the translation of each chapter. We also thank the LSCE for its financial support and Nabil Khelifi for his support from the origin to the end. Last but not least, thank you to all our colleagues—more than 50—who patiently contributed to this book.

Preface

Before taking this journey together into the Earth's paleoclimates, it is important to know what we will be facing. This exploration will bring us into the heart of the 'Earth system': a tangle of interwoven components with very different characteristics and response times, a system in constant interaction.

The first volume is dedicated (Chaps. 1 and 2) to an introduction to climate of the Earth. Chapters 3–9 focus on different time measurement and datation technics. The most important part of this first volume deals with the reconstructions of different climatic parameters from the three major reservoirs (ocean, continent, cryosphere, Chaps. 10–21). The second volume is devoted to modeling the Earth system to better understand and simulate its evolution (Chaps. 1–9). Last but not least, the final chapter (Chap. 10) describes the future climate of the Earth projection from next century to millennia.

The first part of this book (Chaps. 1 and 2) will equip the reader with a 'climate kit' before delving into the study of paleoclimates. This quick overview shows the great diversity in the systems involved. From the microphysics of the clouds that can be seen evolving over our heads by the minute to the huge ice caps that take nearly 100,000 years to reach their peak, the spatiotemporal differences are dizzying (Chap. 1). Yet, it is the same 'Earth system' that, throughout the ages, undergoes various disturbances that we will address. Chapter 2 takes us on a journey through the geological history of our planet. The distribution of continents, oceans, and reliefs changes how energy and heat are transported at the Earth's surface by the ocean and the atmosphere.

The study of paleoclimates requires an understanding of two indispensable concepts in order to describe the past climates of the Earth.

The first is the concept of time. Measuring time is fundamental to our research, and an understanding of the diversity of temporalities particular to paleoclimatic records is essential. The second part (Chaps. 3 to 9) of this book is devoted to the question of the measurement of time. Different techniques may be implemented depending on the timescales considered in Chap. 3. Thus, although carbon-14 (Chap. 4) provides us with reliable measurements going back to 30,000–40,000 years ago, other radioactive disequilibria (Chaps. 5 and 6) need to be used to access longer timescales. But it is not only the radioactivity-based methods that inform us of the age of sediments; the use of magnetism (Chap. 7) is also a valuable way of placing events occurring on the geological timescale into the context of climate. On shorter timescales, the use of tree rings is also a valuable method (Chap. 8). Ice core dating techniques will also be outlined (Chap. 9). This gamut of different methods shows how researchers have succeeded in developing 'paleo-chronometers' which are essential to locate climate archives within a temporal context, but also to establish the connections of cause and effect between the different components of the Earth system during periods of climatic changes.

The second concept is that of climate reconstruction. Indeed, in the same way that there is no single chronometer that allows us to go back in time, there is not one paleothermometer, pluviometer, or anemometer. Just as it was necessary to invent paleo-chronometers based on physical or biological grounds in order to attribute an age and an estimate of its uncertainty to archives, the relevant climatic indicators had to be invented to quantify the variations in temperature, hydrological cycle, and deepwater current. The third part of this book (Chaps. 10

–21) is devoted to the slow and complex work of reconstruction by applying this whole range of indicators. Thus, we can reconstruct the climate of the major components of the climate system: the atmosphere, the ocean, the cryosphere, and the biosphere. But we can also take advantage of the specificities of temperate or tropical lakes, of caves and their concretions (speleothems), of tree rings and even, more recently, of harvesting dates (Chap. 17). How can paleo-winds or, to put it in more scientific jargon, the variations in atmospheric dynamics be reconstructed? Based on the isotopic composition of precipitation (Chap. 10) or of the loess (Chap. 13), not only can the evolution of the surface and deep ocean be reconstructed, but also the geometry and dynamics of large water masses (Chap. 21). For land surfaces, palynology and dendroclimatology enable us to retrace the evolution of vegetation and climate, respectively (Chaps. 12 and 16). Finally, the cores taken from the ice caps of both hemispheres make it possible to reconstruct the polar climate (Chap. 11).

In addition to these two main concepts, we also need to understand how fluctuations in the hydrology of the tropics have caused variations in lakes (Chaps. 18 and 19) and glaciers (Chap. 20); these factors also tell a part of the climate story. Other markers, such as speleothems (Chap. 14) or lake ostracods (Chap. 15) reveal changes in climate in more temperate areas.

Thus, a description of the global climate emerges from the local or regional climate reconstructions. Through coupling these reconstructions with dating, our knowledge of climate evolution progresses constantly. Nevertheless, this image is both fragmentary, because of the strong geographic and temporal disparity of our knowledge, and unclear, because of the uncertainties in the reconstructions that the paleoclimatologist tries to reduce. There is still a long way to go in terms of developing new indicators and improving those widely used in order to complete and refine this description.

The second volume of this book (Chaps. 22–30) focuses on the major processes and mechanisms explaining the evolution of past climate from geological to historical timescales, whereas last Chap. 31 examines future climate projections. First of all, we address, in the very long term, the interactions between tectonics and climate over the timescale of tens to hundreds of millions of years (Chap. 22). Then, we deal with the biogeochemical cycles that govern the concentrations of greenhouse gases in the atmosphere over the last million years (Chap. 23). And finally, we consider the interactions with ice caps (Chap. 24).

We will continue our journey simulating the climate evolution through time from the formation of the Earth (4.6 billion years ago) up to the future climates at scales from a few tens of thousands to a hundred of years. On this journey, it becomes obvious that the dominant processes, those that drive climate change, vary according to timescales: solar power, which increases by about 7% every billion years places its stamp on very long-term evolution, whereas at the scale of tens of millions of years, it is tectonics that sculpt the face of the Earth, from the high mountain ranges to the bathymetry of the ocean floor. Finally, 'the underlying rhythm of Milankovitch,' with a much faster tempo of a few tens of thousands of years can produce, if the circumstances permit, the glacial–interglacial cycles described in the preceding parts. On top of this interconnection of timescales, a broad range of processes and components of the climate system is superimposed. Through these chapters, we would like to highlight the need to model a complex system where different constituents interact at different timescales (Chap. 25). With the development of these models, the scope of investigation is vast. Indeed, ranging from recent Holocene climates (Chap. 30) to geological climates (Chaps. 26 and 27), how they evolve is underpinned by very different processes: from plate tectonics (Chap. 22) to orbital parameters (Chap. 28). The complexity of the system can also be seen in the abrupt reorganizations of the ocean–atmosphere system (Chap. 29). The capacity acquired in recent decades to replicate past climate changes using a hierarchy of models, and to compare these results with different types of data, has demonstrated the relevance of this approach coupling model simulations with data acquisition.

Nevertheless, the field of investigation of the Earth's past climates remains an important area of research with many questions being raised about the causes of climate reorganizations throughout the Earth's history. Even though several chapters clearly show recent breakthroughs in our understanding of past climate changes, and the sensitivity of our models to climate data has undeniably increased the extent to which we can rely on their outputs, we can legitimately question what they contribute to future climate. Chapter 10 addresses these issues. Will the ice caps, which have existed on Earth for only a short time relative to geological time, withstand human disturbance? And can this disturbance, apart from its own duration, have an impact on the rhythm of glacial–interglacial cycles?

At the end of these two volumes, you will have obtained the relevant perspective to project into the Earth's climates of the future. Indeed, by absorbing the most up-to-date knowledge of paleoclimatology in this book, you will be provided with the necessary objectivity to critically assess present and future climate changes. It will also give you the scientific bases to allow you to exercise your critical judgment on the environmental and climatic issues that will be fundamental in the years to come. Indeed, in the context of the Anthropocene, a period where man's influence has grown to become the major factor in climate change, the accumulated knowledge of the climate history of our planet gathered here is precious.

Gif-sur-Yvette, France

Gilles Ramstein
Amaëlle Landais
Nathaelle Bouttes
Pierre Sepulchre
Aline Govin

Contents

Volume 1

1. **The Climate System: Its Functioning and History** 1
   Sylvie Joussaume and Jean-Claude Duplessy

2. **The Changing Face of the Earth Throughout the Ages** 23
   Frédéric Fluteau and Pierre Sepulchre

3. **Introduction to Geochronology** 49
   Hervé Guillou

4. **Carbon-14** 51
   Martine Paterne, Élisabeth Michel, and Christine Hatté et Jean-Claude Dutay

5. **The $^{40}K/^{40}Ar$ and $^{40}Ar/^{39}Ar$ Methods** 73
   Hervé Guillou, Sébastien Nomade, and Vincent Scao

6. **Dating of Corals and Other Geological Samples via the Radioactive Disequilibrium of Uranium and Thorium Isotopes** 89
   Norbert Frank and Freya Hemsing

7. **Magnetostratigraphy: From a Million to a Thousand Years** 101
   Carlo Laj, James E. T. Channell, and Catherine Kissel

8. **Dendrochronology** 117
   Frédéric Guibal and Joël Guiot

9. **The Dating of Ice-Core Archives** 123
   Frédéric Parrenin

10. **Reconstructing the Physics and Circulation of the Atmosphere** 137
    Valérie Masson-Delmotte and Joël Guiot

11. **Air-Ice Interface: Polar Ice** 145
    Valérie Masson-Delmotte and Jean Jouzel

12. **Air-Vegetation Interface: Pollen** 151
    Joël Guiot

13. **Ground-Air Interface: The Loess Sequences, Markers of Atmospheric Circulation** 157
    Denis-Didier Rousseau and Christine Hatté

14 **Air-Ground Interface: Reconstruction of Paleoclimates Using Speleothems** 169
Dominique Genty and Ana Moreno

15 **Air-Interface: $\delta^{18}O$ Records of Past Meteoric Water Using Benthic Ostracods from Deep Lakes** 179
Ulrich von Grafenstein and Inga Labuhn

16 **Vegetation-Atmosphere Interface: Tree Rings** 197
Joël Guiot and Valérie Daux

17 **Air-Vegetation Interface: An Example of the Use of Historical Data on Grape Harvests** 205
Valérie Daux

18 **Air-Ground Interface: Sediment Tracers in Tropical Lakes** 209
David Williamson

19 **Air-water Interface: Tropical Lake Diatoms and Isotope Hydrology Modeling** 213
Florence Sylvestre, Françoise Gasse, Françoise Vimeux, and Benjamin Quesada

20 **Air-Ice Interface: Tropical Glaciers** 219
Françoise Vimeux

21 **Climate and the Evolution of the Ocean: The Paleoceanographic Data** 225
Thibaut Caley, Natalia Vázquez Riveiros, Laurent Labeyrie, Elsa Cortijo, and Jean-Claude Duplessy

Volume 2

22 **Climate Evolution on the Geological Timescale and the Role of Paleogeographic Changes** 255
Frédéric Fluteau and Pierre Sepulchre

23 **Biogeochemical Cycles and Aerosols Over the Last Million Years** 271
Nathaelle Bouttes, Laurent Bopp, Samuel Albani, Gilles Ramstein, Tristan Vadsaria, and Emilie Capron

24 **The Cryosphere and Sea Level** 301
Catherine Ritz, Vincent Peyaud, Claire Waelbroeck, and Florence Colleoni

25 **Modeling and Paleoclimatology** 319
Masa Kageyama and Didier Paillard

26 **The Precambrian Climate** 343
Yves Goddéris, Gilles Ramstein, and Guillaume Le Hir

27 **The Phanerozoic Climate** 359
Yves Goddéris, Yannick Donnadieu, and Alexandre Pohl

28 **Climate and Astronomical Cycles** 385
Didier Paillard

29 Rapid Climate Variability: Description and Mechanisms 405
Masa Kageyama, Didier M. Roche, Nathalie Combourieu Nebout,
and Jorge Alvarez-Solas

30 An Introduction to the Holocene and Anthropic Disturbance 423
Pascale Braconnot and Pascal Yiou

31 From the Climates of the Past to the Climates of the Future 443
Sylvie Charbit, Nathaelle Bouttes, Aurélien Quiquet, Laurent Bopp,
Gilles Ramstein, Jean-Louis Dufresne, and Julien Cattiaux

About the Editors

Gilles Ramstein is a director of research at Laboratoire des Sciences du Climat et de l'Environnement (LSCE, France). His initial degree is in physics and since 1992 he has specialized in climate modeling.

He has been responsible for many French and European research projects on the Pleistocene, Cenozoic, and Precambrian eras. He has also been the advisor of many Ph.D. students who have explored and expanded the frontiers of paleoclimate modeling.

As a climate modeler, he studies very different climate contexts from 'Snowball Earth' episodes (717–635 Ma) to more recent, and occasionally future, climate situations.

The main research topics he focuses on are

- **Geological time from the Precambrian to the Cenozoic**:
– Investigation of relationships between tectonics, the carbon cycle, and the climate with an emphasis on the impact on the climate and the atmospheric CO_2 cycle of major tectonic events such as plate movements, shrinkage of epicontinental seas, mountain range uplift, and the opening/closing of seaways.

– Leading international collaborations on projects on monsoon evolutions and the dispersal of human ancestors during the Neogene periods.

- **From the Pleistocene to future climate**: In this framework, his major interests are interactions between orbital forcing factors, CO_2 and climate. More specifically, his focus is on the response of the cryosphere, an important component of the climate system during these periods, with an emphasis on the development of the Greenland ice sheet at the Pliocene/Pleistocene boundary and abrupt climate changes driven by ice sheet variations.

He has also published several books and co-edited the French version of 'Paleoclimatologie' (CNRS Edition) and contributed to an online masters program devoted to educating journalists on climate change (Understanding the interactions between climate, environment and society ACCES).

Amaëlle Landais is a research director at Laboratoire des Sciences du Climat et de l'Environnement (LSCE, France). Her initial degree is in physics and chemistry and, since her Ph.D. in 2001, she has specialized in the study of ice cores.

She has been responsible for several French and European research projects on ice cores working on data acquisition both in the laboratory and in the field, interacting extensively with modelers. She has been the supervisor of ten Ph.D. students and is deeply committed to supporting and training students in laboratory work.

Her main research interests are the reconstruction of climate variability over the Quaternary and the links between climate and biogeochemical cycles. To improve our understanding of these areas, she develops geochemical tracers in ice cores (mainly isotopes), performs process studies using laboratory and field experiments, and analyzes shallow and deep ice cores from polar regions (Greenland and Antarctica). Through numerous collaborations and improvement of ice core dating methods, she tries to establish connections with other paleoclimatic archives of the Quaternary.

Nathaelle Bouttes is a research scientist at the Laboratoire des sciences du climat et de l'environnement (LSCE/IPSL). Following the completion of her Ph.D. in 2010 on the glacial carbon cycle, she went to the University of Reading (UK) for a five-year postdoc on recent and future sea-level changes. She then spent a year at Bordeaux (France) with a Marie–Curie Fellowship on interglacials carbon cycle before joining the LSCE in 2016. Since then, she has specialized in understanding glacial–interglacial carbon cycle changes using numerical models and model–data comparison.

She is mostly using and developing coupled carbon-climate models to understand past changes of the carbon cycle, in particular the evolution of the atmospheric CO_2. She has been focusing on the period covered by ice core records, i.e., the last 800,000 years. She uses model–data comparison by directly simulating proxy data such as $\delta^{13}C$ to evaluate possible mechanisms for the orbital and millennial changes. She has been involved in several projects covering this topic as well as teaching and supervising Ph.D. students.

Pierre Sepulchre is a CNRS research scientist at the Laboratoire des sciences du climat et de l'environnement (LSCE/IPSL). He completed a Ph.D. on the Miocene climate of Africa in 2007, then went to UC Santa Cruz (USA) for a two-year postdoctoral position working on the links between the uplift of the Andes and atmospheric and oceanic dynamics. His lifelong research project at CNRS is to evaluate the links between tectonics, climate, and evolution at the geological timescales, focusing on the last 100 million years. Through the supervision of Ph.D. students and his collaboration with geologists and evolutionary biologists, he also worked at evaluating paleoaltimetry methods with the use of an isotope-enabled atmospheric general circulation model, as well as linking continental surface deformation, climate, and biodiversity in Africa and Indonesia. In recent years, he led the implementation and validation of a fast version of the IPSL Earth system model that allows running long climate integrations dedicated to paleoclimate studies.

Aline Govin is, since 2015, a research associate at the Laboratoire des Sciences du Climat et de l'Environnement (LSCE, Gif sur Yvette, France). She studied Earth Sciences at the Ecole Normale Supérieure of Paris (France) and obtained in 2008 a Ph.D. thesis in paleoclimatology jointly issued by the University of Versailles Saint Quentin en Yvelines (France) and the University of Bergen (Norway). Before joining the LSCE, she worked for five years as a postdoctoral fellow at the Center for Marine Environmental Sciences (MARUM, University of Bremen) in Germany.

Her research activities focus on the reconstruction of paleoclimatic and paleoceanographic changes by applying various types of geochemical and sedimentological tracers on marine sediment cores. She has mostly worked on the Earth's climatic changes of the last 150,000 years and is an expert of the last interglacial climate, which is an excellent case study to investigate the response of the Earth's climate to past warming conditions that could be encountered in the coming decades. Her research interests include the past variability of the deep North Atlantic circulation, the responses, and drivers of tropical monsoon systems (e.g., South American Monsoon), the development and calibration of paleo-tracers, the development of robust chronologies across archives, and the quantification of related uncertainties, as well as the comparison of paleo-reconstructions to climate model simulations of past climates.

She has authored around 30 scientific publications and has been involved in many French, German, and other international (e.g., Brazilian) projects.

Climate Evolution on the Geological Timescale and the Role of Paleogeographic Changes

Frédéric Fluteau and Pierre Sepulchre

Throughout geological time, major climate changes have marked the history of the Earth (Fig. 22.1). Although paleoclimate markers provide us with the broad outlines of these changes, their causes could be manifold as feedback mechanisms occur between the different compartments of the climate system. In a system where the solid and fluid envelopes are closely linked, understanding the evolution of Earth's climates at the scale of geologic time involves knowing its paleogeographic history. Within this context, climate modelling is presented as a well-adapted tool to help to understand the factors causing climate change over geological time. However, modelling a continuous climate evolution over million-year timescales is out of reach, as it would require a detailed and reliable knowledge of the model boundary conditions, e.g. the location of the continents, the topography, the bathymetry as well as the chemical composition of the atmosphere. Prior to the recent Quaternary period, uncertainties generally tend to increase regarding these conditions, and our knowledge becomes increasingly fragmentary the further into the past we go. Moreover, even with a perfect knowledge of these conditions, several million years simulations are beyond the computing capabilities of the supercomputers and the codes used today to simulate climate. This methodological dilemma is routinely overcome by means of steady-state simulations, called "snapshot experiments", which simulate the response of the climate system to particular boundary conditions, and which require only a few thousand years of simulation, corresponding to the time needed to achieve equilibrium for all compartments of the climate system. The first step is to establish the boundary conditions so as to perform a "contextual" simulation of a large geologic time interval (for example, the late Miocene). One can then study the impact of paleogeographic or geochemical changes within this interval via sensitivity experiments in which, for example, a mountain range is uplifted or lowered, an ocean passage is opened or closed, the chemical composition of the atmosphere is changed or the orbital parameters of the Earth are altered in line with information provided by the geological, geophysical and geochemical data.

In the first part of this chapter, we present the broad outlines of the climate history of the earth and in the second part, we provide examples of how the direct and indirect couplings between the different envelopes, solid, liquid and gaseous, can be studied through modelling and show how they contribute to the explanation of the climate history of the Earth over long time scales.

The Evolution of Climate Over the Past 4.54 Billion Years

Although the Precambrian (4.54–0.54 Ga) represents 88% of the Earth's history we only have a very fragmentary knowledge of the climate during this period. There are very few records of the first 900 million years of the Earth's history (4.5–3.6 Ga). The oldest geological formations discovered in northwestern Canada (Acasta Gneiss) and in Greenland (Isua Greenstone Belt) are dated at 4 Ga and 3.8 Ga respectively (Valley 2006) but do not provide any climate constraints. However, the discovery of zircons in Australia in the Archean metaconglomerates of Mount Narryer and Jack Hills, dated at 4.4 Ga, is evidence of the existence of the first granitic proto-continents (sensu lato). The oxygen isotopic signature ($\delta^{18}O$) of these zircons (5–7‰) confirms the presence of liquid water, and certainly of oceans, 150 million years after the formation of the Earth. Between 4.3 and 2.8 Ga (Archean), the Earth's atmosphere consisted of a mixture of nitrogen and greenhouse gases, notably carbon dioxide and methane. There was no oxygen.

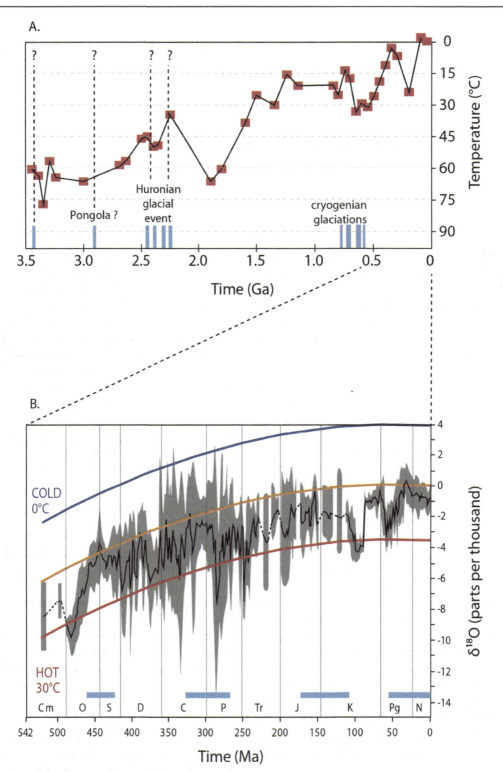

Fig. 22.1 a Estimates of water temperature calculated using maximum $\delta^{18}O$ chert value per 50 Ma age intervals, adapted from Tartèse et al. (2017). The secular trend shows an overall cooling of Earth through geological time. Blue bars indicate episods of widespread glaciation. Dashed lines, and question marks show the apparent contradiction between poorly dated warm temperature estimates and the relatively short glacial events. b Evolution of $\delta^{18}O$ data for low-latitude planktonic foraminifera younger than 118 Ma and brachiopods older than 117 Ma. The temperature curves were calculated using a baseline value for Phanerozoic seawater which mimics the present day 0‰ SMOW. Grey shading indicate 2-sigma intervals. Blue bars indicate the presence of ice caps. Adapted from Veizer and Prokoph (2015)

There are big uncertainties regarding Earth's surface temperature during the Archean, as deduced from isotopic measurements of oxygen in carbonate rocks, with values ranging from 26 to 85 °C (Knauth and Lowe 2003; Robert and Chaussidon 2006; De Wit and Furnes 2016). However, these values may be inaccurate due to our lack of understanding of the average $\delta^{18}O$ of the seawater at that time. Indeed, the average ^{18}O concentration in the oceans is determined by ^{18}O exchanges with oceanic crustal basalts during the circulation of seawater in the hydrothermal systems of the ocean ridges. There is no evidence to indicate that it was within the range of values observed much more recently. Measurements of the isotopic ratio of oxygen and silicon in cherts (siliceous rocks) suggest a paleotemperature of 60–80 °C (Fig. 22.1a). However, these values may be distorted by the contributions of the hydrothermal systems lining the oceanic crust at that time. To overcome these difficulties, new isotopic tools, called the clumped isotopes, have been developed. A promising paleothermometer, based not on the relative abundance of one single isotope to another, but on the abundance ratio of isotopologues with multiple rare isotopes relative to the expected quantity for a stochastic distribution of isotopes within a $(^{13}C^{18}O^{16}O_2)^{2-}$ group (Eiler 2007; Bonifacie et al. 2017). The thermodynamic variable measured and noted Δ_{47} is therefore based on the substitution of carbon and oxygen isotopes within the carbonate ions, which is a function of the temperature at which these carbonates formed. Although uncertainties remain (Daëron et al. 2019), this method does not require knowledge of the average isotopic composition of the ocean and can therefore be applied to the oceans of the Archean. Although the evolution of temperature during the Archean is up for debate, it remained within a range of values able to sustain liquid water on the Earth's surface. This is an essential condition for the development of life which could have started as early as 3.77 Ga (Dodd et al. 2017).

Maintaining a temperature conducive to the development of life on Earth was by no means guaranteed within the context of the "young Sun". In fact, the radiation emitted by a young star is weak, gradually increasing throughout its life. In the case of the Sun, the solar radiation is estimated to have been 30% lower 4 Ga ago than it is today, and 20% lower 3 Ga ago. In these radiation conditions, maintaining a temperature that would support the presence of liquid water on the surface of the globe during the Archean would have been impossible without a powerful greenhouse effect.

Let's examine the data to determine the constraints on the chemical composition of the atmosphere. Precipitation of sodium bicarbonate in the Archean oceans around 3.3 Ga argues in favour of a partial CO_2 pressure of between 1.4 and 15% of the atmospheric pressure of the time (Lowe and Tice 2004), and so lower than the estimates by the models. However, the chemical composition of the Precambrian atmosphere is the subject of intense debate because a high partial pressure of CO_2, necessary to counter the "young sun" effect, would have strongly acidified precipitations, causing the pH to fall to about 3.7, and cause a particularly intense chemical weathering of the rocks, which does not seem to be the case for the 3.5–3 Ga period (Kasting and Howard 2006). Another gas, methane, may therefore have played an important role. In the modern atmosphere, the residence time of methane is around eight years because it is quickly oxidized. However, in an atmosphere devoid of oxygen or at very low partial pressures ($<10^{-5}$ bar) as in the Archean, this residence time is considerably longer. Methane can thus accumulate in the atmosphere and significantly increase the greenhouse effect. We know of two sources of methane at that time: methanogenic organisms and the serpentinization of ultramafic rocks on the ocean floor. Taking the results of the work of Haqq-Mishra et al. (2008), with 1% methane in the atmosphere, the CO_2 partial pressure required to maintain the Earth at 30 °C drops to about 10^{-3} bar. The contribution of methane to the greenhouse effect is therefore extremely effective. Even if atmospheric concentrations remain poorly constrained, a CH_4 to CO_2 ratio greater than 0.2 is impossible (Zerkle et al. 2012).

Traces of several glacial periods, dated between 3.5 and 2.2 Ga, have been discovered in South Africa, Europe, North America and Australia. The oldest glaciation was discovered in South Africa within the Barberton Greenstone Belt. It was dated at about 3.4–3.5 Ga (De Wit and Furnes 2016) and in a latitude band between 20° and 40° (Biggin et al. 2011). It was followed by another glacial episode at 2.9 Ga discovered in units of the Mozaan geological group in South Africa (Young et al. 1998). The end of the Archean and the beginning of the Proterozoic produced the Huron glaciation, originally discovered in the province of Ontario in Canada, but identified in South Africa and Australia. It is actually a succession of three or four glacial events, dated between 2.45 and 2.2 Ga (Caquineau et al. 2018). One of these glacial episodes is demonstrated by the presence of glacial sediment at low latitudes and at low altitude suggesting that the Earth could have been completely frozen at this time. These glaciations are associated with a major event in the history of the Earth: the oxygenation of the atmosphere and of the shallow oceans.

The next billion years (between 1.85 and 0.85 Ga) is often referred to as the 'boring billion' due to the apparent climatic and environmental stability. Indeed, isotopic measurements of carbon ($\delta^{13}C$) show no major disturbance of the carbon cycle and the absence of any known glacial traces during this period suggest (but do not prove) that the Earth's climate had stabilized into a hot configuration. This climate stability ended with the Neoproterozoic. During this time, the Earth experienced three periods of glaciation: the Sturtian glaciation from 717 to 659 Ma, the Marinoan glaciation from 645 to 636 Ma and finally the Gaskiers glaciation around 582 Ma. Marine glacial sedimentary formations at low latitudes indicate that during the Sturtian and Marinoan glaciations, the Earth was completely glaciated: these are the famous episodes of Snowball Earth (Hoffman et al. 2017). Directly on top of these glacial sedimentary formations we find carbonate formations (currently, carbonate production is mainly located in warm tropical seas). The rapidity of this transition between glacial and carbonate formations (on the scale of geological time) is a peculiarity in the climate history of the Earth. Numerous studies have been undertaken in recent years to understand the entry and exit modalities of these glacial phases (Hoffman et al. 2017; and Chap. 5).

The Precambrian/Cambrian boundary (542 Ma) marked a new turning point in the Earth's climate history. According to oxygen isotopic ratio measurements, the mean global temperatures of the Phanerozoic climate became stable within a range comparable to that of the modern day (Fig. 22.1b). This period was punctuated by three major glaciations: one at the end of the Ordovician (around 443 Ma), during the Permo-Carboniferous (between 335 and 260 Ma) and at the end of the Cenozoic (the last 40 Myr). Traces of the Ordovician glaciation are found in Africa, particularly in the Sahara and South Africa but also on the Arabian peninsula. This glaciation is estimated to have lasted a little more than one million years, and is marked by at least three glacial cycles alternating with interglacial periods (Ghienne et al. 2013). The glaciation is associated with a major disruption in the carbon cycle but also with one of the five mass extinctions of the Phanerozoic when close to 86% of the marine benthic and planktonic species died out. During the Silurian and Devonian periods, the Earth experienced a warmer global climate. Carbonate platforms, which develop in warm seas, stretched from 45° S to 60° N. An expansion of this scale would never be seen again (Copper and Scotese 2003). The strong latitudinal expansion of these carbonate platforms suggests weak latitudinal thermal gradients. The widespread evaporite facies suggest a semi-arid to arid climate at subtropical latitudes.

The Devonian period is also marked by the colonization of land surfaces by life. The first forests, made up of *Archeopteris,* appeared at the end of the Devonian at around 370 Ma (Meyer-Berthaud et al. 1999). During the Silurian, some plants, including bryophites, were the pioneers of this land colonization that, until then, had been deserted and barren. The emergence of the continental biosphere affected the carbon cycle and probably influenced the climate of the Earth (Le Hir et al. 2011). The end of the Devonian and the beginning of the Carboniferous mark the return of glacial periods and a more contrasted climate latitudinal gradient. The Earth then underwent a glacial period which started in the Carboniferous (∼335 Ma) and which ended with the Permian around 260 Ma (Montañez and Poulsen 2013). This glaciation, which lasted 70 Ma, was punctuated by several phases of advance and withdrawal of the ice sheets. This is the longest and most important glacial episode of the entire Phanerozoic. Sedimentary formations, striated floors and 'dropstones' (pieces of rock deposited onto unconsolidated marine sediments by icebergs when they melted) are proof of the presence of ice in South America, southern and eastern Africa, on the Arabian peninsula, the Indian subcontinent and Australia, that is, the whole southern part of the Gondwana continent then located in the mid and high latitudes of the southern hemisphere.

At the end of the Carboniferous, while a cold climate developed at the high and mid latitudes of the southern hemisphere, paleoclimatic indicators indicate that there was a tropical and humid climate over a part of Europe and North America, then located close to the equator. At this time, there was a strong contrast between the climates of low and high latitudes.

These conditions disappeared in favor of a warmer and dryer climate during the Late Permian. Wet climates were limited to narrow bands around the equator and in the mid-latitudes. The Paleozoic era ended with two mass extinctions, the first at the end of the "Guadalupian" (∼258 Ma) and the second at the Permo-Trias boundary (∼251 Ma) (Bond et al. 2010; Bond and Wignall 2014). This last crisis is the most important mass extinction of the Phanerozoic with the disappearance of 90% of the fauna and flora (Erwin 1994). During this crisis, climate indicators show significant warming, a disruption of the carbon cycle as well as oceanic anoxia. An exceptionally high level of volcanic activity leading to the contemporaneous formation of the Siberian large igneous province is considered to be the main cause of the extinction. Although the relationship between volcanic activity and this crisis is not fully understood, it should be noted that every ecological crisis of the Phanerozoic, whatever its magnitude, occurred simultaneously with the establishment of a large basaltic province through particularly intense volcanic activity. After the Permo-Trias mass extinction, a warm global climate became established during the Triassic. During the Late Triassic, sedimentary facies in North America suggest a strong seasonal precipitation caused by "mega-monsoon" patterns (Dubiel et al. 1991; Bahr et al. 2020). Subsequently, the Earth experienced a colder global climate during the Jurassic. The presence of ice caps at high latitudes has been suggested, but this hypothesis is based on assumptions that have still to be confirmed (Dromart et al. 2003).

Until the mid-1990s, the Cretaceous (145–65 Ma) was described as a period with a uniformly warm global climate, but the accumulation of data from different climate indicators has completely changed our notions of the climate of this period. The beginning of the Cretaceous period was cold, probably with ice sheets, but experienced a particularly warm period towards the middle of the Cretaceous, before cooling towards the end of the period. On top of this long-term trend, rapid climate variability was superimposed. The isotopic measurements of the oxygen in carbonate tests of planktonic foraminifera and fish teeth provide an estimate of the paleotemperature of the surface waters of the oceans. These data, as well as studies carried out in continental areas, on palynomorphs, for example, reflect the climate variability of this period (Ladant and Donnadieu 2016). Thus, $\delta^{18}O$ measurements show that the Turonian had a particularly hot climate, with sea surface temperatures of between 34 and 37 °C. However, the $\delta^{18}O$ measurements also show the existence of a glaciation event lasting less than 200,000 years (Bornemann et al. 2008). The Cretaceous ended with a rapid cooling just before the major mass extinction of the Cretaceous-Tertiary boundary at 66 Ma, during which nearly 60% of the species on Earth died out. As for the other crises of the Phanerozoic, this crisis is synchronous with the establishment of a large basaltic province, the Deccan traps in India, but also, with another event, the fall of an extraterrestrial body into the Gulf of Mexico.

After this crisis, the Earth entered a warm period during which signs of glaciation disappeared. This period was punctuated by the thermal maximum at the Palaeocene-Eocene boundary (55 Ma) (Zachos et al. 2001). This was a rapid transient event (at the geological time scale), lasting about 300 ka, marked by an abrupt increase in temperatures (considered as the best analogue to current global warming). Tropical flora, but also turtles, crocodiles, and many mammals were discovered on the Ellesmere and Axel Heiberg islands in northern Canada.

As early as the Late Eocene (~ 50 Ma), the Earth experienced gradual cooling with the onset of a new glacial period around 40 Ma marked by a first stage of development of the Antarctic ice cap, corresponding probably to the appearance of glaciers in the Trans-Antarctic chain. This ice cap grew quickly and reached the coast, as evidenced by glacial sediments found off the Antarctic continent, dating from the Eocene-Oligocene boundary (~ 34 Ma). Numerous measurements of the isotopic composition of oxygen in the carbonate tests of benthic foraminifera in sediments collected at different points around the globe confirm a consistent cooling during the Eocene. At the Eocene-Oligocene boundary, a significant and rapid cooling of the ocean bottom waters occurred, estimated at about 6 °C (Hren et al. 2013). Henceforth, the isotopic ratio of strontium increased significantly in response to intensified continental erosion.

During the Lower Miocene (23–15 Ma), the global trend was a slight warming interspersed with brief cooling episodes. Around 14 Ma, rapid cooling led to a new phase of development of the Antarctic ice sheet. This climate trend accelerated during the Upper Miocene and the Pliocene. The development of an ice cap on Greenland probably dates from the Upper Miocene or the Pliocene, but a high probability of sea ice on the Arctic Ocean starting from the climate transition of the Middle Miocene is suggested by the sedimentary facies observed in the Arctic Ocean. This cold climate impacts on the high latitudes of both hemispheres and provides the necessary conditions for the rapid glacial/interglacial fluctuations during the Pleistocene (but does not trigger them).

Paleoclimate indicators can be used to trace the evolution of the Earth's climate, so as to gradually refine its contours and to observe rapid fluctuations superimposed on longer-term trends. However, there are areas of uncertainty (as will probably always be the case), especially for the oldest periods. The causes and mechanisms of these climate changes at the scale of geological time are manifold. Data collected in the field allow us to document this evolution with increasing accuracy, but it is impossible to isolate with certainty the specific cause or causes of these climate disruptions. Since the 1970s, numerical modeling of climates has been used, in addition to data, to test the sensitivity of the climate to different forcings and to try to reproduce numerically the climate changes observed in the field. Over long time scales, paleogeographic changes brought about by plate tectonics have shaped the face of the Earth (Volume 1, Chap. 2) and are a major forcing of the Earth's climate history through their direct effects on atmospheric and oceanic circulation. We will also see the indirect effects of paleogeographic changes induced by feedback mechanisms, in particular on the regulation of the partial pressure of CO_2, another major contributor to the climate system.

Some Consequences of Paleogeographic Changes on the Earth's Climate

Continental Drift

Continental drift leads to changes in the latitudinal and longitudinal distribution of emerged lands with various consequences for climate. The main ones are: changes in the distribution of solar radiation received by the continents; changes in the atmospheric dynamics by the uplift or collapse of mountain ranges or during the formation or break-up of supercontinents; changes in ocean dynamics during the opening or closing of basins or ocean passages (also called seaways), and also indirect effects such as changes in weathering fluxes that impact on the carbon cycle (Donnadieu et al. 2004). In the Upper Permian (~ 260 Ma), paleoclimate indicators suggest that a warm, dry climate developed over a large part of Gondwana (southern hemisphere) located between the narrow rainy equatorial band and the narrow temperate mid-latitude band, below the theoretical subsidence zone of the Hadley cell. Several numerical climate simulations made it possible to develop hypotheses to explain the various mechanisms involved. Fluteau et al. (2001) suggested that high seasonality in Gondwana was typical at mid and high latitudes due to the low heat capacity of the continents, and that monsoon-type atmospheric circulation marked the eastern side of the supercontinent. Other modeling studies (Kiehl and Shields 2005; Shields and Kiehl 2018) have confirmed these results, even suggesting that these characteristics continued despite a high atmospheric concentration of carbon dioxide (3550 ppm). Climate modeling has also looked at the consequences of supercontinent break-up. By simulating the climate response to paleogeographic changes between the Triassic and Cretaceous, it was suggested, for example, that the break-up of Gondwana and Laurasia annihilated the continental effect, preventing the development of large desert bands in the subtropics (Fig. 22.2), and favoring the establishment of wet conditions contemporaneous to the diversification of flowering plants (Angiosperms) (Fluteau et al. 2007; Chaboureau et al. 2014).

Paleogeographic Changes and Ocean Circulation

The oceans are a major component of the climate system. They ensure, in part, the transport of heat from the low to the high latitudes, particularly in modern times, via the Atlantic Meridional Overturning Circulation (AMOC). At the geological time scale, large changes in ocean basin geometry have controlled the dynamics of water bodies and the associated heat and salt fluxes by opening or closing interoceanic connections. This is, for example, the case of the opening of the South Atlantic Ocean which began 135 million years ago. Exchanges between the South Atlantic and Central Atlantic oceans have only happened since 100 Ma (Murphy and Thomas 2013; Granot and Dyment 2015) and led to changes in the global ocean circulation as shown by the isotopic data of neodymium and oxygen (e.g. Donnadieu et al. 2016). Although there is no consensus on the exact evolution of Cretaceous ocean circulation, taking these paleogeographic changes into account in numerical simulations suggests that the establishment of a sea passage between the South Atlantic and Central Atlantic played an important role in the formation and oxygenation of deep waters on a global scale (Poulsen et al. 2003).

The Cenozoic is also characterized by large-scale climate change, including global cooling and reorganization of ocean circulation. Apart from India and Australia, the drift in latitude of most of the continents is small over the last 60 million years, and cannot by itself explain the temperature changes implicit in the data. However, several openings and closings of ocean passages have altered the exchanges between water bodies and the associated heat flows. At the beginning of the Cenozoic, the distribution of the continents meant that the Pacific, Atlantic and Indian basins were connected in the tropical band via three open ocean passages: The Central American seaway (CAS), the east-Tethys seaway, and the Indonesian passage. Numerical simulations suggest that this interoceanic connection operated from east to west at the surface, with the formation of the circum-equatorial current (CEC). Conversely, in the southern hemisphere, the Drake and Tasmanian passages separating South America and Australia from Antarctica, respectively, were closed, preventing the formation of a strong Antarctic Circumpolar Current (ACC). At the Eocene-Oligocene transition, the circumpolar Antarctic maritime passages gradually opened, widened and deepened. This episode would be followed by the gradual closure of tropical passages in the Middle Miocene (15 Ma).

Numerical simulations show that both events contributed to the establishment of deep water formation in the northern hemisphere and to the cooling of the southern hemisphere (see Sijp et al. 2014 for an overview). Since the late 1970s, the establishment of the ACC in response to the opening of the marine passages of the southern hemisphere has been advanced as a cause of the freeze-up of the Antarctic (Kennett 1977). Indeed, the opening of the Tasman Sea and the oceanic exchanges through the Drake Passage in southern South America fostered the thermal isolation of Antarctica. Recent studies show that the paleogeographic configuration,

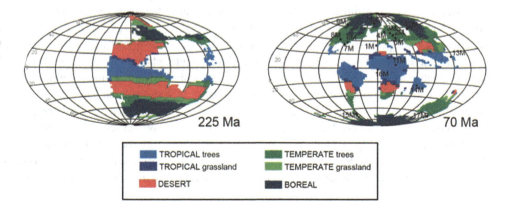

Fig. 22.2 Simulated Jurassic and Cretaceous bioclimatic zones. The numbered points indicate the localisation of angiosperm fossils. Adapted from Chaboureau et al. (2014)

in particular the bathymetry, is decisive to correctly quantify the cooling related to the change of configuration of the seaways, but also that the freeze-up of the Antarctic requires, in addition to paleogeographic changes, a significant decrease of the atmospheric partial pressure of carbon dioxide from 1160 to 560 ppm (Lefebvre et al. 2012).

Numerous modelling studies have also focused on the impact of the central American seaway (CAS) on climate. Early works by Haug and Tiedemann (1998) had suggested that the closure of the isthmus during the early Pliocene led to a reorganization of the Atlantic thermohaline circulation. Specifically, the authors inferred that the shallowing and closure of the CAS intensified the Gulf stream and strengthened deep water formation in the northern Atlantic. More importantly these authors suggested that warmer SST induced by this strengthening of the Gulf stream increased atmospheric moisture content, and ultimately favored the Greenland ice sheet growth. Later, numerous model experiments with open CAS with various depth and width have been carried out (see Zhang et al. 2012; Sepulchre et al. 2014 for reviews). Most of the model indeed showed intensified AMOC with the seaway closure, but models run with explicit ice sheet modelling failed to demonstrate that the closure had a significant impact on the Greenland icecap onset (Lunt et al. 2008; Tan et al. 2017). Our knowledge of the timing of CAS closure also evolved during the last 20 years. Although still very debated (O'Dea et al. 2016), authors have suggested that CAS constriction happened much earlier than previously thought (Montes et al. 2015; Bacon et al. 2015; Jaramillo 2018), with a very restricted seaway by the late Miocene (ca. 10 Ma, Fig. 23.3). This different chronology has many consequences on our understanding of CAS influence on climate, as it involves that its constriction has occurred during other major tectonics events, such as the uplift of the south American cordilleras (Andes).

Other maritime passages have seen their configuration change during the Cenozoic. The drift of Australia towards Indochina has progressively restricted the maritime exchanges between the Pacific Ocean and Indian Ocean through the Indonesian Passage. Numerical simulations show that before closure around 4 Ma, a warm ocean current flowed between the tropical Pacific Ocean and the Indian Ocean (Cane and Molnar 2001; Brierley and Fedorov 2016). After closure, this warm ocean current was blocked and was replaced by a colder current from the north Pacific. Simulated consequences involve a cooling of the surface waters of the Indian Ocean and a marked drying in East Africa, causing tree cover to decrease in favor of savanna vegetation. Although this drying has been confirmed by paleoclimate indicators (Bonnefille 2010), more proximal tectonics changes, namely the uplift of the east African dome might have played a role in this aridification (Sepulchre et al. 2006).

The Influence of Shelf and Epicontinental Seas

Variations in sea level have marked the history of the Earth. Reconstructions by Haq et al. (1987) of eustatic variations show that a high underlying sea level during the Upper Cretaceous (~95 Ma) is responsible for the formation of numerous shelf seas. The functioning and role of these shelf seas is still poorly understood as there is no modern equivalent of these water bodies. From the climate perspective, the answer seems simple: the higher the sea level, the less land surface is exposed, the smoother the seasonal cycle and the more homogeneous the climate becomes. Simulations conducted using a general atmospheric circulation model indicate that the climate response to the formation of shelf seas is

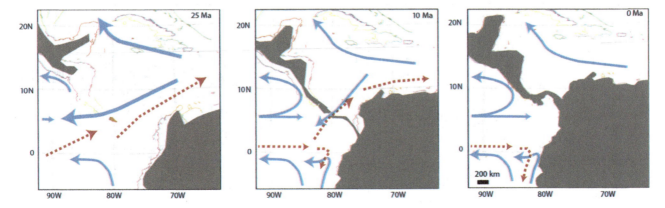

Fig. 22.3 Sketches of putative ocean currents in different paleogeographic configurations for Central America between 25 Ma and present. Reconstructions of continental areas are from Jaramillo (2018). Blues arrows show surface currents, and doted-red arrows show subsurface to mid-depth currents. In these reconstructions, a wide ocean gateway is open between the American continents at 25 Ma, whereas only a very narrow passage allows water exchange during the late Miocene. With this latter configuration, climate models suggest that surface currents flow westward, while undercurrents can bring fresher Pacific waters into the tropical Atlantic ocean

more complex than it might appear (Fluteau et al. 2006). A massive epicontinental sea in North America, the Western Interior Seaway (WIS), stretched across the North American continent from the Arctic ocean to the Gulf of Mexico, due to the sea incursion during the Middle Cretaceous, and to the dynamic topography caused by the subduction of the Farallon Plate under North America. During the Aptian (~20 Ma), this arm of the sea had not yet formed. Simulated atmospheric circulation in winter over North America is driven by a high pressure zone over the northeast of the continent. In summer, the opposite is the case; a zone of low pressure develops over the southwest of North America east of the mountains bordering the Pacific Ocean. The onset of the WIS completely disrupts the atmospheric circulation in summer across the continent (Fig. 22.4). The presence of this sea pushes the low-pressure zone to the east, while in winter temperatures increased by about 4 °C close to the WIS and precipitations, sometimes of snow, intensified along the Arctic Ocean. The warm temperate zones were shifted northward by up to 1000 km along the WIS following its establishment. Although, these results cannot be generalized since they depend on the location of the shelf seas relative to the high and low pressure areas that develop over the continent, they seem to have greatest impact on temperatures in the mid-latitudes and to induce significant increases in precipitation at low latitudes, as the shelf seas act as important reservoirs of water for the atmosphere.

The Impact of Mountains on Climate

As described in Volume 1, Chap. 2, much of the modern topography was formed during the Cenozoic. The paleoaltimetry proxies described earlier and the sedimentological record suggest that the Andes, the East-African dome, as well as the Himalayas and the Tibetan plateau rose mostly over the last twenty million years. These different reliefs form obstacles that alter climate through many mechanisms,

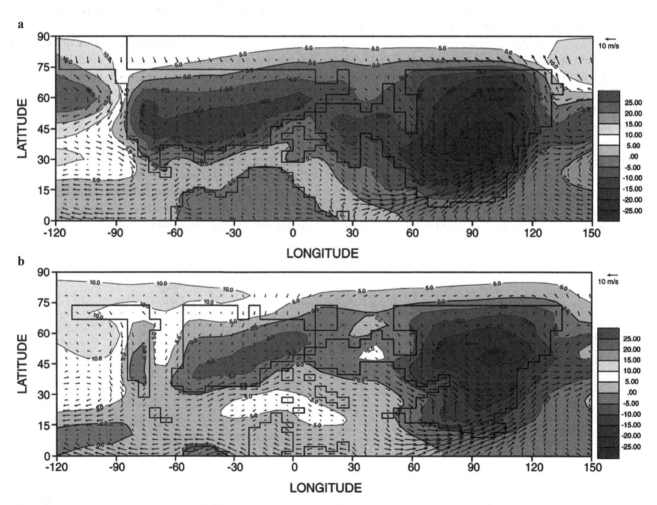

Fig. 22.4 Consequences of the establishment of the Western Interior Seaway (according to Fluteau et al. 2006). **a** Atmospheric circulation in summer during the Aptian (120 Ma). **b** Idem during the Cenomanian (95 Ma). Pressure fields are represented by isobars (P—1000 hPa) (HP = high pressure and BP = low pressure) and by winds (arrows) at 850 hPa (about 1500 m). Shaded areas: exposed land surfaces

at many spatial scales. Here we provide only a short overview of the main interactions between mountain uplift and climate dynamics to understand how the topographic changes of the Cenozoic have changed climate.

First, depending on its velocity, its angle with the mountain range and the size of the obstacle, air arriving towards a mountain range can undergo an orographic ascent that leads to adiabatic cooling until water condensate and precipitate on the windward slope, leading to a rainfall gradient between both sides of the orographic barrier. At the continental scale, this "rainshadow" produces strong heterogeneities in rainfall patterns and associated biomes, such as the ones observed today in Patagonia, for example. The varying temperature lapse rates, i.e. temperature decrease with elevation, that depends on air masses characteristics, also produce thermal heterogeneity in elevated regions. High mountains can also alter the surface radiative balance through snow-induced changes in albedo, creating thermal and pressure gradients that alter air circulation. Lastly, mountain ranges have long been shown to create different kinds of waves that propagate on the lee side and vertically, ultimately disturbing winds and the spatial distribution of precipitation (Warner 2004).

The impact of mountains on climate was considered as early as the middle of the nineteenth century, but it was not until the 1950s that studies quantified this relief-climate relationship. The meteorologists, Charney and Eliassen (1949), followed by Bolin (1950), assessed the impact of mountain ranges on the mid-latitude westerlies, while Flohn (1950) suggested that the Tibetan plateau was a source of sensible heat significant enough to explain the establishment of the Asian monsoon. Between the 1970s and 1990s, many studies benefited from the advances in atmospheric general circulation models to quantify the impact of mountain ranges on the distribution of arid zones globally and the establishment of the Asian monsoon. Broccoli and Manabe (1992) suggested that the aridity of Central Asia and the Great Plains of North America was partly due to the generation of stationary waves downwind of adjacent mountain ranges.

One specific geological event has captured the attention of scientists: the uplift of the Tibetan plateau and the Himalayan range in the context of the India-Asia collision, which occurred at the beginning of the Eocene (Molnar et al. 2010). The first assessment studies on the impact of the uplift of the Tibetan Plateau using climate models were based on sensitivity experiments with varying elevations of the plateau without differentiating between the Himalayan and Tibetan uplifts and without taking other paleogeographic changes into account (e.g. Kutzbach et al. 1993; Ruddiman et al. 1997). These studies suggested that the rise of the Tibetan plateau played a crucial role for atmospheric circulation in general, and for the Asian monsoon in particular. However, the physical mechanisms involved, which use the temperature gradients induced by a large continental area at high altitude, are still in question. Other studies have shown (i) that the thermal contrast between a flat continent and the Indian Ocean was enough to initiate a lower intensity monsoon than the modern monsoon pattern and (ii) that a prerequisite for this was the removal of the Paratethys Sea which enhanced the heating of Central Asia, and therefore the thermal contrast between the Indian Ocean and continental Asia (Ramstein et al. 1997; Fluteau et al. 1999). More recently, the separation of air masses on either side of a mountain range has been mentioned as a major factor in the initiation of convection and associated precipitation (see Boos 2015 for a review) and many modeling studies using various boundary conditions have been undertaken to try to understand the exact role of altitude (e.g., latitudinal position (Zhang et al. 2015, 2018), geographic extension (Chen et al. 2014) and of vegetation cover of the plateau (Hu and Boos 2017) in the establishment and intensification of the Asian monsoon.

Furthermore, the implications of the Himalayan-Tibetan relief for the climate are not limited to the atmosphere. Rind et al. (1997) used a general circulation model coupled with an ocean-atmosphere model (AOGCM) to show that these uplifts caused a rise in sea surface temperatures (~ 2 °C) in the North Atlantic (Norwegian Sea), an increase in heat transport at high latitudes (a crucial parameter for understanding climate change at high latitudes and possibly the freezing-over of Greenland) as well as a reduction of about 10% in deep water production in the Norwegian Sea (linked to the decrease in density of the water mass, due to the warming of the North Atlantic Ocean). More recently, Su et al. (2018) suggested that plateau uplift may have contributed to the establishment of the AMOC by changing the intensity and latitude of the zonal winds, thereby altering sea-ice formation and deep-water formation.

The impact of the reliefs on ocean-atmosphere dynamics are not limited to the Late Cenozoic orogenesis, because the Earth's history is dotted with uplifts creating mountain ranges with different locations, extensions, heights, and orientations. By comparing climate simulations for a world with its current topography and a "flat earth" world, Maffre et al. (2018) showed that the orographic barriers of the Andes and the Rockies constrain the transport of freshwater between the Pacific and the Atlantic, and thus contribute to the high salinity of the latter, favoring the formation of deep water. The Andean uplift is also believed to be responsible for the establishment of convective precipitation in tropical South America (Poulsen et al. 2010) and the strengthening of the Humboldt Current (Sepulchre et al. 2009). The establishment of reliefs in south and east Africa during the Mio-Pliocene probably led to the aridification of East Africa (Sepulchre et al. 2006), to the strengthening of the coastal upwellings of the Benguela current (Jung et al. 2014) and a change in position of the ITCZ in the Atlantic (Potter et al. 2017).

The Indirect Effects of Paleogeographic Changes

The atmospheric partial pressure of carbon dioxide (pCO_2) is driven by the carbon cycle. On the scale of geological time, pCO_2 reflects the balance between the CO_2 emission fluxes from volcanic systems (ocean ridges and aerial volcanism), the degree of magmatic activity from the mantle (plumes), decarbonation in subduction zones, and the CO_2 fluxes consumed by the weathering of the rock silicates on the surface of the Earth and in the oceanic crust, through the burial of the organic matter. Fluctuations in pCO_2 reflect the evolution of one or both of these flows. CO_2 emissions are proportional to the annual rate of production of oceanic crust. We have seen that the variability in this rate of production does not exceed 30% over the last 170 million years (Cogné and Humler 2006). To these rates should be added CO_2 emissions from plumes of mantle volcanism and from decarbonation in subduction zones. Significant magmatic events dating from the late Early Cretaceous and associated with the establishment of some large submarine basalt provinces, such as the Ontong-Java Plateau, increase the production rate of oceanic crust by about 25% (Cogné and Humler 2006) and about the same increase in CO_2 is injected into the ocean-atmosphere system. Contributions from the subduction zones are less constrained. Currently, this process is limited to a few subduction zones in the Pacific or around the Indonesian archipelago, while the main deposition zones are in the Atlantic and Indian Ocean. In the past, the subduction of the Tethysian Ocean could have emitted a significant CO_2 flux (Hoareau et al. 2015).

The intensity of the chemical weathering is a function of climate parameters such as surface temperature and runoff. Moreover, most geochemical models considered runoff as a function of surface temperature: the higher the temperature, the greater the runoff, and consequently, the chemical weathering. The chemical weathering of silicates acts as a climate regulator. We know now that this proportional relationship between temperature and runoff is based on current data and is not transferable to past periods. This is easily understood by analyzing the upper Permian climate. Simulation of chemical weathering is effective only in some areas experiencing a tropical and humid climate. This weak chemical weathering of silicates in the paleogeographic context implies a relatively high pCO_2 equilibrium of about 2500 ppm, and a high global average temperature of about 21 °C (considering that the rate of CO_2 emissions from the ridges is comparable to the current rate). Paleoclimate data confirm the hot and dry climate simulated over a large part of Pangea. The paleogeography of the Triassic maintained the Earth in a relatively stable climate dominated by a high simulated pCO_2 of around 3000 ppm and a high simulated global average temperature (between 21.5 and 23 °C). During the Jurassic, the drift of the Pangea to the north and its break-up brought about an increase in continental surfaces exposed to the hot and humid climate of the equatorial band. This resulted in intensified chemical weathering of silicates thus causing an increase in the consumption of CO_2. The simulated pCO_2 is lower at about 700 ppm as is the average temperature of the globe at a little over 18 °C. The end of the Mesozoic is marked by the final break-up of Gondwana. The arid areas reduced in size, reinforcing global chemical weathering. This resulted in a fairly low simulated pCO_2 of between 300 and 500 ppm for the Cretaceous. The effect of paleogeography on climate (through the regulation of chemical weathering of silicates) is therefore an effective process that can be seen in long-term trends, even if it does not explain every climate variations.

The conditions necessary for the high latitudes to freeze-up were not present at the end of the Permian, but they were in place for a long period from the Lower Carboniferous (340 Ma) to the Lower Permian (280 Ma), already in a supercontinent paleogeographic context. During this glacial period, the Earth experienced a succession of advances and retreats of continental ice over southern Gondwana. What mechanisms would push the Earth into a different climatic state? Although the geographical configuration of Pangea had not changed drastically between the Carboniferous and the Upper Permian, it had been drifting northwards during this period. Indeed, paleomagnetic data show that southern Gondwana was located at the pole during the Carboniferous, favoring a cooler summer in this region, but not yet cold enough for ice to remain outside some high reliefs. Another more effective mechanism was needed. The decrease in pCO_2 was therefore necessary to explain this glacial period. One suggestion was that colonization of emerged lands by plants during the Devonian increased the chemical weathering of silicates leading to a decrease in pCO_2 (Berner 2001). However, this vegetal colonization of continents took place tens of millions of years before the beginning of the glaciation. Goddéris et al. (2017) have shown that the Hercynian orogenesis could have played a major role. This orogenesis, resulting from the collision of Laurussia and Gondwana around 350 Ma, was at the origin of the uplift of a vast chain of mountains, the Hercynian chain, stretching for several thousands of kilometers in the equatorial band. With no mountains, the hot and humid climate of the lower latitudes causes a thick saprolite to form, considerably limiting the weathering of the underlying bedrock. With orogenesis, the presence of relief produces strong mechanical erosion due to the slopes which considerably limit the development of thick saprolites, thus the weathering of silicate rocks is increased, leading to a

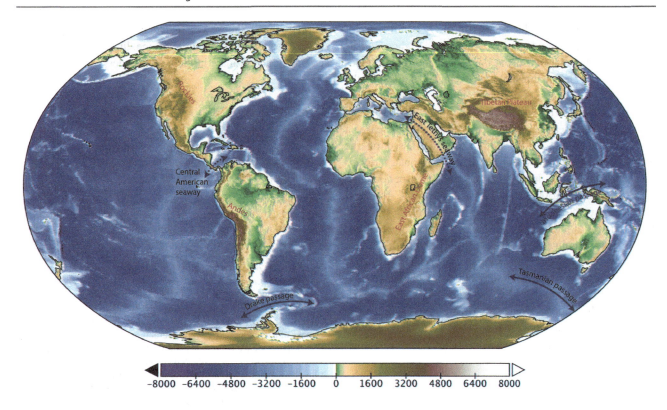

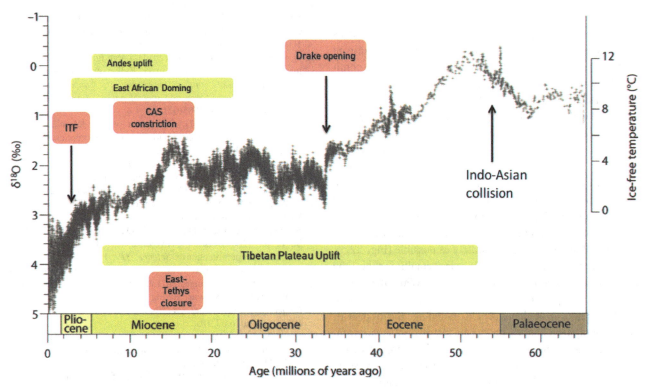

Fig. 22.5 Spatial positions (upper panel) and timing (lower panel) of the modification of the principal mountain ranges and seaways discussed in the text. Bathymetry and topographic are from NOAA. The stable isotope curve in the bottom panel can be interpreted as ice-free temperatures. Adapted from Zachos et al. (2001)

decrease in the pCO_2. This decline in pCO_2 is further reinforced by the increased transport of organic carbon and its burial in sedimentary basins at the foot of the reliefs (Goddéris et al. 2017). The threshold for freezing is defined by the level of atmospheric CO_2 concentration below which freezing occurs. This threshold was reduced during the Carboniferous because of the slow, inexorable increase in insolation produced by the Sun. Estimates, based on the analysis of fossil leaves and pedogenic carbonates, found that in basins at low latitudes pCO_2 fluctuated periodically from about 200 to 300 ppmv during the Upper Carboniferous with a minimum concentration of 160–300 ppmv (Montañez et al. 2016). The periodicity of these fluctuations seems to be dominated by an eccentricity signal of 405 ka. The vegetation cover of these equatorial basins also varies between dry vegetation during interglacials and humid tropical vegetation during interglacials (Montañez et al. 2016) which influence the organic carbon cycle, possibly causing fluctuations in the volume of continental ice on Gondwana and is superimposed on the long-term trend driven by geodynamics.

Conclusion

The impact of paleogeographic changes on climate and its evolution is fundamental. The presence of a supercontinent produces climates characterized by strong contrasts (continental climate at high latitudes, wide arid zone in the subtropics, as in the Permian), while a fragmented continent, combined with vast shelf seas produced as a consequence of a high sea level, as in the early Upper Cretaceous, will be conducive to a climate with relatively low contrasts. The dispersion of landmasses therefore represents an important climate forcing, which does not, however, offer a complete explanation for the climate history of the Earth. Orogenesis is another major forcing of the climate at geological timescales. It is therefore essential to delineate the spatio-temporal evolution of all the orogens as accurately as possible (Fig. 22.5), even if the reconstructed paleoaltitudes remain, for the time being, subject to considerable uncertainty. While the impact of paleogeography on atmospheric circulation has been widely studied through numerical modeling, the impact of paleogeographic changes on ocean circulation is far from well understood. Studies have quantified the impact of the opening of seaways (Panama or Indonesian) or large-scale ocean connections (South Atlantic—Central Atlantic) on the climate through the redistribution of heat carried by ocean currents. The chronology of the openings and closures of the ocean passages, in addition to that of the orogens, is thus essential to a good understanding of the climate system at the scale of geological time. In this framework, the development of new isotopic tools, the improvements of traditional proxies calibration and the rise of paleoclimate modelling should allow a better quantification of the links between paleogeography and climate in the next decades.

References

Bacon, C. D., Silvestro, D., Jaramillo, C., Smith, B. T., Chakrabarty, P., & Antonelli, A. (2015). Biological evidence supports an early and complex emergence of the Isthmus of Panama. *Proceedings of the National Academy of Sciences, 112*, 6110–6115.

Bahr, A., Kolber, G., Kaboth-Bahr, S., Reinhardt, L., Friedrich, O., Pross, J. (2020). Mega-monsoon variability during the late Triassic: Re-assessing the role of orbital forcing in the deposition of playa sediments in the Germanic Basin. *Sedimentology, 67*, 951–970. https://doi.org/10.1111/sed.12668.

Berner, R. A. (2001). GEOCARB III: A revised model of atmospheric CO_2 over Phanerozoic time. *American Journal of Science, 301*(2), 182–204. https://doi.org/10.2475/ajs.301.2.182.

Biggin, A. J., de Wit, M. J., Langereis, C. G., Zegers, T. E., Voûte, S., Dekkers, M. J., & Drost, K. (2011). Palaeomagnetism of Archaean rocks of the Onverwacht Group, Barberton Greenstone Belt (southern Africa): Evidence for a stable and potentially reversing geomagnetic field at ca. 3.5 Ga. *Earth and Planetary Science Letters, 302*, 314–328.

Bolin, B. (1950). On the influence of the Earth's orography on the general character of the westerlies. *Tellus, 2*(3). http://www.tellusa.net/index.php/tellusa/article/download/8547/9993.

Bond, D. P. G., Hilton, J., Wignall, P. B., Ali, J. R., Stevens, L. G., Sun, Y., and Lai, X. (2010). The Middle Permian (Capitanian) mass extinction on land and in the oceans. *Earth-Science Reviews, 102*, 100–116, https://doi.org/10.1016/j.earscirev.2010.07.004.

Bond, D. P. G. & Wignall, P. B. (2014). Large igneous provinces and mass extinctions: An update. In G. Keller, & A. C. Kerr (Eds.), *Volcanism, impacts, and mass extinctions: Causes and effects: Geological society of America special paper, 505* (p. 29). https://doi.org/10.1130/2014.2505(02).

Bonifacie, M., et al. (2017). Calibration of the dolomite clumped isotope thermometer from 25 to 350 °C, and implications for a universal calibration for all (Ca, Mg, Fe) CO3 carbonates. *Geochimica et Cosmochimica Acta, 200*(1), 255–279.

Bonnefille, R. (2010). Cenozoic vegetation, climate changes and hominid evolution in tropical Africa. *Global and Planetary Change, 72*, 390–411.

Boos, W. R. (2015). A review of recent progress on Tibet's role in the South Asian monsoon. *CLIVAR Exchanges, 19*(23), 27.

Bornemann, A., Norris, R. D., Friedrich, O., Beckmann, B., Schouten, S., et al. (2008). Isotopic evidence for glaciation during the Cretaceous supergreenhouse. *Science, 319*, 189–192.

Brierley, C. M., & Fedorov, A. V. (2016). Comparing the impacts of Miocene–Pliocene changes in Inter-Ocean gateways on climate: Central American Seaway, Bering Strait, and Indonesia. *Earth and Planetary Science Letters, 444*(juin), 116–130. https://doi.org/10.1016/j.epsl.2016.03.010.

Broccoli, A. J., & Manabe, S. (1992). The effects of orography on Mid-latitude Northern Hemisphere Dry Climates. *Journal of Climate, 5*(11), 1181–1201. https://doi.org/10.1175/1520-0442(1992)005.

Cane, M. A., & Molnar, P. (2001). Closing of the Indonesian Seaway as a Precursor to East African aridification around 3–4 million years ago. *Nature, 411*, 157–162.

Caquineau, T., Paquette, J.-L., & Philippot, P. (2018). U-Pb detrital zircon geochronology of the Turee Creek Group, Hamersley Basin, Western Australia: Timing and correlation of the Paleoproterozoic glaciations. *Precambrian Research, 307*, 34–50. https://doi.org/10.1016/j.precamres.2018.01.003.

Chaboureau, A. C., Sepulchre, P., Donnadieu, Y., & Franc, A. (2014). Tectonic-driven climate change and the diversification of angiosperms. *Proceedings of the National Academy of Sciences, 111*(39), 14066–14070. https://doi.org/10.1073/pnas.1324002111.

Charney, J. G., & Eliassen, A. (1949). A numerical method for predicting the perturbations of the Middle Latitude Westerlies. *Tellus, 1*(2): 38–54. https://doi.org/10.1111/j.2153-3490.1949.tb01258.x.

Chen, G.-S., Liu, Z., & Kutzbach, J. E. (2014). Reexamining the barrier effect of the Tibetan Plateau on the South Asian Summer Monsoon. *Climate of the Past, 10*(3),1269–1275. https://doi.org/10.5194/cp-10-1269-2014.

Cogné, J. P., & Humler, E. (2006). Trends and rhythms in global seafloor generation rate. *Geochemistry, Geophysics, Geosystems, 7,* Q03011. https://doi.org/10.1029/2005GC001148.

Copper, P., & Scotese, C. R. (2003). Megareefs in Middle Devonain super greenhouse climates. *Special Publications - Geological Society of America, 370*, 209–230.

Daëron, M., Drysdale, R. N., Peral, M., Huyghe, D., Blamart, D., Coplen, T. B., et al. (2019). Most Earth-surface calcites precipitate out of isotopic equilibrium. *Nature Communications, 10*(1), 429. https://doi.org/10.1038/s41467-019-08336-5.

De Wit, M. J., & Furnes, H. (2016). 3.5-Ga hydrothermal fields and diamictites in the Barberton Greenstone Belt—Paleoarchean crust in cold environments. Science Advances, 2, e1500368. https://doi.org/10.1126/sciadv.1500368.

Dodd, M. S., Papineau, D., Grenne, T., Slack, J. F., Rittner, M., Pirajno, F., et al. (2017). Evidence for early life in Earth's oldest hydrothermal vent precipitates. *Nature, 543*(7643), 60–64. https://doi.org/10.1038/nature21377.

Donnadieu, Y., Goddéris, Y., Ramstein, G., Nédélec, A., Meert, J. (2004). A 'snowball Earth'climate triggered by continental break-up through changes in runoff.*Nature, 428*(6980), 303–306.

Donnadieu, Y., Pucéat, E., Moiroud, M., Guillocheau, F., & Deconinc,. J. F. (2016). A better-ventilated ocean triggered by late Cretaceous changes in continental configuration. *Nature Communications, 7* (janvier), 10316. https://doi.org/10.1038/ncomms10316.

Dromart, G., Garcia, J.-P., Picard, S., Atrops, F., Lécuyer, C., & Sheppard, S. M. F. (2003). Ice Age at the Middle–Late Jurassic Transition? *Earth and Planetary Science Letters, 213*(3–4), 205–220. https://doi.org/10.1016/S0012-821X(03)00287-5.

Dubiel, R. F., Parrish, J. T., Parrish, J. M., & Good, S. C. (1991). The Pangaean megamonsoon: Evidence from the Upper Triassic Chinle Formation, Colorado Plateau. *Palaios, 6*, 347–370.

Eiler, J. M. (2007). "Clumped-isotope" geochemistry—the study of naturally-occurring, multiply-substituted isotopologues. *Earth and Planetary Science Letters, 262*, 309–327.

Erwin, D. H. (1994). The Permo-Triassic extinction. *Nature, 367*, 231–236. *https://doi.org/10.1038/367231a0*.

Flohn, H. (1950). Neue Anschauungen über die allgemeine Zirkulation der Atmosphäre und ihre klimatische Bedeutung. *Erdkunde, 4*(3/4), 141–162.

Fluteau, F., Besse, J., Broutin, J., Ramstein, G. (2001). The Late Permian climate. What can be inferred from climate modelling concerning Pangea scenarios and Hercynian range altitude? *Palaeogeography, Palaeoclimatology, Palaeoecology, 167,* 39–71.

Fluteau, F., Ramstein, G., Besse J. (1999). Simulating the evolution of the Asian and African Monsoons during the past 30 Myr using an atmospheric general circulation model. *Journal of Geophysical Research, 104*(D10), 11995–12018.

Fluteau, F., et al. (2006). The impacts of the Paleogeography and sea level changes on the Mid Cretaceous climate. *Palaeogeography, Palaeoclimatology, Palaeoecology, 247*(3–4), 357–381. https://doi.org/10.1016/j.palaeo.2006.11.016

Fluteau, F., Ramstein, G., Besse, J., Guiraud, R., & Masse. J. P. (2007). Impacts of Palaeogeography and sea level changes on the Mid Cretaceous climate. *Palaeogeography, Palaeoclimatology, Palaeoecology, 247,* 357–381.

Ghienne, J-F., Moreau, J., Degermann, L. et al. (2013). Lower Palaeozoic unconformities in an intracratonic platform setting: Glacial erosion versus tectonics in the eastern Murzuq Basin (southern Libya). *International Journal of Earth Sciences, 102*(2), 455–482.

Goddéris, Y., Donnadieu, Y., Carretier, S., Aretz, M., Dera, G., Macouin, M., & Regard, V. (2017). Onset and ending of the late Palaeozoic ice age triggered by tectonically paced rock weathering. *Nature Geoscience, 10*(5): 382–386. https://doi.org/10.1038/ngeo2931.

Granot, R., & Dyment, J. (2015). The Cretaceous opening of the South Atlantic Ocean. *Earth and Planetary Science Letters, 414*(mars), 156–163. https://doi.org/10.1016/j.epsl.2015.01.015.

Haq, B. U., et al. (1987). Chronology of fluctuating sea levels since the Triassic (250 million years ago to present). *Science, 235,* 1156–1166.

Haqq-Misra, J. D., et al. (2008). A revised, Hazy methane greenhouse for the Archean Earth. *Astrobiology, 8,* 1127–1137.

Haug, G. H., & Tiedemann, R. (1998). Effect of the formation of the Isthmus of Panama on Atlantic Ocean Thermohaline circulation. *Nature, 393,* 673–676.

Hoareau, G., Bomou, B., van Hinsbergen, D. J. J., Carry, N., Marquer, D. et al. (2015). Did high Neo-Tethys subduction rates contribute to early Cenozoic warming? *Climate of the Past, 11*(12), 1751–1767.

Hoffman, et al. (2017). Snowball Earth climate dynamics and Cryogenian geology-geobiology. *Science Advances, 3,* e1600983.

Hren, M. T., Sheldon, N. D., Grimes, S. T., Collinson, M. E, Hooker, J. J., Bugler, M., Lohmann, K. C. (2013). Terrestrial cooling in Northern Europe during the Eocene–Oligocene transition, *PNAS, 110,* 7562–7567.

Hu, S., & Boos, W. R. (2017). Competing effects of surface Albedo and Orographic elevated heating on regional climate: Albedo-Elevation compensation. *Geophysical Research Letters, 4*(13), 6966–6973. https://doi.org/10.1002/2016GL072441.

Jaramillo, C. (2018). Evolution of the Isthmus of Panama: Biological, Paleoceanographic and Paleoclimatological implications. In C. Hoorn, A. Perrigo, & A. Antonelli (Eds.), *Mountains, climate and biodiversity* (1st ed.). Wiley.

Jung, G., Prange, M., & Schulz, M. (2014). Uplift of Africa as a potential cause for Neogene Intensification of the Benguela upwelling system. *Nature Geoscience, 7*(10), 741–747. https://doi.org/10.1038/ngeo2249.

Kasting, J., & Howard, M. T. (2006). Atmospheric composition and climate on the early Earth. *Philosophical Transactions of the Royal Society B, 361,* 1733–1742.

Kennett, J. P. (1977). Cenozoic evolution of Antarctic Glaciation, the Circum-Antarctic Ocean, and their impact on global Paleoceanography. *Journal of Geophysical Research, 82*(27), 3843–3860. https://doi.org/10.1029/JC082i027p03843.

Kiehl, J. T., & Shields, C. A. (2005). Climate simulation of the latest Permian: Implications for mass extinction. *Geology, 33*(9), 757–760. https://doi.org/10.1130/G21654.1.

Kutzbach, J. E., Prell, W. L., & Ruddiman, W. F. (1993). Sensitivity of Eurasian climate to surface uplift of the Tibetan Plateau. *The Journal of Geology, 101*(2), 177–190. https://doi.org/10.1086/648215.

Knauth, L. P., & Lowe, D. R. (2003). High Archean climatic temperature inferred from oxygen isotope geochemistry of cherts in the 3.5 Ga Swaziland Supergroup, South Africa. *Geological Society of America Bulletin, 115,* 566–580 (2003).

Ladant, J. B., Donnadieu, Y. (2016). Palaeogeographic regulation of glacial events during the Cretaceous supergreenhous. *Nature communications, 7*, Article number: 12771.

Lefebvre, V., Donnadieu, Y., Sepulchre, P., Swingedouw, D., & Zhang, Z.-S. (2012). Deciphering the role of Southern Gateways and carbon dioxide on the onset of the Antarctic circumpolar current. *Paleoceanography, 27*(4), n/a–n/a. https://doi.org/10.1029/2012PA002345.

Le Hir, G., Donnadieu, Y., Goddéris, Y., Meyer-Berthaud, B., Ramstein, G., & Blakey, R. C. (2011). The climate change caused by the land plant invasion in the Devonian. *Earth and Planetary Science Letters, 310*(3), 203–212. https://doi.org/10.1016/j.epsl.2011.08.042.

Lowe, D. R., & Tice, M. M. (2004). Geologic evidence for Archean atmospheric and climatic evolution: Fluctuating levels of CO_2, CH_4, and O_2 with an overriding tectonic control. *Geology, 32*, 493–496.

Lunt, D. J., et al. (2008). Closure of the Panama Seaway during the Pliocene: Implications for climate and Northern Hemisphere Glaciation. *Climate Dynamics, 3*, 1–18.

Maffre, P., Ladant, J. B., Donnadieu, Y., Sepulchre, P., & Goddéris, Y. (2018). The influence of Orography on modern ocean circulation. *Climate Dynamics, 50*(3–4), 1277–1289. https://doi.org/10.1007/s00382-017-3683-0.

Matte, P. (1986). Tectonics and plate Tectonics model for the Variscan Belt of Europe. *Tectonophysics, 126*, 329–374.

Meyer-Berthaud, B., Scheckler, S., Wendt, J. (1999). Archaeopteris is the earliest known modern tree. *Nature, 398*, 700–701.

Molnar, P., Boos, W. R., & Battisti, D. S. (2010). Orographic controls on climate and Paleoclimate of Asia: Thermal and mechanical roles for the Tibetan Plateau. *Annual Review of Earth and Planetary Sciences, 38* (1), 77–102. https://doi.org/10.1146/annurev-earth-040809-152456.

Montañez, I., & Poulsen, C. (2013). The late Paleozoic ice age: An evolving paradigm. *Annual Review of Earth and Planetary Sciences, 41*, 24.1–24.28.

Montañez, I. P., McElwain, J. C., Poulsen, C. J., White, J. D., DiMichele, W. A., Wilson, J. P., Griggs, G., & Hren, M. T. (2016). Climate, pCO_2 and terrestrial carbon cycle linkages during late Palaeozoic glacial–interglacial cycles. *Nature Geoscience, 9*(11), 824–828. https://doi.org/10.1038/ngeo2822.

Montes, C., Cardona, A., Jaramillo, C., Pardo, A., Silva, J. C., Valencia, V., et al. (2015). Middle Miocene closure of the Central American Seaway. *Science, 348*(6231), 226–229. https://doi.org/10.1126/science.aaa2815.

Murphy, D. P., & Thomas, D. J. (2013). The evolution of late Cretaceous deep-ocean circulation in the Atlantic Basins: Neodymium isotope evidence from South Atlantic Drill Sites for Tectonic controls: Cretaceous Atlantic deep circulation. *Geochemistry, Geophysics, Geosystems, 14*(12), 5323–5340. https://doi.org/10.1002/2013GC004889.

O'Dea, A., Lessios, H. A., Coates, A. G., Eytan, R. I., Restrepo-Moreno, S. A., Cione, A. L., et al. (2016). Formation of the Isthmus of Panama. *Science Advances, 2*(8), e1600883. https://doi.org/10.1126/sciadv.1600883.

Potter, S. F., Dawson, E. J., & Frierson, D. M. W. (2017). Southern African Orography impacts on low clouds and the Atlantic ITCZ in a coupled model: African Orography impacts on the ITCZ. *Geophysical Research Letters, 44*(7), 3283–389. https://doi.org/10.1002/2017GL073098.

Poulsen, C. J., Ehlers, T. A., & Insel, N. (2010). Onset of convective rainfall during gradual late Miocene rise of the Central Andes. *Science, 328*(5977): 490–493. https://doi.org/10.1126/science.1185078.

Poulsen, C. J., Gendaszek, A. S., & Jacob, R. L. (2003). Did the rifting of the Atlantic Ocean Cause the Cretaceous thermal maximum? *Geology, 31*(2), 115–118. https://doi.org/10.1130/0091-7613(2003)031%3c0115:DTROTA%3e2.0.CO;2.

Ramstein, G., Fluteau, F., Besse, J., Joussaume, S. (1997). Effects of orogeny, sea-level change and tectonic drift on the monsoon over the past 30 millions years. *Nature, 386*, 788–795.

Rind, D. G., Russell, G., & Ruddiman, W. F. (1997). The effects of uplift on ocean-atmosphere circulation. In W. F. Ruddiman (Ed.), *Tectonic uplift and climate change* (pp. 123–147). New York: Plenum.

Robert, F., & Chaussidon, M. (2006). A palaeotemperature curve for the Precambrian oceans based on silicon isotopes in cherts. *Nature, 443*, 969–972.

Ruddiman, W. F., et al. (1997). The uplift-climate connection: A synthesis. In W. F. Ruddiman (Ed.), *Tectonic uplift and climate change* (pp. 471–515). New York: Plenum.

Sepulchre, P., Arsouze, T., Donnadieu, Y., Dutay, J-C., et al. (2014). Consequences of shoaling of the Central American Seaway determined from modeling Nd isotopes. *Paleoceanography, 29*(3), 176–189.

Sepulchre, P., Ramstein, G., Fluteau, F., Schuster, M., Tiercelin, J. J., & Brunet, M. (2006). Tectonic Uplift and Eastern Africa Aridification. *Science, 313*(5792), 1419–1423. https://doi.org/10.1126/science.1129158.

Sepulchre, P., Sloan, L. C., Snyder, M., & Fiechter, J. (2009). Impacts of Andean Uplift on the Humboldt current system: A climate model sensitivity study. *Paleoceanography, 24*(4). https://doi.org/10.1029/2008PA001668.

Shields, C. A., & Kiehl, J. T. (2018). Monsoonal precipitation in the Paleo-Tethys warm pool during the latest Permian. *Palaeogeography, Palaeoclimatology, Palaeoecology, 491*(février), 123–136. https://doi.org/10.1016/j.palaeo.2017.12.001.

Sijp, W. P., von der Heydt, A. S. Dijkstra, H. A. Flögel, S., Douglas, P. M. J., & Bijl, P. K. (2014). The role of ocean gateways on cooling climate on long time scales. *Global and Planetary Change, 119*(août), 1–22. https://doi.org/10.1016/j.gloplacha.2014.04.004.

Su, B., Jiang, D., Zhang, R., Sepulchre, P., & Ramstein, G. (2018). Difference between the North Atlantic and Pacific Meridional overturning circulation in response to the uplift of the Tibetan Plateau. *Climate of the Past, 14*(6), 751–762. https://doi.org/10.5194/cp-14-751-2018.

Tan, N., Ramstein, G., Dumas, C., Contoux, C., Ladant, J. B., Sepulchre, P., Zhang, Z., & De Schepper, S. (2017). Exploring the MIS M2 glaciation occurring during a warm and high atmospheric CO_2 Pliocene background climate. *Earth and Planetary Science Letters, 472*, 266–276.

Tartèse, R., Chaussidon, M., Gurenko, A., Delarue, F., & Robert, F. (2017). Warm Archaean Oceans reconstructed from oxygen isotope composition of early-life Remnants. *Geochemical Perspectives Letters*, 55–65. https://doi.org/10.7185/geochemlet.1706.

Valley, J. W. (2006). Early Earth. *Elements 2*(4), 201–204. https://doi-org.insu.bib.cnrs.fr/10.2113/gselements.2.4.201.

Veizer, J., & Prokoph, (2015). Temperatures and oxygen isotopic composition of Phanerozoic oceans. *Earth-Science Reviews, 146*, 92–104. https://doi.org/10.1016/j.earscirev.2015.03.008.

Warner, T. (2004). *Desert Meteorology*. Cambridge: Cambridge University Press. https://doi.org/10.1017/CBO9780511535789.

Young, G. M., Von Brunn, V., Gold, D. J. C., & Minter, W. E. L. (1998). Earth's oldest reported Glaciation: Physical and chemical evidence from the Archean Mozaan Group (2.9 Ga) of South Africa. *Journal of Geology, 106*, 523–538.

Zachos, J., et al. (2001). Trends, rhythms, and aberrations in global climate 65 Ma to present. *Science, 292*, 686–693.

Zerkle, A. L., Claire, M. W., Domagal-Goldman, S. D., Farquhar J., & Poulton, S. W. (2012). A bistable organic-rich atmosphere on the

Neoarchaean Earth. *Nature Geoscience, 5*, 359–363. https://doi.org/10.1038/ngeo1425.

Zhang, R., Jiang, D., Ramstein, G., Zhang, Z., Lippert, P. C. & Yu, E. (2018). Changes in Tibetan Plateau Latitude as an Important factor for understanding East Asian climate since the Eocene: A modeling study. *Earth and Planetary Science Letters, 484*(février), 295–308. https://doi.org/10.1016/j.epsl.2017.12.034.

Zhang, R., Jiang, D., Zhang, Z., & Yu, E. (2015). The impact of regional uplift of the Tibetan Plateau on the Asian Monsoon climate. *Palaeogeography, Palaeoclimatology, Palaeoecology, 417* (janvier), 137–150. https://doi.org/10.1016/j.palaeo.2014.10.030.

Zhang, X., Prange, M., Steph, S., Butzin, M., Krebs, U., Lunt, D. J., et al. (2012). Changes in equatorial Pacific thermocline depth in response to Panamanian seaway closure: Insights from a multi-model study. *Earth and Planetary Science Letters, 317–318*, 76–84.

Biogeochemical Cycles and Aerosols Over the Last Million Years

Nathaelle Bouttes, Laurent Bopp, Samuel Albani, Gilles Ramstein, Tristan Vadsaria, and Emilie Capron

Introduction

The biogeochemical cycles encompass the exchange of chemical elements between reservoirs such as the atmosphere, ocean, land and lithosphere. Those exchanges involve biological, geological and chemical processes, hence the term "biogeochemical cycles". A widely known cycle (which is not a biogeochemical cycle) is the water cycle, whose impact on climate is of major importance and which has been described in Volume 1, Chap. 1. Similarly, chemical elements such as carbon, nitrogen, oxygen and sulphur are also exchanged during cycles. Some of these elements can significantly impact climate through their effect on the atmospheric energy budget when they are in gaseous form (CO_2, CH_4, N_2O). The biogeochemical cycles are also affected by changes in climate, constituting a feedback in the Earth system. Other chemical compounds present in the atmosphere also influence the amount of energy available at the surface and therefore the dynamics of the climate: these are aerosols, small liquid or solid particles in suspension in the air. Because of the effect of climate on these chemical elements and particles, they sometimes record the changes that modified their cycle. It is possible to measure many of these tracers, which provide a valuable insight into past climate changes.

The link between the composition of the atmosphere and climate was discovered in the nineteenth century through the work of Jean-Baptiste Fourier, a French mathematician. He showed that the Earth would be much colder than it currently is, if it was heated by incoming solar radiation alone. To explain the additional heating, he proposed, among other possibilities, that the Earth is insulated by gases present in the atmosphere. This effect of atmospheric gases blocking some of the infrared radiation emitted by the planet is now known as the greenhouse effect. Later, in 1860, John Tyndall, an English chemist, demonstrated that the two main constituents of our atmosphere, dinitrogen and dioxygen, are transparent to infrared radiation and therefore do not contribute to the greenhouse effect. On the other hand, he identified water vapour (H_2O) and carbon dioxide (CO_2) as the two main greenhouse gases in our atmosphere. This led the way for another chemist, Svante Arrhenius, who, in 1896, was the first to estimate the change in the average temperature of the Earth's surface triggered by a change in the concentration of CO_2 in the atmosphere. Lastly, it was only in the second half of the twentieth century that advances in measurement techniques made it possible to measure the impacts of other gases such as methane (CH_4), nitrous oxide (N_2O), ozone (O_3), and chlorofluorocarbons (CFCs), on the greenhouse effect.

The direct effect of aerosols on climate was first suggested by Benjamin Franklin in the eighteenth century to explain the cold winter of 1783–84. Benjamin Franklin noted there was a "constant fog over all Europe, and a great part of North America" which resulted in colder conditions. He suggested volcanic eruptions in Iceland as a possible explanation for this fog. Since then, the big volcanic eruptions that have occurred in the nineteenth and twentieth centuries, such as the Krakatoa (1883), Santa Maria (1902),

Katmai (1912), Agung (1963), El Chichon (1982) and Pinatubo (1991) have provided material to better understand the effects of aerosols that have recently been incorporated into climate models. The indirect effects of aerosols by modifying clouds were discovered more recently. The presence of aerosols can modify cloud characteristics by making them more reflective, or by extending their lifetime before precipitation, for example.

These chemical elements and aerosols are closely linked to climate and climate changes. Their concentration in the atmosphere impacts the Earth's energy budget, either directly or indirectly, while changes in climate modify the exchanges between reservoirs of these compounds and, in fine, their concentration in the atmosphere. Some atmospheric gases have the ability to modify the energy budget of the Earth (Volume 1, Chap. 1). The Earth receives shortwave radiation (ultraviolet, visible, and near-infrared) from the sun. Part of this radiation is reflected back by the surface, clouds and the atmosphere, part of it is absorbed by the atmosphere and clouds, and the last part is absorbed by the surface of the Earth. The Earth's surface emits longwave radiation (infrared) because it is colder than the sun, and also transfers energy to the atmosphere by latent and sensible heat. The longwave radiation from the Earth's surface is partly absorbed by the atmospheric greenhouse gases, which then re-emit radiation in all directions, including towards the surface of the Earth. The latter is then heated, resulting in a higher temperature than on an Earth without greenhouse gases. The radiation absorbed by the gases depends on their properties and in which zones they absorb radiation. Among the gases present in the atmosphere, the main greenhouse gases are, in decreasing order (excluding water vapor) carbon dioxide (CO_2), methane (CH_4) and nitrous oxide (N_2O).

Aerosols have two effects on the energy budget in the atmosphere: direct and indirect. Aerosols are tiny particles—such as sea salt, dust from deserts and fires—in suspension in the atmosphere, either in liquid or solid form. They can absorb and disperse solar radiation, as well as absorb and emit thermal radiation. This is the direct effect. Aerosols also form cloud condensation nuclei and ice nuclei: raindrops and ice develop around these nuclei. This is the indirect effect because it modifies the energy budget through the modification of the microphysics of clouds. Depending on the size of the drops, which is dependent on the type and size of nuclei, the clouds will reflect or absorb radiation. In addition, aerosols deposited back to the surface will alter the amount of solar radiation reflected back to space, and will disperse chemical elements that can influence various biogeochemical cycles.

Some of the changes impacting on the biogeochemical cycles and aerosols are recorded and preserved for thousands of years. The main natural archives that have been used to track such changes are sediment cores extracted from oceans or lakes, and ice cores, drilled from polar ice sheets (Fig. 23.1).

Marine sediment cores, collected from the bottom of the ocean, include various organic and inorganic elements, which can be used to get direct information. For example, examining the type of plankton that lived in a region at a given time can tell us how cold it was. A succession of species that thrive in warm or cold environments will indicate a succession of warm and cold periods. An analysis of the pollen present in sediment will yield information on the proportion of the major plants that lived on the nearby continent. In addition, the material is measured to obtain the ratio of chemical elements such as Pa/Th, and the ratio of isotopes such as oxygen and carbon isotopes ($\delta^{18}O$, $\delta^{13}C$, $\Delta^{14}C$), which can be used as indicators of specific processes such as ocean circulation changes, temperature changes, terrestrial biosphere changes, etc. (see Volume 1, Chap. 21).

The idea that ice from ice sheets could be used to provide information on past changes originated in the 1950s with the work of Willi Dansgaard and others, who hypothesized that the link between temperature and the number of heavy oxygen isotopes in precipitation could be applied to old ice to reconstruct past temperature changes. Ice core drilling began in Antarctica, Alaska and Greenland in the 1950s, but these cores were around 100 m deep and the recovery quality was low. Drilling to extract ice cores was spurred by the International Geophysical Year (1957–1958) and longer ice cores were drilled in Greenland at Site 2 (1956–1957) and in Antarctica at Byrd station (1957–1958) and at Little America V (1958–1959). Many more ice cores have been drilled since then, mainly in Greenland and Antarctica. Past climate and environmental changes are recorded both in the ice and in air bubbles trapped within the ice of the ice cores. For instance, the proportion of hydrogen and oxygen isotopes in the ice provide information on past surface temperatures, while the concentration of greenhouse gases can be directly measured in the air bubbles.

To understand the changes recorded in climate archives and to test various hypotheses on feedback mechanisms, more and more climate models now include biogeochemical cycles, and sometimes isotopes (see Volume 2, Chaps. 25 and 29). Additional mechanisms and elements are added progressively so that simulations and measured data can be compared directly. This continuous comparison helps to increase our knowledge of the climate system resulting in improved models that can be used to evaluate possible future changes. These coupled carbon-climate models are valuable tools to help understand past changes and increase our confidence in future climate projections.

In this chapter we describe the main biogeochemical cycles interacting with the climate: carbon (CO_2 and CH_4),

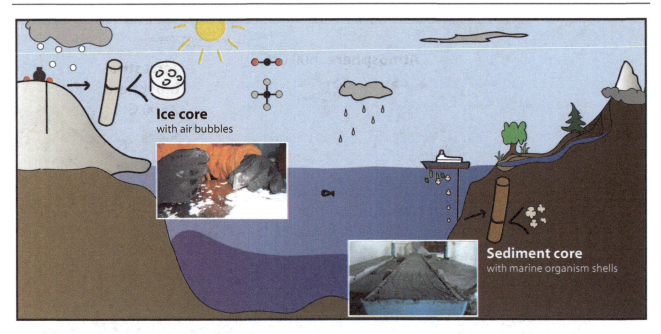

Fig. 23.1 Diagram of the Earth system and schemes showing ice cores and sediment cores extraction

nitrogen, oxygen, sulphur, as well as aerosols. We review existing records from marine sediment cores and ice cores and highlight the current knowns and unknowns.

Carbon Cycle

Two atmospheric gases containing carbon play a major role in the interactions between climate and biogeochemistry: carbon dioxide (CO_2) and methane (CH_4). While CO_2 is the most famous and most studied element, CH_4 is far from negligible and has a rather different cycle. We will first focus on CO_2, then on CH_4, before reviewing past changes.

Carbon Dioxide (CO_2)

On the timescale of a few hundred to a few tens of thousands of years, the main carbon reservoirs are the ocean (including sediments), atmosphere and land (including terrestrial biosphere and permafrost) (Fig. 23.2). On the longer timescale of a few million years the lithosphere (rocks) starts to play a major role (see Volume 1, Chap. 2, Volume 2, Chaps. 22 and 26). Exchanges between these reservoirs involve biotic processes, due to biological activity such as photosynthesis by plants, and abiotic processes.

In the atmosphere, CO_2 is the main form of carbon and is relatively well mixed over a year. Its concentration is currently increasing due to anthropogenic activity and, at the beginning of 2018, the atmospheric CO_2 concentration was over 400 ppm (408 in March 2018, NOAA, https://www.esrl.noaa.gov/gmd/ccgg/trends/monthly.html). At the beginning of the pre-industrial era, it was around 280 ppm, corresponding to a carbon stock of around 600 GtC.

On land, atmospheric CO_2 is taken up by plants during photosynthesis, turning CO_2 into organic carbon. The living biomass is then converted into dead organic carbon matter in litter and soils. Organic carbon is progressively returned to the atmosphere by autotrophic and heterotrophic respiration. Currently, carbon storage in the terrestrial biosphere is around 2500 GtC. If the conditions become cold enough, the soil freezes, locking the carbon into permafrost, i.e. frozen soil. During warming, the permafrost thaws, and carbon is returned to the atmosphere. The current estimate of carbon stored in permafrost is $\sim$1700 GtC (Tarnocai et al. 2009), making it the single largest component of the terrestrial carbon pool. The fluxes of carbon between the atmosphere and land are around 120 GtC/year.

The ocean is the largest carbon reservoir (excluding the lithosphere) with around 38,000 GtC both in organic and inorganic forms. Carbon fluxes to the ocean come from the atmosphere through surface exchanges and from the continent from riverine inputs. In the ocean, dissolved carbon ($CO_{2(aq)}$) gets hydrated into H_2CO_3 (carbonic acid), which then gives bicarbonate ion (or hydrogen carbonate ion, HCO_3^-) and carbonate ion CO_3^{2-}. All these dissolved species are summed up in the term "dissolved inorganic carbon" (DIC). Because the concentration of H_2CO_3 is very small, it is included in $CO_{2(aq)}$. $CO_{2(aq)}$, HCO_3^- and CO_3^{2-} are in equilibrium following the chemical equations:

$$CO_{2(aq)} + H_2O \rightleftarrows HCO_3^- + H^+$$

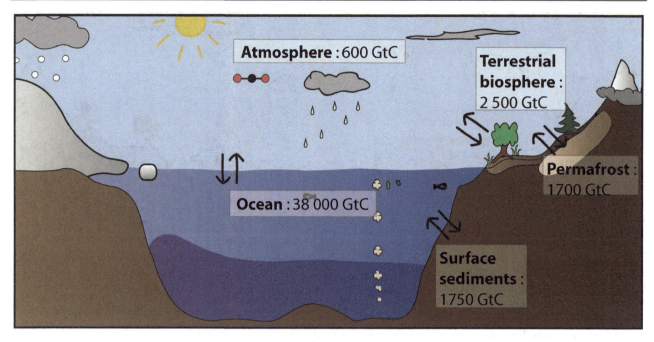

Fig. 23.2 Diagram of the short term carbon cycle (up to a few tens of thousands years)

$$HCO_3^- \rightleftarrows CO_3^{2-} + H^+$$

CO_2 is exchanged between the atmosphere and the surface ocean depending on the difference in partial pressure of CO_2 (pCO_2) between the atmosphere and ocean, on sea ice coverage and on wind. The pCO_2 in the ocean is governed by temperature and salinity. The flux of carbon between atmosphere and dissolved inorganic carbon is around 70 GtC/year in both directions.

In polar regions, CO_2 is more soluble in water because of the colder conditions, resulting in more dissolved inorganic carbon in the surface ocean. These regions are also major sites of deep convection, where surface water becomes denser due to colder and more saline conditions, and can sink to the ocean's depths. The carbon from the surface is thus transported to the deeper ocean. This uptake of carbon from the atmosphere to the ocean is called the solubility pump.

Marine phytoplankton uses dissolved CO_2 during photosynthesis, using nutrients and solar energy, following the simplified equation:

$$106\ CO_2 + 16\ NO_3^- + H_2PO_4^- + 17\ H_3O^+ + 105\ H_2O$$
$$+\ solar\ energy \to (CH_2O)_{106}(NH_3)_{16}(H_3PO_4) + 138\ O_2$$

The ratios C: N: P: O_2 are the Redfield ratios, and the inorganic carbon assimilated by biology constitutes the gross primary production. The difference between gross primary production and the carbon respired by phytoplankton is the net primary production. The phytoplankton that has synthetized organic carbon is then grazed by zooplankton. The depth to which sufficient light penetrates to sustain life is called the euphotic zone and extends to around 100–200 m deep. When plankton dies, it is partially remineralized at the surface, and the remainder sinks to the deeper ocean where it will be progressively remineralized. Carbon can then be found in two forms: particulate organic carbon or dissolved organic carbon. During remineralisation, carbon and nutrients are returned to the solution in dissolved form and can be used again in the euphotic zone for photosynthesis when it is brought back to the surface. Hence, biology also transfers carbon from the surface to the deep ocean leading to carbon being taken up from the atmosphere by the ocean: this is the biological pump. The flux of carbon between dissolved inorganic carbon and marine biology is around 50 GtC/year.

Marine biology not only produces organic carbon, but some organisms also create a shell made of calcium carbonate ($CaCO_3$). Most shells are made of calcite (coccolithophores and foraminifera) or aragonite (pteropods), two forms of calcium carbonate. The equation for calcium carbonate production is:

$$Ca^{2+} + CO_3^{2-} \to CaCO_3$$

In deep areas in the ocean where water is undersaturated with respect to calcium carbonate, the inverse equation takes place and the shells are progressively dissolved. If they reach the bottom of the ocean, they are buried in sediments. At the ocean surface, the production of calcium carbonate shells leads to a decrease in CO_3^{2-}, which increases CO_2 due to the following equation:

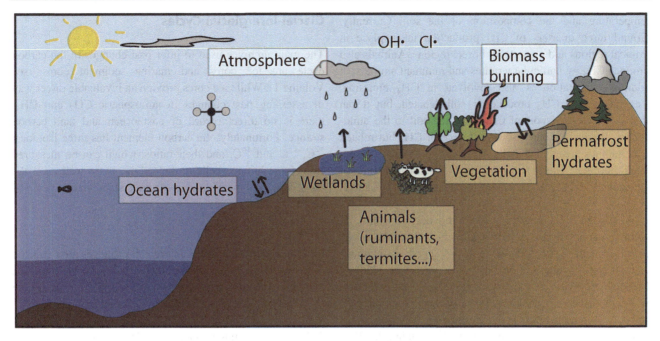

Fig. 23.3 Diagram of the methane cycle

$$CO_2 + CO_3^{2-} + H_2O \rightleftharpoons 2HCO_3^-$$

Hence this process, called the carbonate pump, counteracts the two carbon pumps described above.

Finally, ocean circulation impacts on the carbon cycle in the ocean by modifying the distribution of carbon and nutrients, the latter modifying primary production and thus the carbon distribution. Primary production is limited by several nutrients such as nitrate (N), phosphorus (P) and iron (Fe). The concentration of these nutrients mainly depends on transport from the nutrient-rich deeper ocean layers to the surface which is nutrient-depleted due to nutrient absorption by marine biology. Hence, in upwelling zones where lots of nutrients are brought to the surface primary production is high. When deep convection is active, it also provides important nutrient transport to the surface favouring production. In the opposite, in low latitude gyres nutrients are lacking and primary production is low. Some regions in the North Pacific and in the Southern Ocean display low productivity despite high nitrate concentrations (HNLC regions for high nitrate low chlorophyll). This is due to the lack of iron which limits production.

Methane (CH₄)

Methane is the second most important greenhouse gas after CO_2. Discovered in 1976 by Wang et al., its capacity to absorb infrared radiation is around 28 times more efficient than CO_2, over a time period of 100 years. Hence, although the concentration of CH_4 is 20 times lower than CO_2, it still plays a crucial role as a greenhouse gas, with a radiative forcing of around one third that of CO_2. The natural sinks and sources of CH_4 are different from those of CO_2, yielding a distinctly different—although often with some common features—evolution over time as described in Sect. "Glacial-Interglacial Cycles".

In the atmosphere, CH_4, like CO_2, is well mixed over a year. However, while CO_2 stays around 100 years in the atmosphere, CH_4 has a shorter lifetime of around 9 years. This is because the main sink of CH_4 is in the atmosphere: CH_4 is oxidized by the hydroxyl radical, OH. Oxidation by OH, which is photochemically produced in the atmosphere, takes place mainly in the troposphere, but also, to a lesser extent, in the stratosphere, and depends on several parameters. First, it depends on the speed of the reaction with OH, hence on temperature. Second, it depends on the quantity of free OH, which itself, depends on other compounds reacting with OH such as volatile organic compounds (VOCs) and ozone. In addition, the reaction of CH_4 with OH produces the CH_4 feedback effect: if CH_4 decreases, OH increases, which in turn reduces even more the concentration of CH_4 (Prather 2007). Another smaller sink of CH_4 in the atmosphere is the reaction of CH_4 with chlorine gas.

Contrary to CO_2, the main natural sources of CH_4 are from the continents (Fig. 23.3). The main contributors are wetlands, areas saturated with water such as marshes and swamps. CH_4 is produced by microbes (methanogenic archaea) in anoxic conditions in wetlands. Locally, CH_4 production strongly depends on oxygen availability, temperature and the composition of the soil. Currently, around three quarters of CH_4 production takes place in tropical regions and a quarter in boreal regions. Animals also produce CH_4, in particular termites and ruminant animals. In addition, vegetation is also involved in CH_4 emission. Vegetation as a CH_4 producer is still debated, but it can impact on the transport of CH_4 from the soil to the atmosphere. Permafrost can also be a source of CH_4 but related uncertainties are high. Finally, CH_4 is also emitted by biomass burning. There is also a small continental sink of atmospheric CH_4: the oxidation by methanotrophic bacteria in soils.

A smaller CH_4 source is from the ocean, with CH_4 coming mainly from coastal regions. However, the ocean could contain large CH_4 quantities trapped as methane clathrate in sediments, mainly on continental shelves. Methane clathrate, or methane hydrate, is a compound in which methane is trapped in a crystal of water. Clathrates are stable at low temperature and high pressure. The size of this reservoir is poorly constrained and could contain 500–2500 GtC, a smaller amount than the very high quantities suggested in the 1970s (Milkov 2004).

While the atmospheric CH_4 mixing ratio impacts on climate by modifying the atmospheric radiative balance, climate impacts on the sources and sinks of methane. In particular, wetlands are very dependent on the hydrological cycle and microbial activity, and thus the emission of methane, is strongly dependent on temperature. In addition, in the atmosphere, methane oxidation by OH radicals is one of the most temperature-sensitive reactions.

Glacial-Interglacial Cycles

The main archives used to infer past changes in the carbon cycle are ice cores and marine sediment cores (see Volume 1). While ice cores provide an invaluable direct way of assessing past changes in atmospheric CO_2 and CH_4, there is no direct record of past ocean and land carbon storage. Fortunately, the carbon element has three isotopes: ^{12}C, ^{13}C and ^{14}C, and their ratios, which can be measured from foraminifera shells in the sediments, as well as in the air in ice cores, provide clues to understand changes in past carbon reservoirs.

Ice Cores

Ice cores recording glacial-interglacial changes are extracted from the Greenland ice sheet in the Northern Hemisphere, and the Antarctic ice sheet in the Southern Hemisphere (Volume 1, Chap. 9). The oldest ice obtained from Greenland is ~130,000 years old (NEEM Community Members 2013), much younger than the oldest ice from Antarctica which is ~800,000 years old (EPICA Community Members 2004). This age could be extended further and future expeditions are planned to find and drill Antarctic ice older than a million years (Dahl-Jensen 2018).

Ice cores provide a direct record of past atmospheric gas concentrations, such as CO_2 and CH_4, thanks to air bubbles trapped in ice (Fig. 23.4). The air is trapped only at the bottom of the firn, a 60–120 m permeable layer below the surface where snow progressively densifies into ice (see Volume 1, Chap. 9). It results in an age difference between

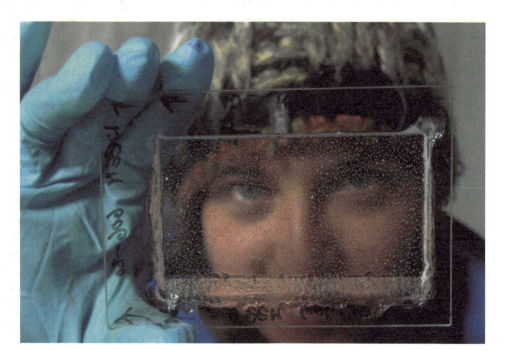

Fig. 23.4 Air bubbles trapped in ice taken from an ice core. *Credit* Sepp Kipfstuhl (Alfred Wegener Institute)

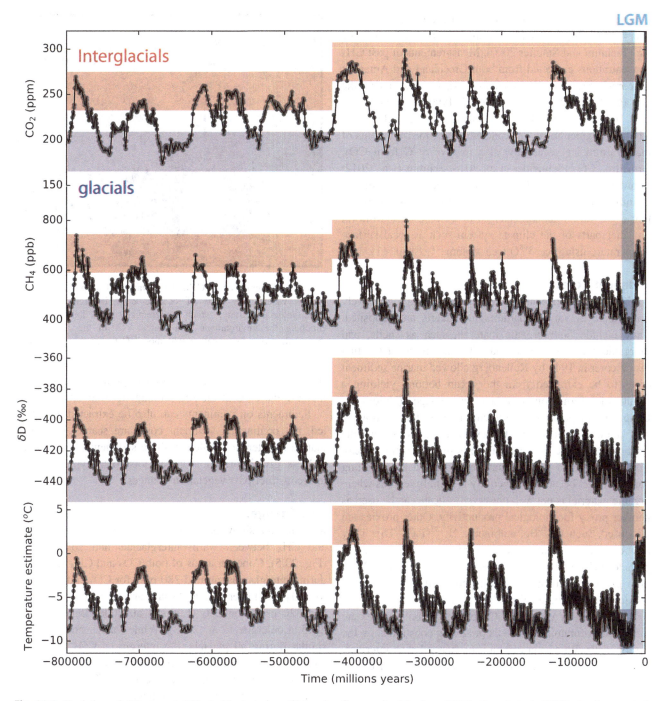

Fig. 23.5 Evolution of CO_2 (ppm), CH_4 (ppb), deuterium (‰) and temperature (°C) over the last 800,000 years. The Last Glacial Maximum (LGM) is indicated by the blue bar. Data from EPICA Community Members (2004), Jouzel et al. (2007), Loulergue et al. (2008), Bereiter et al. (2015)

the ice and the trapped air at a given depth level. This age difference that may be as much as several hundred years and even up to several thousands of years in the case of central Antarctic sites (Schwander and Stauffer 1984). Estimates of the age difference between the ice and the entrapped air are possible but there are associated uncertainties which are detailed in Volume 1, Chap. 8. The first reliable measurements of past CO_2 changes recorded in Antarctic ice were published in the 1980s, initially covering the last 30,000 years (Delmas et al. 1980), then the last 160,000 years (Barnola et al. 1987), and now as much as the last 800,000 years (Lüthi et al. 2008; Bereiter et al. 2015, Fig. 23.5). Note that only past atmospheric CO_2 concentrations determined from Antarctic ice cores are reliable, as

Greenland ice is prone to in situ production of CO_2, which alters the atmospheric CO_2 concentration already within the ice (Tschumi and Stauffer 2000). Measurements of past CH_4 concentrations extracted from both Greenland and Antarctic ice cores are reliable. While the first CH_4 records covered the last 160,000 years (Chappellaz et al. 1990), they also now date back to 800,000 years (Loulergue et al. 2008). In addition to greenhouse gas concentrations, measurements of the air from ice cores now also involve $\delta^{13}C$ from CO_2, termed $\delta^{13}CO_2$ (Schneider et al. 2013; Schmitt et al. 2012; Lourantou et al. 2010). Ice cores offer a unique possibility to quantify the sequence of events occurring between greenhouse gas variations and changes in ice core tracers inform on other parts of the climate system such as local surface temperature using ice $\delta^{18}O$ (see Volume 1, Chap. 11).

Sediment Cores

At the bottom of oceans and lakes, sediments progressively accumulate as various particles and debris are deposited. They include both organic material, such as shells, and inorganic material, such as clay. The invention of the first piston corer in 1947 by Kullenberg allowed marine sediment cores to be extracted from the ocean bottom, yielding a wealth of information on past ocean changes (Volume 1, Chap. 21). In particular, sediments provide information on past marine productivity, for example by measuring the fraction of organic material, calcite or opal (biogenic silica). The proportion of organic material measured in upwelling zones is large, because productivity is high. In regions where organic material is not well preserved, silicate can be used as another proxy for biological productivity. Other proxies are also used, such as ^{10}Be, authigenic U, $^{231}Pa/^{230}Th$). They rely on the fact that some elements preferentially fix to particles (Th) while others remain in solution (U, Pa). Their ratio gives an indication of past particle flux in the water column, hence biological productivity.

As detailed in Volume 1, Chap. 21, carbon isotopes are measured in foraminifera shells providing constraints on the carbon cycle. Whenever carbon is exchanged at an interface, fractionation takes place, which modifies $\delta^{13}C$ defined as:

$$\delta^{13}C = \left(\frac{\left(\frac{^{13}C}{^{12}C}\right)_{sample}}{\left(\frac{^{13}C}{^{12}C}\right)_{standard}} - 1 \right) \times 1000$$

The standard is the PDB (Peedee belemnite) carbon isotope standard, which corresponds approximately to average limestone (Craig 1957).

For example, biological activity preferentially uses the light ^{12}C over ^{13}C, so that plants or plankton are enriched in ^{12}C, and the environment (atmosphere for terrestrial biosphere, surface ocean for plankton) has higher $\delta^{13}C$ values

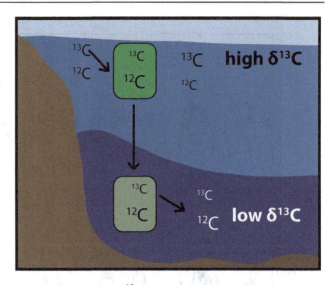

Fig. 23.6 Diagram of $\delta^{13}C$ in the surface and deep ocean. In the surface waters, the value of oceanic $\delta^{13}C$ is high because of photosynthetic activity, which preferentially uses light ^{12}C, thus enriching the environment with heavy ^{13}C. In the deep ocean, remineralisation releases carbon with more ^{12}C, therefore lowering $\delta^{13}C$

(Fig. 23.6). A more active biological productivity will thus increase the $\delta^{13}C$ in the surrounding environment.

Sediments on continents can also be extracted and studied, but continental sediment cores are scarcer than their marine counterpart. Pollen in these sediments, for example in loess, give information on past local plant types and help to reconstruct past vegetation changes.

Past Changes

Ice core data show large changes in concentrations of CO_2 and CH_4 between warm interglacials and cold glacials (Fig. 23.5). Concentrations of both CO_2 and CH_4 are higher during interglacials: around 280 ppm for CO_2 and 780 ppbv for CH_4, and lower during glacials: around 190 ppm for CO_2 and 320 ppbv for CH_4 (Lüthi et al. 2008; Bereiter et al. 2015; Loulergue et al. 2008). While values are relatively similar during all glacials, the interglacial CO_2 values are around 20 ppm lower for the older interglacials (before 430 ka) compared to the more recent ones. The concomitant records of climate and air composition demonstrate the strong link between climate and greenhouse gases, both for CO_2 (Barnola et al. 1987; Petit et al. 1999; Siegenthaler et al. 2005; Lüthi et al. 2008) and CH_4 (Chappellaz et al. 1990; Petit et al. 1999; Spahni et al. 2005; Loulergue et al. 2008). Recent research has shown that CO_2 and Antarctic temperature changed synchronously at the start of the two last deglaciations (within 200 years; Parrenin et al. 2013; Landais et al. 2013).

Several explanations have been put forward to explain the atmospheric CO_2 waxing and waning during glacial-interglacial cycles. Atmospheric CO_2 changes are driven by changes in carbon storage in other reservoirs, particularly on land and in the ocean.

On land, the lower sea level, by 120 m, during glacial time increased the surface area where plants could develop, although the larger ice sheets covering part of North America and Eurasia reduced the available area. Overall, due to the colder and drier climate, as well as reduced atmospheric CO_2 concentration, photosynthesis by land vegetation is reduced and the terrestrial biosphere tends to represent a smaller carbon reservoir during glacial period, which would increase atmospheric CO_2, not lower it. This is indicated by ocean $\delta^{13}C$ which decreased by around 0.03–0.04‰ during the Last Glacial Maximum compared to the pre-industrial level. This is explained by the transfer of continental carbon with low $\delta^{13}C$ values (due to fractionation during photosynthesis) to the ocean (Shackleton 1977; Bird et al. 1994), causing a reduction in continental carbon of around 270–720 GtC. An understanding of the changes in the terrestrial biosphere can also be obtained from paleo-biomes, using pollen from sediment cores (Adams et al. 1990; Crowley 1995; Maslin et al. 1995) which show a carbon loss of around 750–1350 GtC. Pollen data have also been used to reconstruct maps of vegetation types during the LGM (Fig. 23.7).

However, frozen soils, i.e. permafrost, may have increased in glacial times, potentially storing more carbon, which could partly explain the lower CO_2 (Ciais et al. 2012) and help resolve the $\delta^{13}CO_2$ signal recorded in ice cores, which strongly depends on land carbon changes (Crichton et al. 2016).

Although permafrost probably played a role, most of the change is likely to have come from the ocean, which is a much bigger carbon reservoir. In addition, $\delta^{13}C$ measurements from sediment cores indicate large changes in the ocean (Curry and Oppo 2005; Marchal and Curry 2008; Hesse et al. 2011). In the ocean, the carbon cycle changes could originate from modifications of biological activity and physical or chemical changes. Known processes include temperature change as colder temperatures lead to more carbon being stored in the ocean. The sea level drop of $\sim$120 m during glacial maxima results in higher concentrations of salinity, which causes a reduction in the storage capacity of carbon in the ocean, and an increase in nutrients, which increases biological activity and thus increases ocean carbon storage. However, these processes are not sufficient to fully explain the decrease in CO_2 decrease (see review by Sigman and Boyle 2000; Archer et al. 2000), and additional mechanisms are needed.

Four main hypotheses have been proposed to explain the CO_2 lowering by increased ocean carbon storage: increased biological pump, isolation of the ocean from the atmosphere due to sea ice coverage, changes in ocean dynamics, and carbonate compensation.

The biological pump causes more carbon to be stored when it is stimulated, for example when more nutrients are delivered to the ocean (Broecker and Peng 1982). In regions of high nutrients low chlorophyll (HNLC), biological activity is limited due to the lack of iron. An influx of iron to these zones during glacial periods would increase biological activity. Alternatively, the biological pump could also store more carbon if it becomes more efficient, for example with a greater carbon to nutrient ratio (Broecker and Peng 1982) or a switch of plankton species with higher productivity (Archer and Maier-Reimer 1994). However, both data and model simulations have shown that changes in biological activity are not sufficient to sufficiently account for the decrease in atmospheric CO_2 (Kohfeld et al. 2005; Bopp et al. 2003a; Tagliabue et al. 2009; Lambert et al. 2015).

Increased sea ice coverage has also been proposed, as this could isolate the ocean, preventing carbon from getting to the atmosphere, hence lowering atmospheric CO_2 (Stephens and Keeling 2000). But such an impact has only been simulated in very simple models, more complex models do not show such an effect on CO_2 (Archer et al. 2003).

Most current theories involve changes in ocean dynamics, and point to the Southern Ocean (Fischer et al. 2010). A larger ocean volume occupied by AABW, or slower overturning, could result in more carbon stored in the deep ocean, reducing atmospheric CO_2. Ocean circulation changes are supported by data generally indicating a reduced NADW and a more stratified Southern Ocean (Adkins 2013). In particular, $\delta^{13}C$ measurements show lower $\delta^{13}C$ values in the deep glacial ocean, especially around Antarctica, and higher values near the surface (Curry and Oppo 2005; Marchal and Curry 2008; Hesse et al. 2011). In addition, very salty water has been measured in the deep Southern Ocean (Adkins et al. 2002). Complementary data, such as from neodymium isotopes (Basak et al. 2018), B/Ca ratio (Yu et al. 2016) and $\Delta^{14}C$ (Skinner et al. 2010), also point towards changes in the circulation in the Southern Ocean.

Comparison of model simulations over the last decade have shown that models simulate a large range of ocean circulation changes, which are generally opposite to those deduced from data. In PMIP3, most models simulate a strengthening and deepening of the NADW with LGM boundary conditions (Muglia and Schmittner 2015). Yet simulations have shown that better agreement with $\delta^{13}C$ and CO_2 data requires lower NADW intensity and/or shoaling of NADW (Tagliabue et al. 2009; Tschumi et al. 2011; Menviel et al. 2017). This is also seen in terms of water mass volume with a smaller volume of NADW and a larger volume of AABW filling the ocean. The latter has a larger DIC content,

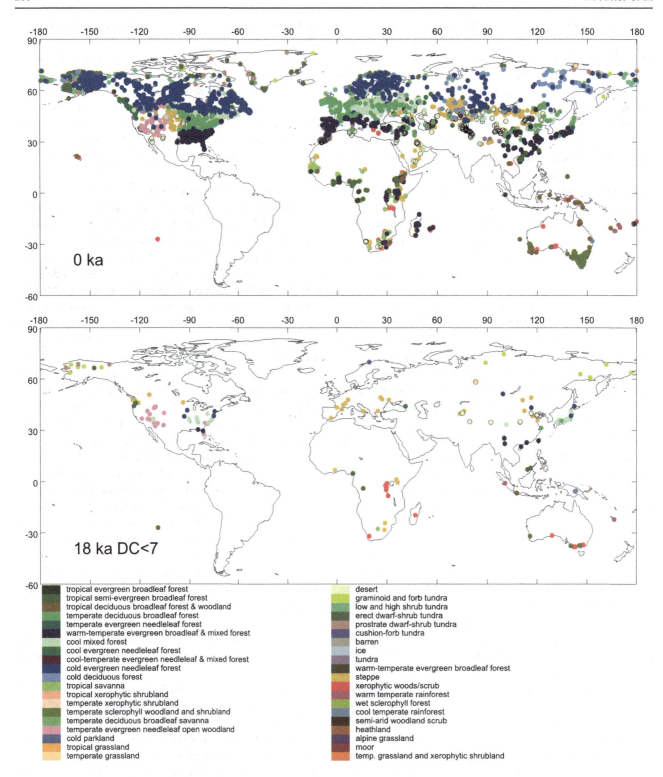

Fig. 23.7 Land vegetation maps for the modern era and the LGM, data from Prentice et al. (2000), Harrison et al. (2001), Bigelow et al. (2003) and Pickett et al. (2004). This figure with homogenised nomenclature has been downloaded from http://www.bridge.bris.ac.uk/resources/Databases/BIOMES_data

Fig. 23.8 Diagram of the carbonate compensation mechanism

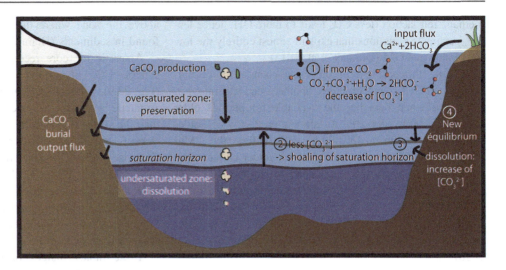

resulting in more carbon in the ocean when the AABW occupies relatively more volume than the NADW (Brovkin et al. 2012). This is referred to as the "standing volume effect" (Skinner 2009).

Changes of ocean circulation could be due to changes of winds (Anderson et al. 2009; Toggweiler et al. 2006). But data show no clear evidence of large wind changes (Kohfeld et al. 2013) and model simulations have shown that this is unlikely to have had a strong effect on the carbon cycle (Menviel et al. 2008a). It could also be linked to ocean diffusion (Bouttes et al. 2009, 2011) and particularly to bottom topography induced diffusion (De Boer and Hogg 2014). Alternatively, it could be linked to sea ice changes and modifications in bottom water formation (Ferrari et al. 2014; Bouttes et al. 2010). Indeed, sea ice formation around Antarctica was probably enhanced, especially during winter (Gersonde et al. 2005). Yet models usually fail to represent glacial sea ice extent (Roche et al. 2012; Goosse et al. 2013; Marzocchi and Jansen 2017). Improving sea ice formation in models during the LGM and the sinking of dense water around Antarctica should help towards explaining the glacial atmospheric CO_2 concentration.

Finally, on longer timescales of a few thousand years, carbonate compensation amplifies the increase of oceanic carbon storage, by maintaining a balance between inputs and outputs of alkalinity fluxes (Fig. 23.8). For example, (1) if $[CO_{2(aq)}]$ increases (for example due to the solubility pump), the equation below is displaced towards more $[HCO_3^-]$, reducing $[CO_3^{2-}]$.

$$CO_2 + CO_3^{2-} + H_2O \rightleftarrows 2HCO_3^-$$

(2) When $[CO_3^{2-}]$ is reduced, the saturation horizon, the limit between oversaturated and undersaturated water, is shifted upwards and a larger volume of water is undersaturated. (3) The larger undersaturated zone results in more $CaCO_3$ dissolution, which increases $[CO_3^{2-}]$, counteracting the initial reduction and leading to a lowering of $[CO_2]$ as the previous equation is shifted to the right, allowing the ocean to take up more CO_2 from the atmosphere. (4) When $[CO_3^{2-}]$ is increased, the oversaturated zone increases and the saturation horizon is shifted down until a new equilibrium is reached. Overall, the ocean takes up more carbon with this mechanism through dissolution of $CaCO_3$.

The concomitant lowering of CH_4 during the glacial period could be due to either a decrease in CH_4 sources, mainly wetlands, or an increase in sinks, mainly more oxidation by increased OH.

During the LGM, the colder climate, larger ice sheets and reduced hydrological cycle all led to a reduction of wetlands and reduced emissions. The first hypothesis to explain the lower CH_4 concentration during the LGM has thus focused on reduced emissions, possibly from low latitude wetlands (Chappellaz et al. 1993). Later, process-based models were developed and used to evaluate emissions (Valdes et al. 2005; Kaplan et al. 2006; Weber et al. 2010). But the resulting reduction of emissions was not enough to account for the low CH_4 concentration detected in ice cores. It was then hypothesized that the oxidizing atmospheric capacity had changed, for instance through a reduction of emissions of volatile organic compounds (VOCs) from forests (Valdes et al. 2005). The VOCs react with OH in the same way as CH_4, thus constituting an OH sink, which increases the lifetime of CH_4. If more VOCs are produced, OH concentration is reduced and the concentration of CH_4 is increased. However, more complex chemical models show that certain processes compensate for the reduced atmospheric oxidizing capacity, such as temperature, humidity, lightening (Levine et al. 2011; Murray et al. 2014). In conclusion, it now appears that the atmospheric oxidizing capacity is probably of secondary importance, as process-based models and the methane retroaction (less CH_4 leads to more OH, hence less CH_4) estimate emissions that explain almost entirely the low CH_4 value at LGM (Quiquet et al. 2015).

Beyond the problem of the large changes in glacial-interglacial GHG concentrations, another issue was raised in the 2000s when data became available for periods older than $\sim 430,000$ years BP. As shown on Fig. 23.5, older interglacials before $\sim 430,000$ ka BP (before the "Mid Brunhes Event") are characterized by a colder climate than more recent interglacials, associated to lower GHG concentrations. The colder interglacial climate can be attributed to different orbital configurations and lower CO_2 (Yin and Berger 2010, 2012) but the reason for the lower GHG concentrations still remains to be explained (Bouttes et al. 2018).

Abrupt Changes

At the centennial to millennial scale, climate variability is superimposed onto the orbital-scale glacial-interglacial cycles and is recorded by specific expressions at different latitudes and in different climate archives (Clement and Peterson 2008; see Volume 2, Chap. 29). Indeed, Greenland ice cores unveiled a succession of events called Dansgaard-Oeschger (D-O) events during the last glacial period (Dansgaard et al. 1993; North Greenland Ice Core Project Members 2004). Typically, a DO event is depicted as an abrupt warming of 5–16 °C of the mean annual surface temperature within a few decades toward a relatively mild phase. This phase is then usually characterised by a gradual cooling over several centuries and its end is marked by a rapid cooling leading to a relatively stable cold phase persisting over several centuries or even up to a thousand years. The signature of DO events is recorded in continental and marine records in the Northern Hemisphere. In the Southern Hemisphere, there are counterparts to these DO events. Ice cores indicate more gradual (millennial-scale) warming in Antarctica during the Greenland cold phases (EPICA Community Members 2006; Barker et al. 2009; WAIS Divide Project Members 2015). The antiphase relationship between the two hemispheres is attributed to the thermal bipolar seesaw, a mechanism whereby heat is redistributed in the Atlantic Ocean (Stocker and Johnsen 2003).

In addition to the succession of Dansgaard-Oeschger events, another prominent feature identified in marine sediments from the North Atlantic is the occurrence of the Heinrich events. Heinrich events are identified by the presence of debris in sediments, and were first discovered by Ruddiman in 1977. These debris are too big to be transported by oceanic currents, and in 1988 Heinrich proposed that they could have been brought by icebergs which melted above the zone where these ice rafted debris (IRD) were found in sediments (Hemming 2004).

Antarctic ice core records show a rapid rise in atmospheric CO_2 of around 15 ppm over 2000–4000 years (Fig. 23.9), generally synchronous with the millennial-scale Antarctic warming, followed by a more gradual decrease than the Antarctic temperature drop (Ahn and Brook 2008; Bereiter et al. 2012). Measurements have shown that the CO_2 rise was not steady, but punctuated by events with a rapid increase (Ahn et al. 2012). Other data measurements have resulted in several carbon sources being suggested to explain these increases in atmospheric CO_2, such as the Southern Ocean (Gottschalk et al. 2016) or the North Atlantic (Ezat et al. 2017).

These rapid changes of atmospheric CO_2 have been studied only relatively recently as previously, the temporal resolution in the records was not sufficient. Temperature changes have been studied for longer since more high resolution data were available. To replicate these rapid climate changes in simulations, modellers have found that artificially adding freshwater to the North Atlantic, for example theoretically due to the melting of numerous icebergs, could slow down or even stop the Atlantic meridional overturning circulation. This then generally leads to warming in the North Hemisphere and cooling in the South Hemisphere, in line with changes observed in the data (see Volume 2, Chap. 29).

More recently, the impact of such hosing experiments on the carbon cycle and atmospheric CO_2 evolution has also been tested in carbon-climate models, to evaluate the role of the terrestrial biosphere and the ocean, in particular, in coupled ocean-atmosphere-terrestrial biosphere models. The model response to the freshwater input appears to be very dependent on the type, duration and amplitude of the freshwater input, on the background climate (glacial vs pre-industrial) and the model. For example, the LOVECLIM model simulates a 15 ppmv increase in the context of a pre-industrial climate, but a 10 ppm CO_2 decrease in a glacial climate in response to a decrease in the AMOC driven by the same freshwater input (Menviel et al. 2008b). In both cases, the ocean takes up more carbon and the terrestrial biosphere loses carbon, but the balance between the two outcomes results in opposite effects on the atmospheric CO_2. This balance also depends on the different time reactions of the carbon reservoirs: vegetation reacts more rapidly than the ocean.

In general, most models simulate an overall increase in atmospheric CO_2 ranging from a few ppm up to more than 20 ppm, depending on the model and the size of the freshwater flux (Obata 2007; Schmittner and Galbraith 2008; Menviel et al. 2008b; Bozbiyik et al. 2011; Bouttes et al. 2012; Matsumoto and Yokoyama 2013), but the causes are different: in some models the ocean gains carbon and the

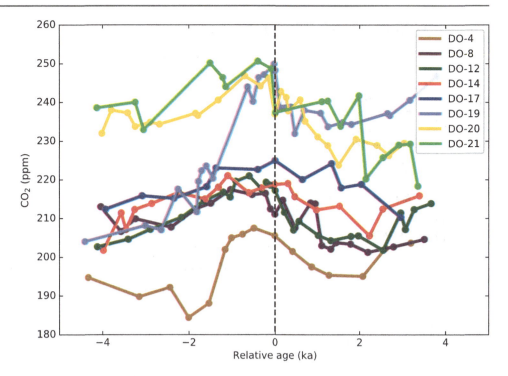

Fig. 23.9 CO_2 variations relative to abrupt warming in Greenland as presented in Ahn and Brook (2008)

terrestrial biosphere loses carbon (Obata 2007; Menviel et al. 2008b; Bozbiyik et al. 2011) while in others it is the opposite with the ocean losing carbon and terrestrial biosphere gaining carbon (Schmittner and Galbraith 2008; Bouttes et al. 2012; Matsumoto and Yokoyama 2013). Alternatively, simulations with artificially increased salinity in the Southern Ocean have been tested, resulting in a strengthening of the AABW and a loss of ocean carbon, yielding an overall CO_2 increase of around 20 ppm (Menviel et al. 2015). In general, simulations with atmosphere-ocean-terrestrial biosphere models in glacial background climate produce a large range of CO_2 changes due to the different processes, which are summed up in Fig. 23.10.

In addition, rapid CO_2 changes during the last deglaciation have been recently highlighted, such as the rapid CO_2 rise concomitant to the warming in the Northern Hemisphere at the Bølling Allerød around 14,600 years ago (Marcott et al. 2014). On top of changes in the Atlantic meridional overturning circulation, permafrost thawing, releasing large quantities of CO_2 trapped in frozen soil, has been suggested as a potential driver of the CO_2 rise (Köhler et al. 2014).

Changes in atmospheric methane (CH_4) concentration measured in ice cores are linked closely to the rapid surface temperature variations in Greenland during the last glacial period. In particular, CH_4 increases of 100–200 ppbv are associated with the abrupt DO warming events. CH_4 concentration is a global signal that reflects the response of the terrestrial biosphere, mainly wetlands, to hydroclimate changes (Brook et al. 2000) and its close link with Greenland temperature is classically interpreted as reflecting changing CH_4 emissions from tropical and boreal wetlands in phase with Greenland temperature (Chappellaz et al. 1993). A detailed study of the abrupt Bølling warming, the penultimate warming in the series of abrupt climate changes during the last glacial, suggests that changes in Greenland temperatures and atmospheric CH_4 emissions occurred essentially synchronously (within 20 yr; Rosen et al. 2014). CH_4 concentrations measured in Greenland ice cores are higher than those measured in Antarctic ice cores primarily because of enhanced CH_4 emissions in the Northern Hemisphere due to its larger land area (Chappellaz et al. 1997). However, rapid CH_4 changes are seen in both hemispheres and this feature is commonly used to synchronise Antarctic and Greenland ice core chronologies (Blunier et al. 1998; Buizert et al. 2015).

Nitrogen Cycle

Natural Nitrogen Cycle

Nitrogen (N) interacts with climate in two ways. First, like carbon dioxide and methane, nitrous oxide N_2O is a greenhouse gas. It is in fact more powerful than CO_2 or CH_4, but its atmospheric concentration is less, currently 325 ppb (NOAA, http://esrl.noaa.gov/gmd/). Second, nitrogen is also

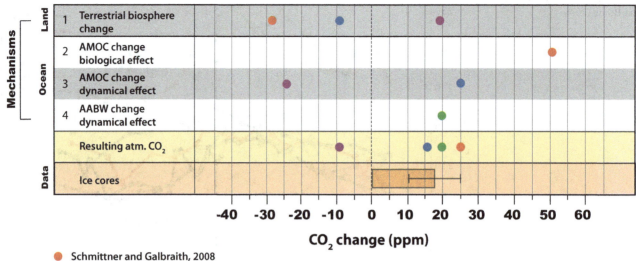

Fig. 23.10 Changes of atmospheric CO_2 due to different processes in atmosphere-ocean-terrestrial biosphere model simulations with glacial background climate. Figure modified from Mariotti (2013). Model data are taken from the simulations with maximum CO_2 change and at the time when atmospheric CO_2 for ocean and land carbon changes is at a peak

a limiting nutrient for the growth of vegetation on land and in the ocean, along with other nutrients such as phosphate. Nitrogen is essential for photosynthesis, which produces organic carbon. This connects the nitrogen cycle to the carbon cycle and atmospheric CO_2 concentration, and ultimately connects the nitrogen cycle to climate.

The nitrogen cycle is governed by biochemical reactions oxidising and reducing nitrogen which can be divided into three main processes: N_2-fixation, nitrification and denitrification (Fig. 23.11). These processes take place both in the ocean and on land.

Although N_2 is very abundant in the atmosphere it cannot be used in this form by most organisms. N_2 has to be transformed to a bioavailable—or reactive—form (ammonia NH_3 or ammonium NH_4^+) in order to be usable, a conversion, which requires a lot of energy to break the strong triple bond of the N_2 molecule. Most N_2 fixation is done by bacteria called diazotrophs, which have a specific enzyme called nitrogenase which combines gaseous nitrogen with hydrogen to produce ammonia. In the ocean, the main source of bioavailable nitrogen comes from N_2-fixation by marine diazotrophs (cyanobacteria and proteobacteria) which are mainly present in warm waters in the low latitudes. The efficiency of N_2-fixation depends on the environment, in particular, radiation, temperature, the presence of other nutrients (such as phosphate and iron), and O_2 concentration. Smaller marine inputs of bioavailable nitrogen include atmospheric nitrogen deposition and riverine inputs.

Nitrification by soil or marine bacteria is the oxidation of ammonia (NH_3) into nitrate (NO_3^-). This oxidation is done in two separate steps: first the oxidation from NH_3 to nitrite (NO_2^-) by ammonia-oxidizing bacteria and archaea. Then the oxidation of NO_2^- into NO_3^- by nitrite-oxidizing bacteria. Nitrification in the ocean takes place at the lower boundary of the euphotic zone where photosynthesis is limited by the low penetration of light, preventing the assimilation of nitrate by phytoplankton, and where remineralisation of organic matter increases nitrate concentration. In addition to this aerobic oxidation of ammonia into nitrate, anaerobic nitrification can also take place, called anammox (anaerobic ammonia oxidation).

Denitrification is the process that reduces NO_3^- to N_2 gas, releasing it back to the atmosphere. It happens during respiration by anaerobic bacteria in low O_2 conditions, and removes bioavailable nitrogen from the environment. In the process, intermediate gases are produced such as N_2O,

Fig. 23.11 Diagram of the nitrogen cycle

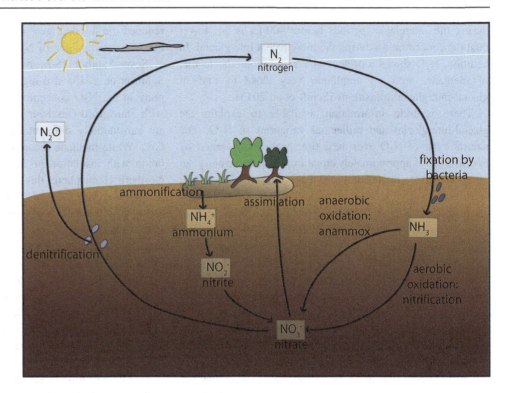

which is a powerful GHG. In conditions where O_2 is completely depleted, N_2O instead of NO_3^- is respired into N_2.

Ammonification is the production of NH_4^+ by bacteria and fungi from organic nitrogen originating from dead plants or animals, or animal waste.

During nitrification and denitrification, N_2O, a greenhouse gas, is also produced. The sources of N_2O come from both ocean and land. The sink of N_2O is due to photochemical reaction with ozone in the stratosphere.

Changes in the Nitrogen Cycle During Glacial Interglacial Cycles

Atmospheric nitrous oxide (N_2O) has a large glacial-interglacial amplitude (Fig. 23.12), with values of ~ 200 ppbv during glacial maxima and ~ 270 ppbv during interglacials including the Holocene and up to 280 ppb during the interglacial $\sim 400{,}000$ years ago (MIS 11) (Schilt et al. 2010). Unlike CO_2 and CH_4, the concentration of N_2O

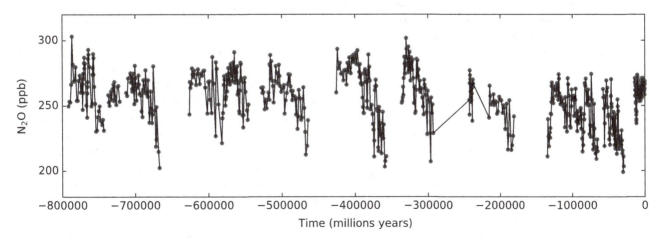

Fig. 23.12 Evolution of N_2O (ppb), over the last 800,000 years. Data from Schilt et al. (2010)

during the interglacial periods before 400 ka is not lower than the concentration during the more recent interglacial. In addition, N_2O shows a millennial variability similar to that of methane, with an amplitude very close to that of glacial-interglacial transitions (Schilt et al. 2013).

There is little information available to explain the glacial-interglacial and millennial variations in N_2O. The natural cycle of N_2O over these time scales is governed by two sources of approximately equal importance currently, an ocean source (estimated at 4 ± 2 TgN/year) and a terrestrial source (the natural part estimated at 7 TgN/year), and by an atmospheric sink mainly linked to the photolysis of the N_2O molecule in the stratosphere (12 Tg/yr). Modeling studies simulating atmospheric chemistry over the last few thousand years suggest that the N_2O sink during the LGM was similar to now (Crutzen and Brühl 1993), resulting in a lifetime of ~ 120 years for N_2O in the atmosphere. Therefore, it appears that variations in the sources are responsible for the changes in its atmospheric concentration during the Quaternary.

Isotopic data, in particular measurements of the ^{15}N of N_2O, are used to separate the ocean source from the terrestrial source, as the former generally shows enriched isotopic signatures compared to the latter (probably due to a relatively stronger denitrification process in N_2O production in an ocean environment). These data suggest that the ratio of ocean to land sources did not change over the last 33,000 years (Sowers et al. 2003). These two sources appear to vary in phase with each other, with an increase of nearly 40% during warm periods compared to cold periods.

The mechanisms for explaining these variations are still largely hypothetical. The production of N_2O in the ocean is generally linked to the presence of suboxic zones, low in oxygen, which are found directly below some of the major productive regions (east of the tropical Pacific, the Arabian Sea). In these zones deficient in dissolved oxygen, nitrates are used by the microorganisms as a source of oxygen during the denitrification reactions leading to the remineralization of the organic matter; nitrous oxide (N_2O) is a by-product of these reactions. A decrease in the source of N_2O during the glacial period could be the consequence of a shrinking of these zones, which would itself be a result of changes in ocean dynamics or local marine productivity leading to an increase in dissolved oxygen at the sub-surface. However, this hypothesis is only partially supported by reconstructions of paleoproductivity.

Another theory concerning the ocean source is related to the expansion of flooded surfaces on the continental shelves. Recent estimates suggest that a significant portion (between 0.6 and 2.7 TgN/year) of marine N_2O production comes from the continental shelves. The significant drop in sea level (-120 m) during cold periods would have greatly reduced the flooded areas at the continental edges and thus the associated source of N_2O.

Several authors (Sowers et al. 2003; Flückiger et al. 2004; Schilt et al. 2013) also highlighted variations of around 40 ppbv in the N_2O concentration in the atmosphere in phase with Dansgaard-Oeschger events. These variations in N_2O are substantially different in amplitude from variations in CH_4. While methane recordings show a fairly strong correlation with insolation in the low and mid latitudes of the Northern Hemisphere, this is not the case for N_2O.

The concentration of N_2O begins to increase before methane (during the warming phase in the Southern Hemisphere), and then the concentrations of the two gases reach their maximum at the peak of the hot phase of the D-O event. This information is compatible with the idea that both marine and terrestrial sources play an important part in the evolution of N_2O: the marine source is stronger in the Southern Hemisphere, in phase with the warming period in the south, and the terrestrial source is stronger in the northern hemisphere, in phase with the warming period in the north.

So far, very little modelling work has focused on changes in the concentration of N_2O in the atmosphere over the last hundreds of thousands of years. A simulation for the Younger Dryas episode suggests a combination of changes in the marine and terrestrial sources of N_2O to explain the variations measured during this event. Simulations run by Schmittner and Galbraith (2008) show that changes in ocean circulation play a major role N_2O variations. A reduction of the AMOC leads to decreased productivity and better ventilation resulting in increased subsurface oxygen concentrations, which explains the decrease in N_2O production.

Oxygen Cycle

The Oxygen Cycle and Its Ocean Component

Atmospheric free oxygen does not directly impact climate since it does not absorb infrared radiations. Despite this fact, the cycle of oxygen has gained lots of attention primarily because of its tight relationship with life on Earth. Indeed, the main source of free oxygen comes as a waste product of photosynthesis by plants on land and phytoplankton in the ocean. The main oxygen sink is due to respiration and/or remineralization of organic matter by almost all living organisms, which consume di-oxygen and release carbon dioxide. Other minor sources include the photolosyis of N_2O and H_2O in the atmosphere, whereas oxygen sinks are numerous and include a number of oxidation and chemical weathering pathways (see Walker 1980, for a review of the global oxygen cycle).

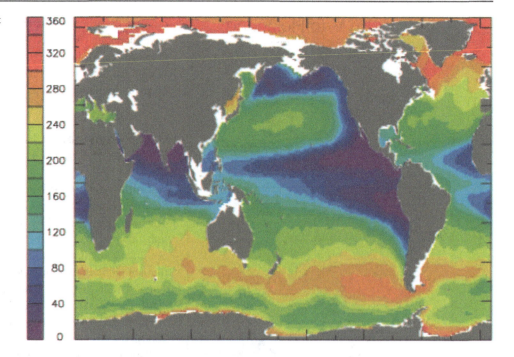

Fig. 23.13 O$_2$ concentrations at mid ocean depth (averaged over 200–600 m, in micromole/L) from the World Ocean Atlas (2009) (Garcia et al. 2010)

Many studies have examined the oxygenation of the Earth's atmosphere over geological time scales, focusing for example on the Great Oxidation Event 2.45 billion years ago (e.g., Holland 1994) or on the variations of atmospheric O$_2$ over the Phanerozoic (Berner et al. 2003). Over the last million years, the consensus is that the atmospheric concentration of O$_2$ has been very stable due to the very long residence time of O$_2$ in the atmosphere-ocean system (on the order of two million years, Catling and Claire 2005).

The oceanic component of the oxygen cycle has been however much more variable over glacial-interglacial cycles (Jaccard and Galbraith 2012). This is due to the fact that the oceanic reservoir of oxygen is much smaller than the atmospheric one (225 Tmol O$_2$ in the ocean vs. 3.8 × 10^7 Tmol of O$_2$ in the atmosphere), and that dissolved oxygen concentrations in the ocean are very heterogeneous, with O$_2$ concentrations ranging from 0 to almost 400 micromol/L. A map of O$_2$ concentrations at mid ocean depth illustrates this heterogeneity (Fig. 23.13), with O$_2$-enriched waters at high latitudes and O$_2$-depleted waters in the Eastern Tropical Pacific and in the Northern Indian Basin (depicting the so-called Oxygen Minimum Zones).

The oxygen content of the ocean results from a fine balance between the consumption of oxygen by respiring organisms feeding on organic matter sinking from the surface, and the supply of O$_2$-rich waters coming from the surface of the ocean through ocean ventilation.

Because oxygen is a fundamental resource for aerobic organisms, the distribution of oxygen in the ocean has a large imprint on marine life, shaping for example the habitat of large fish such as tunas or billfishes (Stramma et al. 2012).

Over the past decades, observations have shown that oxygen concentrations have decreased in the open ocean in many ocean regions and that the tropical oxygen minimum zones (OMZs) have likely expanded (Rhein et al. 2013). The mechanisms involved are a decrease in the oxygen solubility due to ocean warming and the combination of reduced ocean ventilation and increased stratification that prevents the penetration of oxygen into the interior of the ocean. These mechanisms are very consistent with the recent global warming trend, suggesting that deoxygenation will continue with future anthropogenic climate change. Indeed, climate models do simulate a clear deoxygenation trend with global warming, with an oceanic loss of oxygen of a few percent at the end of the twenty-first century (Bopp et al. 2013). At the regional scale however, there is yet no consensus on the evolution of subsurface oxygen levels, with very large model uncertainties.

Ocean Oxygenation at the Last Glacial Maximum

The past record of ocean oxygenation during glacial-interglacial cycles provides a complementary perspective on how the oceanic oxygen content may respond to climate change or climate variability. The reconstruction of past ocean oxygenation relies on sedimentary proxies of bottom water oxygenation. The most common proxies for ocean oxygenation are based on the presence of sediment laminations (that testify very low levels of bottom water oxygen levels), on redox sensitive trace metals (such as uranium and molybdenum) and on benthic foraminifera assemblages.

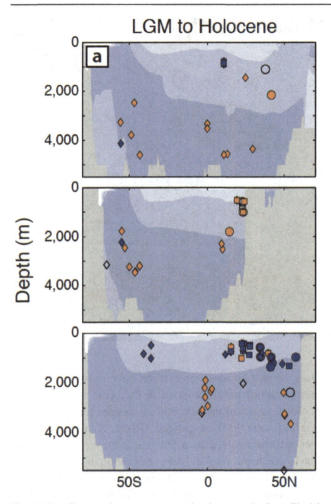

Fig. 23.14 Changes in ocean oxygenation between the Last Glacial Maximum and the Early Holocene for the Atlantic (top), Indian (middle) and Pacific (bottom) basins. The symbols refer to the different proxies used to infer past oxygenation (squares corresponding to laminations, circles to foraminifera species assemblages, and diamonds to redox-sensitive trace metals). Blue shadings indicate a relative decrease in oxygenation and orange shadings a relative increase in oxygenation from the LGM to the early Holocene. From Jaccard et al. (2014)

Despite recent progress on the developments of new proxies and the multiplication of the application of such methods to a growing number of sediment cores, the indications of past ocean oxygenation over the last glacial-interglacial cycles portrayed by these records stay very qualitative and mostly apply to the last deglaciation (Jaccard and Galbraith 2011). Overall, marine sediments indicate an oxygenation of the deep ocean throughout the last deglaciation and an expansion of low-oxygen waters in the upper ocean. These changes seem quite consistent across all ocean basins (Fig. 23.14).

The mechanisms responsible for such changes are still debated (Jaccard et al. 2014; Bopp et al. 2017). In the sub-surface ocean for example, the decreasing trend in oxygen levels during the deglaciation could be due to an increase in carbon export from the surface layers that would have increased oxygen consumption in the subsurface ocean (Jaccard and Galbraith 2012), but also to a decreased ventilation of sub-surface layers in response to changes in oceanic circulation (Bopp et al. 2017).

Ocean Oxygenation in the Mediterranean Sea: The Case of Sapropels

Definition and overview

Sapropel events are clearly identified in marine sediments. They are characterized by an organic rich layer sedimentation, mainly found in the eastern basin (Fig. 23.15). The peculiarity of this sedimentation suggests an anoxic environment allowing the organic matter preservation-due to the shutdown of the thermohaline ventilation (Möbius et al. 2010) and/or an enhanced biological productivity leading to an increased oxygen consumption (Martinez-Ruiz et al. 2000).

Sapropel events have been largely described and investigated since they have been identified in the middle of the twentieth century (Kullenberg 1952). Since the closure of East Tethys seaway 14 million years ago (Hamon et al. 2013), the only connection between the Mediterranean basin and the global ocean has been the Gibraltar straight. This semi-enclosed configuration favored sapropel events that occurred with a frequency of 21,000 years (Emeis et al. 2003).

Large variability in contexts, developments and processes associated with sapropels occurrences

There is a large variability in the imprints (strength, extension, duration) of these sapropels (see Rohling et al. (2015) for a review). For example the carbon organic content levels in sapropels typically range between 1 and 10% with a large variability, for instance sapropel S5 occurring during the last interglacial reached values from 7–15% (Grant et al. 2012) and Pliocene sapropel may reach 30% (Nijenhuis and De Lange 2000).

Whereas pacing of sapropel is strongly correlated to precessional cycles, there are major differences in preconditioning water masses that favor the occurrence of sapropels. Superimposed to this variability, the triggering of sapropel has been shown to be several thousand years after the maximum summer insolation in the northern hemisphere. Ziegler et al. (2010) inferred a recurrent lag between the northern hemisphere insolation maximum and sapropel deposition.

Sapropels typically lasted between ∼3 and ∼8 kyrs. For instance, durations of ∼4.4, ∼4.0, ∼6.2, and ∼7.4 kyr for S1, S3, S4, and S5, respectively has been attributed by different methods (Grant et al. 2012) Despite a large

Fig. 23.15 Left, sapropels within a sediment core recovered in 2001 during *RV Meteor* cruise M51-3 (Hemleben et al. 2003) (photograph by Eelco. J. Rohling and Kristian.C. Emeis). Right, late Pliocene sapropel layers outcropping at Punta Piccola (Plancq et al. 2015)

variability in time and space of benthic recolonization at the end of a sapropel. The end of enhance monsoon seems to mark a sharp and widespread (basinwide) onset of deep water oxygenation due to resumption of strong convective deep-water formation.

Sapropels are more common in the eastern Mediterranean (east of the Strait of Sicily) than in the western Mediterranean, where they are also known as Organic Rich Layers (ORLs) (Rogerson et al. 2008).

As suggested by Rohling et al. (2015), the eastern Mediterranean is more sensitive to development of deep-sea anoxia than the western Mediterranean, because of differences in the efficiency of deep-water renewal.

For a long time, the relationship between insolation changes associated with precession cycles and impacts on hydrological changes through fresh water inputs has been invoked as causal link to explain sapropel occurrences (Rossignol-Strick et al. 1982). Nevertheless, there are many different contexts in, which sapropels have occurred since the Miocene (Rohling et al. 2015). For instance, the role of the cryosphere was certainly different during Pliocene and Pleistocene. For Quaternary, it has been shown that, superimposed to the major effect of African monsoon and enhancement of freshwater from the Nile river, other forcing factors have to be accounted for. The imprint of glacial-interglacial cycle and associated sea level changes has been shown to contribute strongly for sapropel occurring during deglaciation as S1 (Rohling et al. 2015; Grimm et al. 2015).

Most recent advance on S1 sapropel modeling

Modeling represents also a unique tool to investigate the responses of the Mediterranean basin to different external forcing factors from insolation changes to associated hydrological perturbations that may produce sapropel events.

Two important developments have been down recently concerning sapropel modeling: much longer simulation and much higher spatial resolution.

Superimposed to precession cycles, since one million years the 100 ky glacial-interglacial cycle has also affected sapropel occurrence (Köng et al. 2017). Recent modeling studies (Grimm et al. 2015) aimed to simulate the S1 from its onset. A more specific scenario, involving a preconditioning of cold and poorly salted water coming from the last Heinrich event, was suggested as a possible cause for the S1 formation.

Another important issue is to reach high resolution to capture convection patterns in the Mediterranean basin. Using a coupled AOGCM (Atmospheric-Ocean Global Circulation Model) including a regional Mediterranean Sea model (1/8° much higher than previously used), Vadsaria et al. (2019) have revisited the impact of Nile hosing fresh water increase on triggering sapropel S1. This improvement allows for better simulating the intermediate and deep convection occurring in winter (Adloff et al. 2015).

Moreover, the simulation of oceanic tracer as Nd allows one to validate changes in ocean dynamics (Ayache et al. 2016).

Sulphur

The sulfur cycle is of interest to climatologists because it leads to the formation of sulfuric acid (H_2SO_4), a submicronic aerosol that reflects solar radiation efficiently (direct effect) and which, due to its highly hygroscopic nature, has a physical influence on clouds (indirect effect). It is formed in the atmosphere by the oxidation of SO_2 whose emission level through the combustion of fossil reserves is 35–45 Tg per year. The IPCC estimates that due to the increase in sulfate aerosols in the atmosphere, anthropogenic emissions of SO_2 could be responsible for a radiative forcing of -0.4 W/m^2, an opposite forcing but equivalent to about one third of the radiative forcing linked to the increase of CO_2 in the atmosphere ($+1.2$ W/m^2). However, as noted above, the comparison between these two values is of limited

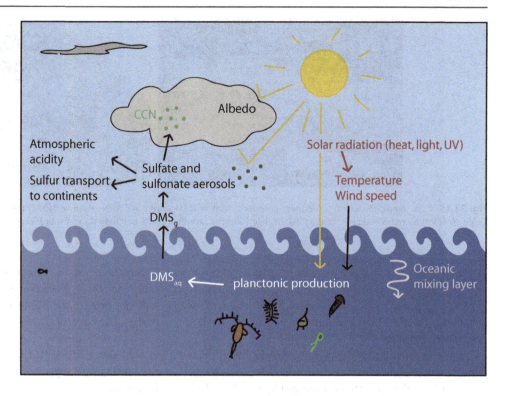

Fig. 23.16 Feedback mechanisms linking our planet's climate to DMS emissions from the ocean (adapted from Charlson et al. 1987)

significance since the forcing of the aerosol may, for example, be much greater over heavily anthropized regions of the northern hemisphere. Moreover, the temporal evolution of anthropogenic SO_2 emissions varies greatly from one continent to another: emissions reached a peak in the 1980s in Western Europe and the USA, while an acceleration of their emissions is underway in India, for example. The sulfur cycle in the atmosphere also involves two major natural sources, volcanic activity and marine DMS biogenic emissions. An important dimension of the latter source is its potential response to climate conditions, a hypothesis proposed by Charlson, Lovelock, Andreae and Warren, four authors of a well-known article in Nature in 1987, proposing that DMS emissions produced by plankton may act as a climate regulator (Fig. 23.16). Responses by the marine biosphere (changes in primary productivity, phytoplankton speciation, stratification of surface ocean layers, wind patterns, etc.) to the current climate change are thus possible. Its direction and magnitude remain uncertain, but some ocean-atmosphere models incorporating an ocean biology module, coupling sulfur and carbon, suggest that a 40% increase in DMS emissions could occur in the 40 °S zone in response to a doubling of CO_2 content and to the associated climate change (Bopp et al. 2003b).

The idea of examining the relationship between marine biogenic sulfur emissions and past climate is the motivation behind many ice core studies. Given that sulphate in the atmosphere has several origins, the first attempt at reconstituting marine biogenic emissions was made by examining the MSA content of ice (Saigne and Legrand 1987). This study of the last climate cycle in the Vostok ice core indicated an increase in MSA deposition fluxes by a factor of almost three between an interglacial period and the final stage of an ice age, despite the fact mentioned previously that sulphate flux has shown little variation in the past (Legrand et al. 1991). Since isotopic studies of sulfur undertaken on sulphate have subsequently clearly established that DMS emissions are the major source of sulphate in Antarctica, a contradiction therefore appeared between the two proxies of DMS emissions, with the MSA suggesting an increase in DMS emissions, but not in sulphate. The hypothesis of an increase in DMS emissions during the glacial period seemed reasonable given that the phytoplankton species that emits a lot of DMS (Phaeocystis) has a particular affinity with sea ice.

The difference between the sulphate and MSA records remains unclear to date. The atmospheric studies carried out over the last few years in Antarctica demonstrate the complexity of the problem. Although the atmospheric levels of the three sulfur species coincide well over time with the DMS concentrations in the Southern Ocean, the interannual variability observed in the Southern Ocean does not show any straightforward connection with annual sea ice cover (Preunkert et al. 2007). These measurements also reveal intricate processes involving photochemistry and atmospheric dynamics which makes the link between the two sulfur species and DMS very complex. These studies continue to be pursued actively due to the fact that, to date, only from ice can past variability in marine biogenic emissions potentially be reconstructed, since organisms like Phaeocystis do not leave any traces in marine sediments. The issue is important because it contributes to our understanding of the feedback taking place in high latitudes (regions highly sensitive to global changes in climate), involving complex processes between marine biology, sea ice, and climate.

Aerosols and Dust

Aerosols are small liquid or solid particles, ranging from a few nm to 100 μm, in suspension in the atmosphere. Natural aerosols include desert dust, sea salt, carbonaceous, sulphur and nitrogen species, largely emitted from dry and vegetated landscapes, the oceans, and volcanoes (Carslaw et al. 2010). Primary aerosols are emitted directly from the surface of the Earth, whereas secondary aerosols are formed from gaseous precursors in the atmospheric environment. Aerosols are washed out by precipitation, or removed by gravitational settling and dry deposition, so that their lifetime in the atmosphere is short, a few days only—except when they reach the stratosphere, where they can stay for a few years, as can happen during giant volcanic eruptions. Therefore, unlike well-mixed GHGs, aerosols are considered to be short-lived climate forcing agents, and their impacts are characterized by a strong regional component (Boucher et al. 2013). Aerosol emissions vary depending on surface climate conditions; on the other hand, aerosols impact the climate system themselves, though direct and indirect (cloud-mediated) interactions with the atmospheric radiation budget, by changing the surface albedo, as well as by means of indirect impacts on global biogeochemical cycles (Mahowald et al. 2017).

Size, shape and composition define the specific interactions of aerosols with radiation, including absorption and scattering of shortwave (solar) and longwave (terrestrial) radiation. The scattering of shortwave radiation results in cooling, but absorption can lead to warming when it is above a highly reflective surface. Through their absorption of outgoing longwave radiation, aerosols also behave like GHGs. For some aerosol species, one particular effect is dominant, in other cases, opposing effects coexist. Considering the variability of aerosol spatial distribution, the coexistence and mixing of different aerosol species, and the diversity of aerosol-radiation interactions, it is clear that direct impacts of aerosols on climate constitute a complex problem (Boucher et al. 2013).

Aerosols also interact with clouds. Changes in relative humidity linked to the vertical stability of the atmospheric column and the surface evapotranspiration balance, occurring as a rapid adjustment to direct aerosol forcing, can influence cloud formation. This is called the semi-direct effect. The indirect effects, on the other hand, involve aerosols acting as cloud condensation (CCN) or ice nuclei (IN), which means that water or ice aggregates around them. This modifies the type, extent and lifetime of clouds. For example, the presence of aerosols leads to smaller but more numerous droplets, which yields a more reflective cloud than it would be without aerosols. This also means that clouds formed with aerosols will have a longer lifetime since the droplets are smaller and won't reach the critical size for precipitation. The indirect effect can result in warming or cooling depending on the altitude where clouds are formed. Because the effect of aerosols on radiation and on clouds is complex and depends on many parameters, it remains one of the main sources of uncertainties in models (Boucher et al. 2013).

In addition, aerosols depositing back to the surface can also modify the albedo. This is the case of dust, and especially black carbon; they can cause snow and ice to darken, which reduces the albedo and leads to warming.

In virtue of their composition, aerosols also act as carriers of specific elements, such as nitrogen, phosphorus, sulphur, and iron, which are linked to important biogeochemical cycles, including the carbon cycle (Mahowald et al. 2017). In particular, phosphorus and iron are linked to the dust cycle, and the peculiarity is that windblown inputs can be fundamental to the mass budgets of those elements in remote regions, far from the dust sources. For instance, dust-borne phosphorus from North Africa replenishes the pool of this element in the Amazon, where the loss by fluvial erosion would otherwise deplete it, with implications for the rainforest. Iron, on the other hand, is a micronutrient for marine ecosystems. Because its sources are the continents, remote marine areas are depleted in this element. Where the abundance of macronutrients such as nitrogen and phosphorus is accompanied by a relative scarcity of iron, which limits the primary production at the ecosystem level, i.e. in High-Nutrient Low-Chlorophyll (HNLC) areas, dust-borne inputs of iron become of great importance in sustaining algal blooms—this is notably the case of the Southern Ocean (Jickells et al. 2005).

Natural Aerosols: Overview

In this section we will briefly describe the main natural aerosol types. Note that mineral dust and sea salt are still the most abundant primary aerosol species by mass in the present day atmosphere.

Mineral (desert) dust (Fig. 23.17) is emitted into the atmosphere in response to wind erosion of the surface, in dry and semi-dry areas, with low vegetation cover. Far-travelled dust particles are mostly clays and fine silts below 10 μm in diameter, and are composed mainly of silicates, along with carbonates, gypsum, and metal oxides. Dust aerosols interact

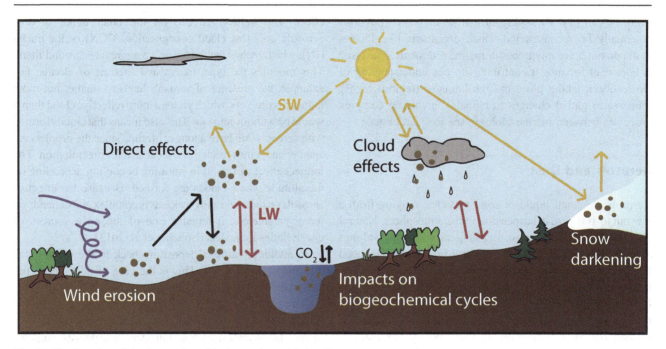

Fig. 23.17 Overview of the dust cycle and its interactions with the climate system

with both solar and terrestrial radiation, and act as ice nuclei in mixed phase clouds. As already mentioned, dust deposition on the surface can reduce snow albedo and impact on the carbon cycle through phosphorus and iron mediated interactions with global biogeochemical cycles.

Land vegetation is a source of primary biogenic aerosol particles (PBAP), in the form of fungal spores, viruses, bacteria, pollens and plant debris, as well as aerosol precursors such as isoprene and other volatile organic compounds (VOC), which result in the formation of secondary organic aerosols (SOA). Wildfires are responsible for the emissions of particulate organic matter and black carbon. Particulate carbonaceous aerosols have a slight net cooling effect in the atmosphere, and play a role as cloud condensation and ice nuclei. Black carbon is a strongly absorbing aerosol, associated with snow darkening and has a warming effect on the atmosphere.

Wetlands emissions of ammonia, and NO_x produced by biomass burning, biogenic soil emissions, stratospheric injection and by the interaction of lightning with atmospheric N_2, are the main precursors to nitrogen aerosol species, which overall have a moderate net cooling effect in the atmosphere, and may impact on biogeochemical cycles.

Volcanic emissions of SO_2 are one of the primary sources of precursors for the formation of natural sulphate aerosols, which are characterized by a strong cooling impact on the atmospheric radiation, and they are very efficient CCN. Volcanic eruptions can also eject glass shards and ash, which can leave a mark by depositing as tephra layers—very useful stratigraphic markers.

Wind stress on the surface of the oceans drives aerosol emissions at the sea-air interface, where bubbles bursting within breaking waves eject sea sprays composed mostly of sea salt (dominating the super-micron fraction), along with organic particles (concentrated in the sub-micron fraction). Sea salt and some organic particles are also emitted from the surface of sea ice, where these form frost flowers during the process of brine rejection that accompanies the freezing of seawater. Sea salts tend to cool the atmosphere, and because of their high hydroscopicity, they act as CCN (O'Dowd and de Leeuw 2007). Oceans are also a major source of precursors of sulphate aerosols, in the form of biogenic emissions of dimethylsulfide (DMS) that is then oxidised to sulphuric acid and methane sulfonic acid (MSA) in the atmospheric environment (Legrand and Mayewski 1997).

Dust Variability and Impacts on Past Climates

The past history of the dust cycle is imprinted in natural archives such as ice, marine sediments, loess/paleosol sequences, and peat bogs. Dust has the greatest preservation potential among all aerosol species, because it is essentially insoluble, and traces of it are present in a variety of environments all around the globe. In general, we can obtain a paleodust record from natural archives when the following conditions are met: there is preservation of the deposition signal; we can establish a chronology; and we are able to separate eolian contributions from the sedimentary matrix.

The fifty-meter thick loess deposits of China and North America are probably the most spectacular evidence of how the dust cycle is capable of shaping immense landscapes—loess deposits cover 10% of the emerged landmasses. Loess accumulation in China has been an ongoing process for over 20 million years, since the uplift of the Tibetan plateau caused widespread aridification of central-eastern Asia. Beyond that, little is known, although isolated information on deep paleoclimate conditions at least dating back to the Paleozoic (∼500 million years ago) can be derived, based on the analysis of geologic formations whose origin can be ultimately linked to eolian sedimentation.

We have a better picture of the global dust cycle on Quaternary time scales, especially since the late Pleistocene. From polar ice core records we know that a strong dust-climate coupling was a persistent feature at least over the last eight glacial-interglacial cycles; colder climate states are characterized by increased dustiness, as shown by the milestone paleoclimate records from the Vostok and EPICA Dome C (EPICA Community Members 2004) ice cores from Antarctica (Fig. 23.18). Preservation of stratigraphy and chronologies based on numerical (absolute) dating methods allow for a more detailed reconstruction of the last glacial-interglacial cycle. In particular, global compilations of paleodust records based on dust mass accumulation rates provide a quantitative metric to compare paleodust records from different natural archives, and constitute a benchmarking tool for models (Kohfeld and Harrison 2001).

During the LGM, global dust emissions were enhanced by a factor of 2–4, and the increase in dust deposition in high latitudes was even by a factor of 10 or more. A combination of changes in dust source areas and transport patterns, driven

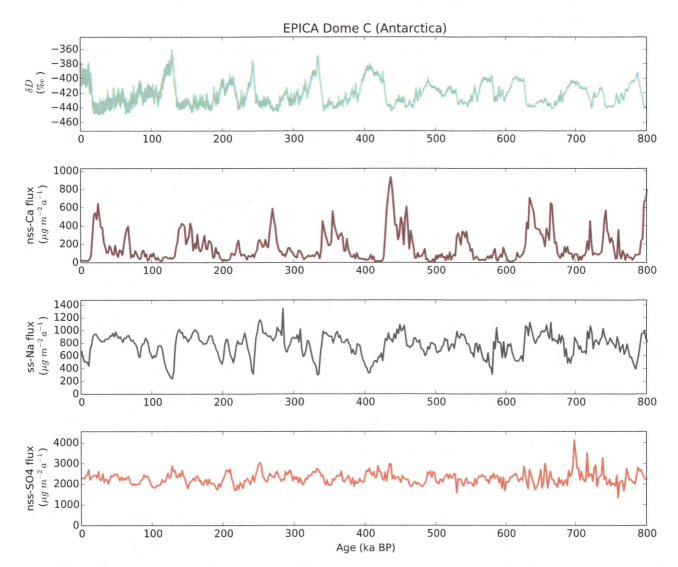

Fig. 23.18 Overview of aerosol (nss-SO_4^{2-} as a proxy for sulfate aerosols, ss-Na for sea salt, and nss-Ca for dust) deposition fluxes from the EPICA Dome C ice core (Wolff et al. 2006), Antarctica, along with indicators of global climate, i.e. deuterium excess (Jouzel et al. 2007)

by the large-scale atmospheric circulation, shaped the geographical variability in dust cycle dynamics. The sharp decline in dust deposition to Antarctica paralleling the rise in atmospheric CO_2 levels, led oceanographer John Martin ("Give me a tanker of iron, and I will give you an ice age"), in the late eighties, to formulate the famous iron hypothesis: the glacial increase in dust-borne iron inputs to the Southern Ocean may have stimulated the ocean biological pump, resulting in an increased productivity in HNLC areas and subsequent carbon sequestration in the deep ocean (Jickells et al. 2005). State of the art ESMs suggest that this mechanism may have been responsible for a decrease of ~ 20 ppmv of atmospheric CO_2, out of 80/100 ppmv—the overall decrease in CO_2 concentrations during the LGM measured in ice cores. On the other hand, the impact of the increased dust burden on atmospheric radiation is estimated by models to be at least -1 W/m^2 globally, compared to -6 W/m^2 forcing by reduced GHGs atmospheric concentrations and changes in surface albedo from decreased sea levels and the growth of ice sheets in the Northern Hemisphere. Nonetheless, this figure might hide much larger dust impacts of the opposite sign, with strong cooling downwind of the major dust sources, and warming over the bright, glaciated Arctic regions (Albani et al. 2018).

During the course of the entire glacial period, millennial scale variability expressed by Dansgaard-Oeschger and Heinrich events in the $\delta^{18}O$ record in Greenland ice cores, is almost paralleled by variations in dust deposition rates, with colder phases associated to dustier conditions (Rasmussen et al. 2014). The same picture emerges from the corresponding alternation of stadial/interstadial periods in Antarctica (EPICA Community Members 2006).

The deglaciation followed, characterized by a non-monotonic increase in global temperatures and CO_2, mirrored by decreasing dust levels in both hemispheres. While the Holocene was initially described as a relatively flat period in terms of dust, based on polar records, in the last three decades, a few studies of North Atlantic sediment cores highlighted the large variability in North African dust emissions; a reduction by a factor 2–5 corresponded to the "Green Sahara" phase of the Early and mid-Holocene, characterized by an enhanced summer monsoon, compared to the drier late Holocene after ~ 5 ka BP. The possible (positive or negative) feedbacks between the monsoon system and variations in the North African dust cycle are a subject of study by the modelling community (Albani et al. 2015).

Ironically, we know relatively less about the more recent past. Marine sediments and loess profiles generally can't achieve a temporal resolution fine enough to resolve the last millennia, and often surface layers of loess/paleosol deposits are disturbed by agricultural practices, as they tend to be very fertile soils. Ice cores do provide this kind of temporal resolution, but we still have very little data. The top meters of polar cores, corresponding to this time frame, are actually made of firn, which complicates the analysis because of the risk of contamination of the samples. Alpine ice cores, on other hand, allow the analysis of dust concentration and can have good chronologies for the last few decades, but reliable, quantitative estimates of dust mass accumulation rates are hampered by the extreme spatial variability of snow accumulation and post-depositional processes. Therefore, we do not have a clear pre-industrial reference state for dust (Carslaw et al. 2010). A few studies trying to address dust trends during the twentieth century yield contrasting results; a generally increasing trend over parts of the last century suggested by some authors may have "masked" a fraction of global temperature increase, due to the net cooling effect of dust.

Other Aerosol Species in Past Climates

There is much less information about other aerosol species in the past; unlike dust, solubility and volatility limit the preservation of most species in many environmental settings, and pose additional analytical challenges. Most of the information we have is from ice cores (Wolff et al. 2006; Preunkert and Legrand 2013).

Na^+ is a stable proxy for sea salts. Yet, we do not really have information from lower latitudes; sea salt records from polar ice cores (Fig. 23.18) show, similar to dust, increased deposition rates in glacial climates, by a factor 3–5 in Antarctica and 1.5–3 in Greenland. It is not fully clear to what extent this was due to an increase in emissions or transport from open waters, rather than to the expanded sea ice source, although the latter seems to be the dominant factor. Increases in sea salts could have had a negative forcing, via direct and indirect effects, on the atmospheric radiation budget.

Sulphur and nitrogen aerosol species present more important issues with preservation, and are more difficult to interpret. The baseline sulphate records from polar ice cores (Fig. 23.18), mainly derived from DMS emissions, show a flat signal in Antarctica and some variability in Greenland. This has dampened some early enthusiasm for the idea that DMS could be a dominant feedback in driving glacial-interglacial variability, given the strong cooling effects associated with sulphate aerosols. Sulphate concentration spikes are associated with inputs from large volcanic eruptions, and are sometimes associated with the presence of tephra. In certain cases, a lag of one or two years was observed in the peaking of the two signals, suggesting a longer stratospheric residence time of sulphates. There are no clear trends in volcanic activity on late Quaternary time scales, although several volcanic spikes marked the deglaciation.

There is very little direct information on carbonaceous aerosols from ice cores on late Quaternary time scales. Some data is available on paleofire proxies, such as ammonium or levoglucosan, which can be combined with information from charcoal, preserved in paleosols, peat bogs, and lakes—on the other hand, an indicator of the source of the aerosols and precursors, rather than the sink. Finally, the variation of pollen assemblages archived in lakes and peat can provide information on the state of past vegetation, which is a source of many aerosols and precursors. The general picture is that land biogenic and fire emissions during the last glacial period were lower than today, at least at high latitudes, which is consistent with a drier and colder climate, and the reduction of vegetated areas linked to the growth of the large ice sheets in North America and Eurasia. More data on the recent past can be found in snow and ice, showing the influence of anthropic activities on carbonaceous, sulphate and nitrogen aerosols (Preunkert and Legrand 2013).

Conclusion

In this chapter, we have looked at the major biogeochemical cycles that interact with climate: carbon (CO_2 and CH_4), nitrogen (N_2O), sulphur (SO_2) as well as dust. These biogeochemical cycles are crucial for climate mainly because of their role in the atmospheric radiative budget. In turn, climate evolution impacts these cycles by changing their sources and sinks. Past changes are documented in archives such as sediment and ice cores, which have been described in more details in Volume 1. To better understand past changes of the carbon, nitrogen, sulphur and dust cycles and evaluate their impact on climate, they have been included in numerical climate models. Going back and forth between data and model simulations is crucial to unravel the causes of past changes, and in the future, both more data and improved modelling should help to better understand the changes in the biogeochemical cycles over the last million years.

References

Adams, J. M., Faure, H., Faure-Denard, L., McGlade, J. M., & Woodward, F. I. (1990). Increases in terrestrial carbon storage from the Last Glacial Maximum to the present. *Nature, 348,* 711–714.

Adkins, J. F., McIntyre, K., & Schrag, D. P. (2002). The salinity, temperature, and $\delta^{18}O$ of the glacial deep ocean. *Science, 298,* 1769–1773. https://doi.org/10.1126/science.1076252.

Adkins, J. F. (2013). The role of deep ocean circulation in setting glacial climates. *Paleoceanography, 28,* 539–561. https://doi.org/10.1002/palo.20046.

Adloff, F., Somot, S., Sevault, F., Jordà, G., Aznar, R., Déqué, M., et al. (2015). Mediterranean Sea response to climate change in an ensemble of twenty-first century scenarios. *Climate Dynamics, 45* (9–10), 2775–2802. https://doi.org/10.1007/s00382-015-2507-3.

Ahn, J., & Brook, E. J. (2008). Atmospheric CO_2 and climate on millennial time scales during the last glacial period. *Science, 322,* 83. https://doi.org/10.1126/science.1160832.

Ahn, J., Brook, E. J., Schmittner, A., & Kreutz, K. (2012). Abrupt change in atmospheric CO_2 during the last ice age. *Geophysical Reseach Letters, 39,* L18711. https://doi.org/10.1029/2012GL053018.

Albani, S., Mahowald, N. M., Winckler, G., Anderson, R. F., Bradtmiller, L. I., Delmonte, B., et al. (2015). Twelve thousand years of dust: The Holocene global dust cycle constrained by natural archives. *Climate of the Past, 11,* 869–903. https://doi.org/10.5194/cp-11-869-2015.

Albani, S., Balkanski, Y., Mahowald, N., Winckler, G., Maggi, V., & Delmonte, B. (2018). Aerosol-climate interactions during the Last Glacial Maximum. *Current Climate Change Reports, 4,* 99–114. https://doi.org/10.1007/s40641-018-0100-7.

Anderson, R. F., Ali, S., Bradtmiller, L. I., Nielsen, S. H. H., Fleisher, M. Q., Anderson, B. E., et al. (2009). Wind-driven upwelling in the Southern Ocean and the deglacial rise in atmospheric CO_2. *Science, 323,* 1443. https://doi.org/10.1126/science.1167441.

Archer, D., & Maier-Reimer, E. (1994). Effect of deep-sea sedimentary calcite preservation on atmospheric CO_2 concentration. *Nature, 367,* 260–263.

Archer, D., Winguth, A., Lea, D., & Mahowald, N. (2000). What caused the glacial/interglacial pCO_2 cycles? *Reviews of Geophysics, 38,* 159–189.

Archer, D. E., Martin, P. A., Milovich, J., Brovkin, V., Plattner, G.-K., & Ashendel, C. (2003). Model sensitivity in the effect of Antarctic sea ice and stratification on atmospheric pCO_2. *Paleoceanography, 18*(1), 1012.

Ayache, M., Dutay, J.-C., Arsouze, T., Révillon, S., Beuvier, J., & Jeandel, C. (2016), High resolution neodymium characterization along the Mediterranean margins and modeling of εNd distribution in the Mediterranean basins. Biogeosciences, (April), 1–31. http://doi.org/10.5194/bg-2016–109.

Barker, S., Diz, P., Vautravers, M. J., Pike, J., Knorr, G., Hall, I. R., et al. (2009). Interhemispheric Atlantic seesaw response during the last deglaciation. *Nature, 457,* 1097–1102. https://doi.org/10.1038/nature07770.

Barnola, J. M., Raynaud, D., Korotkevich, Y. S., & Lorius, C. (1987). Vostok ice core provides 160,000-year record of atmospheric CO_2. *Nature, 329,* 408–414.

Basak, C., Fröllje, H., Lamy, F., Gersonde, R., Benz, V., Anderson, R. F., et al. (2018). Breakup of last glacial deep stratification in the South Pacific. *Science, 359*(6378), 900–904. https://doi.org/10.1126/science.aao2473.

Bereiter, B., Lüthi, D., Siegrist, M., Schüpbach, S., Stocker, T. F., & Fischer H. (2012). Mode change of millennial CO_2 variability during the last glacial cycle associated with a bipolar marine carbon seesaw. *Proceedings of the National Academy of Sciences, 109*(25), 9755–9760. https://doi.org/10.1073/pnas.1204069109.

Bereiter, B., Eggleston, S., Schmitt, J., Nehrbass-Ahles, C., Stocker, T. F., Fischer, H., et al. (2015). Revision of the EPICA Dome C CO_2 record from 800 to 600 kyr before present. *Geophysical Reseach Letters, 42,* 542–549. https://doi.org/10.1002/2014GL061957.

Berner, R. A., Beerling, D. J., Dudley, R., Robinson, J. M., & Wildman, R. A. (2003). Phanerozoic atmospheric oxygen. *Annual Review of Earth and Planetary Sciences, 31,* 105–134.

Bigelow, N. H., Brubaker, L. B., Edwards, M. E., Harrison, S. P., Prentice, I. C., Anderson, P. M., Andreev, A. A., Bartlein, P. J., Christensen, T. R., Cramer, W., Kaplan, J. O., Lozhkin, A. V.,

Matveyeva, N. V., Murray, D. V., McGuire, A. D., Razzhivin, V. Y., Ritchie, J. C., Smith, B., Walker, D. A., Gajewski, K., Wolf, V., Holmqvist, B. H., Igarashi, Y., Kremenetskii, K., Paus, A., Pisaric, M. F. J., & Vokova, V. S. (2003). Climate change and Arctic ecosystems I. Vegetation changes north of 55 °N between the Last Glacial Maximum, mid-Holocene and present. *Journal of Geophysical Research, 108*(D19), 8170.

Bird, M. I., Lloyd, J., & Farquhar, G. D. (1994). Terrestrial carbon storage at the LGM. *Nature, 371,* 566.

Blunier, T., Chappellaz, J., Schwander, J., Dällenbach, A., Stauffer, B., Stocker, T. F., et al. (1998). Asynchrony of Antarctic and Greenland climate change during the last glacial period. *Nature, 394,* 739–743.

Buizert, C., Cuffey, K. M., Severinghaus, J. P., Baggenstos, D., Fudge, T. J., Steig, E. J., et al. (2015). The WAIS Divide deep ice core WD2014 chronology—Part 1: Methane synchronization (68–31 ka BP) and the gas age–ice age difference. *Climate of the Past, 11,* 153–173. https://doi.org/10.5194/cp-11-153-2015.

Bopp, L., Kohfeld, K. E., Quéré, C. L., & Aumont, O. (2003a). Dust impact on marinebiota and atmospheric CO_2 during glacial periods. *Paleoceanography, 18*(2), 1046.

Bopp, L. O., Aumont, S. Belviso, & Monfray, P. (2003b). Potential impact of climate change on marine dimethyl sulfide emissions. *Tellus B: Chemical and Physical Meteorology, 55*(1), 11–22. https://doi.org/10.3402/tellusb.v55i1.16359.

Bopp, L., et al. (2013). Multiple stressors of ocean ecosystems in the 21st century: Projections with CMIP5 models. *Biogeosciences, 10,* 6225–6245. https://doi.org/10.5194/bg-10-6225-2013.

Bopp, L., Resplandy, L., Untersee, A., Mezo, P. L., & Kageyama, M. (2017). Ocean (de)oxygenation from the Last Glacial Maximum to the twenty-first century: insights from Earth system models. *Philosophical Transactions of the Royal Society A, 375,* 20160323.

Boucher, O., Randall, D., Artaxo, P., Bretherton, C., Feingold, G., Forster, P., Kerminen, V.-M., Kondo, Y., Liao, H., Lohmann, U., Rasch, P., Satheesh, S. K., Sherwood, S., Stevens, B., & Zhang, X. Y. (2013). Clouds and aerosols. In T. F. Stocker, D. Qin, G.-K. Plattner, M. Tignor, S. K. Allen, J. Boschung, A. Nauels, Y. Xia, V. Bex, & P. M. Midgley (Eds.), *Climate Change 2013: The Physical Science Basis. Contribution of Working Group I to the Fifth Assessment Report of the Intergovernmental Panel on Climate Change.* Cambridge, UK and New York, NY, USA: Cambridge University Press.

Bouttes, N., Roche, D. M., & Paillard, D. (2009). Impact of strong deep ocean stratification on the glacial carbon cycle. *Paleoceanography, 24,* PA3203. https://doi.org/10.1029/2008pa001707.

Bouttes, N., Paillard, D., & Roche, D. M. (2010). Impact of brine-induced stratification on the glacial carbon cycle. *Climate of the Past, 6,* 575–589. https://doi.org/10.5194/cp-6-575-2010.

Bouttes, N., Paillard, D., Roche, D. M., Brovkin, V., & Bopp, L. (2011). Last Glacial Maximum CO_2 and $\delta^{13}C$ successfully reconciled. *Geophysical Reseach Letters, 38,* L02705. https://doi.org/10.1029/2010GL044499.

Bouttes, N., Roche, D., & Paillard, D. (2012). Systematic study of the impact of fresh water fluxes on the glacial carbon cycle. *Climate of the Past, 8,* 589–607.

Bouttes, N., Swingedouw, D., Roche, D. M., Sanchez-Goni, M. F., & Crosta, X. (2018). Response of the carbon cycle in an intermediate complexity model to the different climate configurations of the last nine interglacials. *Climate of the Past, 14,* 239–253. https://doi.org/10.5194/cp-14-239-2018.

Bozbiyik, A., Steinacher, M., Joos, F., Stocker, T. F., & Menviel, L. (2011). Fingerprints of changes in the terrestrial carbon cycle in response to large reorganizations in ocean circulation. *Climate of the Past, 7,* 319–338.

Broecker, W. S., & Peng, T.-H. (Eds.). (1982). *Tracers in the Sea.* Palisades, New York: Lamont-Doherty Geological Observatory of Columbia University.

Brook, E. J., Harder, S., Severinghaus, J., Steig, E. J., & Sucher, C. M. (2000). On the origin and timing of rapid changes in atmospheric methane during the last glacial period. *Global Biogeochemical Cycles, 14*(2), 559–572. https://doi.org/10.1029/1999GB001182.

Brovkin, V., Ganopolski, A., Archer, D., & Munhoven, G. (2012). Glacial CO_2 cycle as a succession of key physical and biogeochemical processes. *Climate of the Past, 8,* 251–264. https://doi.org/10.5194/cp-8-251-2012.

Carslaw, K. S., Boucher, O., Spracklen, D. V., Mann, G. W., Rae, J. G. L., Woodward, S., et al. (2010). A review of natural aerosol interactions and feedbacks within the Earth system. *Atmospheric Chemistry and Physics, 10*(4), 1701–1737.

Catling, D. C., & Claire, M. W. (2005). How Earth's atmosphere evolved to an oxic state: A status report. *Earth and Planetary Science Letters, 237,* 1–20.

Chappellaz, J., Barnola, J. M., Raynaud, D., Korotkevich, Y. S., & Lorius, C. (1990). Ice-Core Record of Atmospheric Methane over the Past 160,000 Years. *Nature, 345,* 127–131.

Chappellaz, J. A., Fung, I. Y., & Thompson, A. M. (1993). The atmospheric CH_4 increase since the Last Glacial Maximum. *Tellus, 45B,* 228–241. 1. Source Estimates.

Chappellaz, J., Brook, E., Blunier, T., & Malaizé, B. (1997). CH_4 and $\delta^{18}O$ of O_2 records from Antarctic and Greenland ice: A clue for stratigraphic disturbance in the bottom part of the Greenland Ice Core Project and the Greenland Ice Sheet Project 2 ice cores. *Journal Geophysical Research, 102,* 26547–26557.

Charlson, R. J., Lovelock, J. E., Andreae, M. O., & Warren, S. G. (1987). Oceanic phytoplankton, atmospheric sulphur, cloud albedo and climate. *Nature, 326,* 655–661.

Ciais, P., Tagliabue, A., Cuntz, M., Bopp, L., Scholze, M., Hoffmann, G., et al. (2012). Large inert carbon pool in the terrestrial biosphere during the Last Glacial Maximum. *Nature Geoscience, 5,* 74–79. https://doi.org/10.1038/ngeo1324.

Clement, A. C., & Peterson, L. C. (2008). Mechanisms of abrupt climate change of the last glacial period. *Reviews of Geophysics, 46,* RG4002. https://doi.org/10.1029/2006rg000204.

Craig, H. (1957). Isotopic standards for carbon and oxygen and correction factors for massspectrometric analysis of carbon dioxide. *Geochimica et Cosmochimica Acta, 12,* 133–149.

Crichton, K. A., Bouttes, N., Roche, D. R., Chappellaz, J., & Krinner, G. (2016). Permafrost carbon as a missing link to explain CO_2 changes during the last deglaciation. *Nature Geoscience, 9,* 683–686. https://doi.org/10.1038/ngeo2793.

Crowley, T. (1995). Ice age terrestrial carbon changes revisited. *Global Biogeochemical Cycles, 9*(3), 377–389.

Crutzen, P. J., & Brühl, C. (1993). A model study of atmospheric temperatures and the concentrations of ozone, hydroxyl, and some other photochemically active gases during the glacial, the pre-industrial Holocene and the present. *Geophysical Reseach Letters, 20*(11), 1047–1050.

Curry, W. B., & Oppo, D. W. (2005). Glacial water mass geometry and the distribution of $\delta^{13}C$ of ΣCO_2 in the western Atlantic Ocean. *Paleoceanography, 20,* PA1017. https://doi.org/10.1029/2004pa001021.

Dahl-Jensen, D. (2018). Drilling for the oldest ice. *Nature Geoscience, 11,* 703–704.

Dansgaard, W., Johnsen, S. J., Clausen, H. B., Dahl-Jensen, D., Gundestrup, N. S., Hammer, C. U., et al. (1993). Evidence for general instability of past climate from a 250-kyr ice-core record. *Nature, 364,* 218–220.

De Boer, A. M., & Hogg, A. M. C. (2014). Control of the glacial carbon budget by topographically induced mixing. *Geophysical Research Letters, 41*. doi:10.1002/2014GL059963.

Delmas, R., Ascencio, J. M., & Legrand, M. (1980). Polar ice evidence that atmospheric CO_2 20,000 year BP was 50% of present. *Nature, 284*, 155–157.

Emeis, K.-C., Schulz, H., Struck, U., Rossignol-Strick, M., Erlenkeuser, H., Howell, M. W., Kroon, D., Mackensen, A., Ishizuka, S., Oba, T., Sakamoto, T., & Koizumi, I. (2003). Eastern Mediterranean surface water temperatures and 18 O composition during deposition of sapropels in the late Quaternary. *Paleoceanography, 18*(1), n/a–n/a. https://doi.org/10.1029/2000pa000617.

EPICA Community Members. (2004). Eight glacial cycles from an Antarctic ice core. *Nature, 429*, 623–628. https://doi.org/10.1038/nature02599.

EPICA Community Members. (2006). One-to-one coupling of glacial climate variability in Greenland and Antarctica. *Nature, 444*(7116), 195–198. https://doi.org/10.1038/nature05301.

Ezat, M. M., Rasmussen, T. L., Hönisch, B., Groeneveld, J., & deMenocal, P. (2017). Episodic release of CO_2 from the high-latitude North Atlantic Ocean during the last 135 kyr. *Nature Communications, 8*, 14498.

Ferrari, R., Jansen, M. F., Adkins, J. F., Burke, A., Stewart, A. L., & Thompson, A. F. (2014). Antarctic sea ice control on ocean circulation in present and glacial climates. *PNAS, 111*(24), 8753–8758. https://doi.org/10.1073/pnas.1323922111.

Fischer, H., Schmitt, J., Lüthi, D., Stocker, T. F., Tschumi, T., Parekh, P., et al. (2010). The role of Southern Ocean processes on orbital and millennial CO_2 variations—A synthesis. *Quaternary Science Reviews, 29*(1–2), 193–205. https://doi.org/10.1016/j.quascirev.2009.06.007.

Flückiger, J., Blunier, T., Stauffer, B., Chappellaz, J., Spahni, R., Kawamura, K., Schwander, J., Stocker, T. F., & Dahl-Jensen D. (2004). N_2O and CH_4 variations during the last glacial epoch: Insight into global processes. *Global Biogeochemical Cycles, 18*, GB1020. https://doi.org/10.1029/2003gb002122.

Garcia, H. E., Locarnini, R. A., Boyer, T. P., Antonov, J. I., Baranova, O. K., Zweng, M. M., & Johnson, D. R. (2010). World Ocean Atlas 2009, Volume 3: Dissolved oxygen, apparent oxygen utilization, and oxygen saturation. In: S. Levitus (Ed.), NOAA Atlas NESDIS 70, Washington, D.C.: U.S. Government Printing Office, 344 pp.

Gersonde, R., Crosta, X., Abelmann, A., & Armand, L. (2005). Sea-surface temperature and sea ice distribution of the Southern Ocean at the EPILOG Last Glacial Maximum—A circum-Antarctic view based on siliceous microfossil records. *Quat. Sci. Rev., 24*, 869–896.

Goosse, H., Roche, D. M., Mairesse, A., & Berger, M. (2013). Modeling past sea ice changes. *Quaternary Science Reviews, 79*, 191–206. https://doi.org/10.1016/j.quascirev.2013.03.011.

Gottschalk, J., Skinner, L. C., Lippold, J., Vogel, H., Frank, N., Jaccard, S. L., et al. (2016). Biological and physical controls in the Southern Ocean on past millennial-scale atmospheric CO_2 changes. *Nature Communications, 7*, 11539.

Grant, K. M., Rohling, E. J., Bar-Matthews, M., Ayalon, A., Medina-Elizalde, M., Bronk Ramsey, C., et al. (2012). Rapid coupling between ice volume and polar temperature over the past 150 kyr. *Nature, 491*, 744–747.

Grimm, R., Maier-Reimer, E., Mikolajewicz, U., Schmiedl, G., Müller-Navarra, K., Adloff, F., Grant, K. M., Ziegler, M., Lourens, L. J., & Emeis, K.-C. (2015). Late glacial initiation of Holocene eastern Mediterranean sapropel formation. *Nature Communications, 6*(7099), 12 pp. https://doi.org/10.1038/ncomms8099.

Hamon, N., Sepulchre, P., Lefebvre, V., & Ramstein, G. (2013). The role of eastern Tethys seaway closure in the Middle Miocene Climatic Transition (ca. 14 Ma). *Climate of the Past, 9*(6), 2687–2702. https://doi.org/10.5194/cp-9-2687-2013.

Harrison, S. P., Yu, G., Takahara, H., & Prentice, I. C. (2001). Palaeovegetation—Diversity of temperate plants in east Asia. *Nature, 413*, 129–130.

Hemleben, C., Hoernle, K., Jørgensen, B. B., & Roether, W. (2003). Ostatlantik-Mittelmeer- Schwarzes Meer, Cruise 51, 12 September–28 December 2001. Meteor Ber. 03-1, 213 pp.

Hemming, S. R. (2004). Heinrich events: Massive late Pleistocene detritus layers of the North Atlantic and their global climate imprint. *Reviews of Geophysics, 42*, RG1005. https://doi.org/10.1029/2003rg000128.

Heinrich, H. (1988). Origin and consequences of cyclic ice rafting in the Northeast Atlantic Ocean during the past 130,000 years. *Quaternary Research, 29*(2), 142–152. https://doi.org/10.1016/0033-5894(88)90057-9.

Hesse, T., Butzin, M., Bickert, T., & Lohmann, G. (2011). A model-data comparison of $\delta^{13}C$ in the glacial Atlantic Ocean. *Paleoceanography, 26*, PA3220. https://doi.org/10.1029/2010pa002085.

Holland, H. D. (1994). *Early Life on Earth*. In S. Bengston (Ed.) (pp. 237–244). New York: Columbia University Press.

Jaccard, S. L., & Galbraith, E. D. (2011). Large climate-driven changes of oceanic oxygen concentrations during the last deglaciation. *Nature Geoscience, 5*, 151–156.

Jaccard, S. L., & Galbraith, E. D. (2012). Large climate-driven changes of oceanic oxygen concentrations during the last deglaciation. *Nature Geoscience, 5*, 151–156.

Jaccard, S., Galbraith, E., Frölicher, T., & Gruber, N. (2014). Ocean (de)oxygenation across the last deglaciation: Insights for the future. *Oceanography, 27*, 26–35.

Jickells, T. D., An, Z. S., Andersen, K. K., Baker, A. R., Bergametti, G., Brooks, N., et al. (2005). Global iron connections between desert dust, ocean biogeochemistry, and climate. *Science, 308*, 67. https://doi.org/10.1126/science.1105959.

Jouzel, J., Masson-Delmotte, V., Cattani, O., Dreyfus, G., Falourd, S., Hoffmann, G., et al. (2007). Orbital and Millennial Antarctic climate variability over the past 800,000 years. *Science, 317*(5839), 793–797.

Kaplan, J. O., Folberth, G., & Hauglustaine, D. A. (2006). Role of methane and biogenic volatile organic compound sources in late glacial and Holocene fluctuations of atmospheric methane concentrations. *Global Biogeochemical Cycles, 20*, GB2016. https://doi.org/10.1029/2005gb002590.

Kohfeld, K. E., & Harrison, S. P. (2001). DIRTMAP: The geological record of dust. *Earth-Science Reviews, 54*, 81–114. https://doi.org/10.1016/S0012-8252(01)00042-3.

Kohfeld, K. E., Quéré, C. L., Harrison, S. P., & Anderson, R. F. (2005). Role of marine biology in glacial-interglacial CO_2 cycles. *Science, 308*, 74–78.

Kohfeld, K. E., Graham, R. M., de Boer, A. M., Sime, L. C., Wolff, E. W., Le Quéré, C., et al. (2013). Southern Hemisphere westerly wind changes during the Last Glacial Maximum: Paleo-data synthesis. *Quaternary Science Reviews, 68*, 76–95. https://doi.org/10.1016/j.quascirev.2013.01.017.

Köhler, P., Knorr, G., & Bard, E. (2014). Permafrost thawing as a possible source of abrupt carbon release at the onset of the Bølling/Allerød. *Nature Communications*. https://doi.org/10.1038/ncomms6520.

Köng, E., Zaragosi, S., Schneider, J. L., Garlan, T., Bachèlery, P., Sabine, M., et al. (2017). Gravity-driven deposits in an active margin (Ionian Sea) over the last 330,000 years. *Geochemistry, Geophysics, Geosystems, 18*(11), 4186–4210. https://doi.org/10.1002/2017GC006950.

Kullenberg, B. (1952). On the salinity of the water contained in marine sediments. *Meddelanden från Oceanografiska institutet i Göteborg, 21,* 1–38.

Lambert, F., Tagliabue, A., Shaffer, G., Lamy, F., Winckler, G., Farias, L., et al. (2015). Dust fluxes and iron fertilization in Holocene and Last Glacial Maximum climates. *Geophysical Research Letters, 42,* 6014–6023. https://doi.org/10.1002/2015GL064250.

Landais, A., Dreyfus, G., Capron, E., Jouzel, J., Masson-Delmotte, V., Roche, D. M., et al. (2013). Two-phase change in CO_2, Antarctic temperature and global climate during termination II. *Nature Geoscience, 6,* 1062–1065. https://doi.org/10.1038/NGEO1985.

Legrand, M., Feniet-Saigne, C., Sattzman, E. S., Germain, C., Barkov, N. I., & Petrov, V. N. (1991). Ice-core record of oceanic emissions of dimethylsulphide during the last climate cycle. *Nature, 350,* 144–146.

Legrand, M., & Mayewski, P. (1997). Glaciochemistry of polar ice cores: A review. *Reviews of Geophysics, 35*(3), 219–243.

Levine, J. G., Wolff, E. W., Jones, A. E., Sime, L. C., Valdes, P. J., Archibald, A. T., et al. (2011). Reconciling the changes in atmospheric methane sources and sinks between the Last Glacial Maximum and the pre-industrial era. *Geophys. Res. Lett., 38,* L23804. https://doi.org/10.1029/2011GL049545.

Loulergue, L., Schilt, A., Spahni, R., Masson-Delmotte, V., Blunier, T., Lemieux, B., et al. (2008). Orbital and millennial-scale features of atmospheric CH_4 over the past 800,000 years. *Nature, 453,* 383–386. https://doi.org/10.1038/nature06950.

Lüthi, D., Le Floch, M., Bereiter, B., Blunier, T., Barnola, J.-M., Siegenthaler, U., et al. (2008). High-resolution carbon dioxide concentration record 650,000–800,000 years before present. *Nature, 453,* 379–382. https://doi.org/10.1038/nature06949.

Lourantou, A., Lavrič, J. V., Köhler, P., Barnola, J.-M., Paillard, D., Michel, E., Raynaud, D., & Chappellaz, J. (2010). Constraint of the CO_2 rise by new atmospheric carbon isotopic measurements during the last deglaciation. *Global Biogeochemical Cycles, 24,* GB2015. https://doi.org/10.1029/2009gb003545.

Mahowald, N. M., Scanza, R., Brahney, J., Goodale, C. L., Hess, P. G., Moore, J. K., et al. (2017). Aerosol deposition impacts on land and ocean carbon cycles. *Current Climate Change Reports, 3*(1), 16–31.

Marcott, S. A., Bauska, T. K., Buizert, C., Steig, E. J., Rosen, J. L., Cuffey, K. M., et al. (2014). Centennial-scale changes in the global carbon cycle during the last deglaciation. *Nature, 514,* 616–619. https://doi.org/10.1038/nature13799.

Mariotti, V. (2013). *Le cycle du carbone en climat glaciaire: état moyen et variabilité,* Ph.D. thesis. 252 p.

Marchal, O., & Curry, W. B. (2008). On the Abyssal circulation in the Glacial Atlantic. *Journal of Physical Oceanography, 38,* 2014–2037. https://doi.org/10.1175/2008JPO3895.1.

Martinez-Ruiz, F., Kastner, M., Paytan, A., Ortega-Huertas, M., & Bernasconi, S. (2000). Geochemical evidence for enhanced productivity during S1 sapropel deposition in the eastern Mediterranean. *Paleoceanography, 15*(2), 200–209. https://doi.org/10.1029/1999PA000419.

Marzocchi, A., & Jansen, M. F. (2017). Connecting Antarctic sea ice to deep-ocean circulation in modern and glacial climate simulations. *Geophysical Research Letters, 44,* 6286–6295. https://doi.org/10.1002/2017GL073936.

Maslin, M., Adams, J., Thomas, E., Faure, H., & Haines-Young, R. (1995). Estimating the carbon transfer between the ocean, atmosphere and the terrestrial biosphere since the Last Glacial Maximum. *Terra Nova, 7*(3), 358–366.

Matsumoto, K., & Yokoyama, Y. (2013). Atmospheric $\Delta^{14}C$ reduction in simulations of Atlantic overturning circulation shutdown, *Global Biogeochemical Cycles, 27.* https://doi.org/10.1002/gbc.20035.

Menviel, L., Timmermann, A., Mouchet, A., & Timm, O. (2008a). Climate and marine carbon cycle response to changes in the strength of the Southern Hemispheric westerlies. *Paleoceanography, 23,* PA4201.

Menviel, L., Timmermann, A., Mouchet, A., & Timm, O. (2008b). Meridional reorganizations of marine and terrestrial productivity during Heinrich events. *Paleoceanography, 23,* PA1203. https://doi.org/10.1029/2007pa001445.

Menviel, L., Spence, P., & England, M. H. (2015). Contribution of enhanced Antarctic bottom water formation to Antarctic warm events and millennial-scale atmospheric CO_2 increase. *Earth and Planetary Science Letters, 413,* 37–50. https://doi.org/10.1016/j.epsl.2014.12.050.

Menviel, L., Yu, J., Joos, F., Mouchet, A., Meissner, K. J., & England, M. H. (2017). Poorly ventilated deep ocean at the Last Glacial Maximum inferred from carbon isotopes: A data-model comparison study. *Paleoceanography, 32,* 2–17. https://doi.org/10.1002/2016PA003024.

Milkov, A. V. (2004). Global estimates of hydrate-bound gas in marine sediments: How much is really out there? *Earth-Science Reviews, 66*(3–4), 183–197. https://doi.org/10.1016/j.earscirev.2003.11.002.

Möbius, J., Lahajnar, N., & Emeis, K. C. (2010). Diagenetic control of nitrogen isotope ratios in Holocene sapropels and recent sediments from the Eastern Mediterranean Sea. *Biogeosciences, 7*(11), 3901–3914. https://doi.org/10.5194/bg-7-3901-2010.

Muglia, J., & Schmittner, A. (2015). Glacial Atlantic overturning increased by wind stress in climate models. *Geophysical Research Letters, 42.* https://doi.org/10.1002/2015GL064583.

Murray, L. T., Mickley, L. J., Kaplan, J. O., Sofen, E. D., Pfeiffer, M., & Alexander, B. (2014). Factors controlling variability in the oxidative capacity of the troposphere since the Last Glacial Maximum. *Atmospheric Chemistry and Physics, 14,* 3589–3622. https://doi.org/10.5194/acp-14-3589-2014.

NEEM Community Members. (2013). Eemian interglacial reconstructed from a Greenland folded ice core. *Nature, 493,* 489–494. https://doi.org/10.1038/nature11789.

Nijenhuis, I. A., & De Lange, G. J. (2000). Geochemical constraints on Pliocene sapropel formation in the eastern Mediterranean. *Mar. Geol., 163,* 41–63.

North Greenland Ice Core Project Members. (2004). High-resolution record of Northern Hemisphere climate extending into the last interglacial period. *Nature, 431,* 147–151.

Obata, A. (2007). Climate–carbon cycle model response to freshwater discharge into the North Atlantic. *Journal of Climate, 20,* 5962–5976. https://doi.org/10.1175/2007JCLI1808.1.

O'Dowd, C. D., & de Leeuw, G. (2007). Marine aerosol production: A review of the current knowledge. *Philosophical Transactions of the Royal Society A, 365*(1856), 1753–1774.

Parrenin, F., Masson-Delmotte, V., Köhler, P., Raynaud, D., Paillard, D., Schwander, J., Barbante, C., Landais, A., Wegner, A., & Jouzel, J. (2013). Synchronous change of atmospheric CO_2 and Antarctic temperature during the last deglacial warming. *Science, 339,* 1060. https://doi.org/10.1126/science.1226368.

Petit, J. R., Jouzel, J., Raynaud, D., Barkov, N. I., Barnola, J. M., Basile, I., et al. (1999). Climate and atmospheric history of the past 420,000 years from the Vostok Ice Core, Antarctica. *Nature, 399,* 429–436.

Pickett, E. J., Harrison, S. P., Hope, G., Harle, K., Dodson, J. R., Peter Kershaw, A., et al. (2004). Pollen-based reconstructions of biome distributions for Australia, Southeast Asia and the Pacific (SEAPAC region) at 0, 6000 and 18,000 14C yr BP. *Journal of Biogeography, 31,* 1381–1444. https://doi.org/10.1111/j.1365-2699.2004.01001.x.

Plancq, J., Grossi, V., Pittet, B., Huguet, C., Rosell-Melé, A., & Mattioli, E. (2015). Multi-proxy constraints on sapropel formation during the late Pliocene of central Mediterranean (southwest Sicily). *Earth and Planetary Science Letters, 420,* 30–44.

Prather, M. J. (2007). Lifetimes and time scales in atmospheric chemistry. *Philosophical Transactions of the Royal Society A: Mathematical, Physical and Engineering Sciences, 365.* http://doi.org/10.1098/rsta.2007.2040.

Prentice, I. C., Jolly, D., & BIOME 6000 Participants. (2000). Mid-Holocene and glacial-maximum vegetation geography of the northern continents and Africa. *Journal of Biogeography 27,* 507–519.

Preunkert, S., Legrand, M., Jourdain, B., Moulin, C., Belviso, S., Kasamatsu, N., et al. (2007). Interannual variability of dimethylsulfide in air and seawater and its atmospheric oxidation by-products (methanesulfonate and sulfate) at Dumont d'Urville, coastal Antarctica (1999–2003). *Journal of Geophysical Research, 112,* D06306. https://doi.org/10.1029/2006JD007585.

Preunkert, S., & Legrand, M. (2013). Towards a quasi-complete reconstruction of past atmospheric aerosol load and composition (organic and inorganic) over Europe since 1920 inferred from Alpine ice cores. *Climate of the Past, 9*(4), 1403–1416.

Quiquet, A., Archibald, A. T., Friend, A. D., Chappellaz, J., Levine, J. G., Stone, E. J., et al. (2015). The relative importance of methane sources and sinks over the Last Interglacial period and into the last glaciation. *Quaternary Science Reviews, 112,* 1–16. https://doi.org/10.1016/j.quascirev.2015.01.004.

Rasmussen, S. O., Bigler, M., Blockley, S. P., Blunier, T., Buchardt, S. L., Clausen, H. B., et al. (2014). A stratigraphic framework for abrupt climatic changes during the last glacial period based on three synchronized Greenland ice-core records: Refining and extending the INTIMATE event stratigraphy. *Quaternary Science Reviews, 106,* 14–28.

Rhein, M., et al. (2013). Observations: Ocean. In T. F. Stocker et al. (Eds.), *Climate Change 2013: The Physical Science Basis. Contribution of Working Group I to the Fifth Assessment Report of the Intergovernmental Panel on Climate Change* (pp. 255–310). Cambridge, UK: Cambridge University Press.

Roche, D. M., Crosta, X., & Renssen, H. (2012). Evaluating Southern Ocean sea ice for the Last Glacial Maximum and pre-industrial climates: PMIP-2 models and data evidences. *Quaternary Science Reviews, 56,* 99–106. https://doi.org/10.1016/j.quascirev.2012.09.020.

Rogerson, M., Cacho, I., Jimenez-Espejo, F., Reguera, M. I., Sierro, F. J., Martinez-Ruiz, F., Frigola, J., & Canals, M. (2008). A dynamic explanation for the origin of the western Mediterranean organic rich layers. *Geochemistry, Geophysics, Geosystems, 9,* Q07U01. https://doi.org/10.1029/2007gc001936.

Rohling, E. J., Marino, G., & Grant, K. M. (2015). Mediterranean climate and oceanography, and the periodic development of anoxic events (sapropels). *Earth-Science Reviews, 143*(2015), 62–97.

Rosen, J. L., Brook, E. J., Severinghaus, J. P., Blunier, T., Mitchell, L. E., Lee, J. E., et al. (2014). An ice core record of near-synchronous global climate changes at the Bølling transition. *Nature Geoscience, 7,* 459–463.

Rossignol-Strick, M., Nesteroff, V., Olive, P., & Vergnaud-Grazzini, C. (1982). After the deluge; Mediterranean stagnation and sapropel formation. *Nature, 295,* 105–110.

Saigne, C., & Legrand, M. (1987). Measurements of methanesulphonic acid in Antarctic ice. *Nature, 330,* 240–242.

Schilt, A., Baumgartner, M., Blunier, T., Schwander, J., Spahni, R., Fischer, H., et al. (2010). Glacial–interglacial and millennial-scale variations in the atmospheric nitrous oxide concentration during the last 800,000 years. *Quaternary Science Reviews, 29,* 182–192.

Schilt, A., Baumgartner, M., Eicher, O., Chappellaz, J., Schwander, J., Fischer, H., et al. (2013). The response of atmospheric nitrous oxide to climate variations during the last glacial period. *Geophysical Research Letters, 40,* 1888–1893. https://doi.org/10.1002/grl.50380.

Schmitt, J., Schneider, R., Elsig, J., Leuenberger, D., Lourantou, A., Chappellaz, J., Köhler, P., Joos, F., Stocker, T. F., Leuenberger, M., & Fischer, H. (2012). Carbon isotope constraints on the deglacial CO_2 rise from ice cores. *Science, 336,* 711. https://doi.org/10.1126/science.1217161.

Schmittner, A., & Galbraith, E. D. (2008). Glacial greenhouse-gas fluctuations controlled by ocean circulation changes. *Nature, 456* (7220), 373–376. https://doi.org/10.1038/nature07531.

Schneider, R., Schmitt, J., Köhler, P., Joos, F., & Fischer, H. (2013). A reconstruction of atmospheric carbon dioxide and its stable carbon isotopic composition from the penultimate glacial maximum to the last glacial inception. *Climate of the Past, 9,* 2507–2523. https://doi.org/10.5194/cp-9-2507-2013.

Schwander, J., & Stauffer, B. (1984). Age difference between polar ice and the air trapped in its bubbles. *Nature, 311,* 45–47.

Shackleton, N. J. (1977). Carbon-13 in Uvegerina: Tropical rainforest history and the equatorial Pacific carbonate dissolution cycles. In N. R. Andersen & A. Malako (Eds.), *The fate of fossil fuel CO_2 in the oceans* (pp. 401–428). New York: Plenum.

Siegenthaler, U., Stocker, T. F., Monnin, E., Lüthi, D., Schwander, J., Stauffer, B., et al. (2005). Stable carbon cycle-climate relationship during the Late Pleistocene. *Science, 310*(5752), 1313–1317. https://doi.org/10.1126/science.1120130.

Sigman, D. M., & Boyle, E. A. (2000). Glacial/interglacial variations in atmospheric carbon dioxide. *Nature, 407,* 859–869.

Skinner, L. C., Fallon, S., Waelbroeck, C., Michel, E., & Barker, S. (2010). Ventilation of the deep Southern Ocean and deglacial CO_2 rise. *Science, 328*(5982), 1147–1151. https://doi.org/10.1126/science.1183627.

Skinner, L. C. (2009). Glacial-interglacial atmospheric CO_2 change: A possible "standing volume" effect on deep-ocean carbon sequestration. *Climate of the Past, 5,* 537–550. https://doi.org/10.5194/cp-5-537-2009.

Sowers, T., Alley, R. B., & Jubenville, J. (2003). Ice core records of atmospheric N_2O covering the last 106,000 years. *Science, 301,* 945–948.

Stephens, B. B., & Keeling, R. F. (2000). The influence of Antarctic sea ice on glacial-interglacial CO_2 variations. *Nature, 404,* 171–174.

Stramma, L., Prince, E. D., Schmidtko, S., Luo, J., Hoolihan, J. P., Visbeck, M., et al. (2012). Expansion of oxygen minimum zones may reduce available habitat for tropical pelagic fishes. *Nature Climate Change, 2,* 33–37.

Spahni, R., Chappellaz, J., Stocker, T. F., Loulergue, L., Hausammann, G., Kawamura, K., et al. (2005). Atmospheric methane and nitrous oxide of the late Pleistocene from Antarctic ice cores. *Science, 310,* 1317–1321.

Stocker, T. F., & Johnsen, S. J. (2003). A minimum thermodynamic model for the bipolar seesaw. *Paleoceanography, 18*(4), 1087. https://doi.org/10.1029/2003PA000920.

Tagliabue, A., Bopp, L., Roche, D. M., Bouttes, N., Dutay, J.-C., Alkama, R., et al. (2009). Quantifying the roles of ocean circulation and biogeochemistry in governing ocean carbon-13 and atmospheric carbon dioxide at the Last Glacial Maximum. *Climate of the Past, 5,* 695–706.

Tarnocai, C., Canadell, J. G., Schuur, E. A. G., Kuhry, P., Mazhitova, G., & Zimov, S. (2009). Soil organic carbon pools in the northern circumpolar permafrost region. *Global Biogeochemical Cycles, 23,* GB2023. https://doi.org/10.1029/2008gb003327.

Toggweiler, J. R., Russell, J. L., & Carson, S. R. (2006). Midlatitude westerlies, atmospheric CO_2, and climate change during the ice ages. *Paleoceanography, 21,* PA2005.

Tschumi, J., & Stauffer, B. (2000). Reconstructing past atmospheric CO_2 concentration based on ice-core analyses: Open questions due to in situ production of CO_2 in the ice. *Journal of Glaciology, 46* (152), 45–53. https://doi.org/10.3189/172756500781833359.

Tschumi, T., Joos, F., Gehlen, M., & Heinze, C. (2011). Deep ocean ventilation, carbon isotopes, marine sedimentation and the deglacial CO_2 rise. *Climate of the Past, 7,* 771–800. https://doi.org/10.5194/cp-7-771-2011.

Vadsaria, T., Ramstein, G., Dutay, J. C., Li, L., Ayache, M., & Richon, C. (2019). Simulating the occurrence of the last sapropel event (S1): Mediterranean basin ocean dynamics simulations using Nd isotopic composition modeling. *Paleoceanography and Paleoclimatology.* http://doi.org/10.1029/2019PA003566.

Valdes, P. J., Beerling, D. J., & Johnson, C. E. (2005). The ice age methane budget. *Geophysical Research Letters, 32,* L02704. https://doi.org/10.1029/2004GL021004.

WAIS Divide Project Members. (2015). Precise interpolar phasing of abrupt climate change during the last ice age. *Nature, 520,* 661–668. https://doi.org/10.1038/nature14401.

Walker, J. C. G. (1980). The oxygen cycle. In O. Hutzinger (Ed.), *The natural environment and the biogeochemical cycles* (pp. 87–104). Berlin, Heidelberg: Springer.

Wang, W., Yung, Y., Lacis, A., Mo, T., & Hansen, J. (1976). Greenhouse effects due to man made perturbations of trace gases. *Science, 194*(4266), 685–690.

Weber, S. L., Drury, A. J., Toonen, W. H. J., & van Weele, M. (2010). Wetland methane emissions during the Last Glacial Maximum estimated from PMIP2 simulations: Climate, vegetation, and geographic controls. *Journal Geophysical Research, 115,* D06111. https://doi.org/10.1029/2009JD012110.

Wolff, E. W., Fischer, H., Fundel, F., Ruth, U., Twarloh, B., Littot, G. C., Mulvaney, R., Röthlisberger, R., de Angelis, M., Boutron, C. F., Hansson, M., Jonsell, U., Hutterli, M. A., Lambert, F., Kaufmann, P., Stauffer, B., Stocker, T. F., Steffensen, J. P., Bigler, M., Siggaard-Andersen, M. L., Udisti, R., Becagli, S., Castellano, E., Severi, M., Wagenbach, D., Barbante, C., Gabrielli P., & Gaspari V. (2006). Southern Ocean sea-ice extent, productivity and iron flux over the past eight glacial cycles. *Nature 440,* 23. https://doi.org/10.1038/nature04614.

Yin, Q. Z., & Berger, A. (2010). Insolation and CO_2 contribution to the interglacial climate before and after the Mid-Brunhes Event. *Nature Geoscience, 3,* 243–246. https://doi.org/10.1038/ngeo771.

Yin, Q. Z., & Berger, A. (2012). Individual contribution of insolation and CO_2 to the interglacial climates of the past 800,000 years. *Climate Dynamics, 38*(3–4), 709–724.

Yu, J., Menviel, L., Jin, Z. D., Thornalley, D. J. R., Barker, S., Marino, G., Rohling, E. J., Cai, Y., Zhang, F., Wang, X., Dai, Y., Chen P., & Broecker W. S. (2016). Sequestration of carbon in the deep Atlantic during the last glaciation. *Nature Geoscience, 9,* 319–324.

Ziegler, M., Tuenter, E., & Lourens, L. J. (2010). The precession phase of the boreal summer monsoon as viewed from the eastern Mediterranean (ODP Site 968). *Quaternary Science Reviews, 29,* 1481–1490.

The Cryosphere and Sea Level

Catherine Ritz, Vincent Peyaud, Claire Waelbroeck, and Florence Colleoni

Introduction

Several times during the history of the Earth extensive ice sheets covered part of the continents. As a result, a significant proportion of freshwater was stored in solid form, which caused a drop in sea level.

Because of their impact on other components of the Earth system (atmosphere, ocean, land), the dynamics of these ice masses must be taken into account in order to understand the evolution of the climate over the time scale of the last glacial-interglacial cycles. This topic can be addressed in different ways, depending to the various scientific disciplines and tools. One approach is to characterize these ice sheets according to the traces they have left behind, whether on land or in marine records. Marine sediments contain a record of changes in the overall volume of ice over time through changes in the oxygen isotopic composition of calcareous fossils. Due to isotopic fractionation that takes place during the evaporation of water, the drop in sea level during cold periods has been accompanied by an enrichment of seawater, not only in salt, but also in heavy isotopes of water (water molecules containing the ^{18}O isotope of oxygen rather than the most widespread isotope, ^{16}O. See Chap. 20, Volume 1). This enrichment leads to variations in the isotopic composition of the calcareous shells of the foraminifera preserved in the sediments. However, the isotopic composition of foraminifera also depends on the temperature at which the calcite was formed, so the benthic signal must be corrected in order to deduce the variations in sea level.

Another approach is to try to understand the physical mechanisms governing the formation and evolution of these ice masses. In both cases, observations of the two large remaining ice sheets, Antarctica and Greenland are pertinent. Finally, numeric simulation uses all of the information gathered (mechanisms, data) to develop models to calculate the evolution of the polar ice caps as they interact with the climate. These tools (referred to later as 'ice sheet models') allow us to study, for example, the role of the ice sheets in the climate system, in particular the non-linear effects that can amplify the forcings caused by variations in the Earth's orbital parameters. These models are also indispensable tools to assess the rise of sea levels in the context of global warming.

What Is an Ice Sheet?

Some definitions of glaciological terms

An ice cap is a mass of freshwater ice which rests on the ground. A notable difference between an ice cap and a mountain glacier is that the highest point of an ice cap, usually centered on a massif, is made of ice and called a dome, while a glacier flows down from a mountain (or from an ice cap). Ice caps can be of different sizes such as mountain peaks completely covered in ice in almost all the great mountain ranges of the world as well as significantly larger ice caps on the Arctic archipelagos (Fig. 24.1), in Iceland, on the Antarctic Peninsula and on some islands in the Southern Ocean. Finally, when an ice mass covers the land on a continental scale it is called an ice sheet. Two large ice sheets, in Greenland and Antarctica, cover virtually the whole continent on which they lie. North America and Northern Eurasia were partially covered by such ice sheets during glacial periods (Fig. 24.1). Although ice caps and ice sheets are driven by the same processes, the discussion below focuses on ice sheets that are large enough to significantly modify sea level.

C. Ritz (✉) · V. Peyaud
Institute of Engineering, Univ. Grenoble Alpes, CNRS, IRD, Grenoble INP, IGE, 38000 Grenoble, France
e-mail: catherine.ritz@univ-grenoble-alpes.fr

C. Waelbroeck
Laboratoire d'Océanographie et du Climat : Expérimentation et Approches Numériques, LOCEAN/IPSL, Sorbonne Université-CNRS-IRD-MNHN, UMR7159, Paris, France

F. Colleoni
OGS (National Institute of Oceanography and Applied Geophysics), Borgo Grotta Gigante 42/c, Viale Aldo Moro 44, 34010 Sgonico (TS), Italy

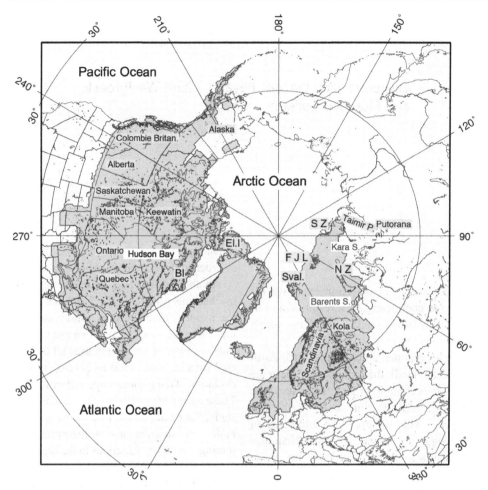

Fig. 24.1 Map of the northern hemisphere during the last glacial maximum. Ice caps and ice sheets are in blue (Peltier Ice4G). The ice sheet over North America is called Laurentide ice sheet. Sometimes, the ice cap over the Rocky Mountains is referred to as the Cordilleran ice cap. The Greenland ice sheet was not much larger than at present, but was probably connected to Ellesmere Island. The ice sheet over Eurasia was called the Fennoscandian ice sheet. The existence of a connection with the ice cap over the British islands is still disputed. It can be seen that a good part of the Fennoscandian ice sheet extended as far as Svalbard, occupying what is currently the sea of Barents. *Abbreviations* Sval—Svalbard; FJL—François Joseph Land; SZ—Severnaia Zembla; El.I.—Ellesmere Island; BI—Baffin Island. From the thesis of Peyaud (2007)

When the ice from an ice sheet flows to the coast, it begins to float on the sea, forming either tongues of ice or ice-shelves. Currently, the largest of these are the Ross and Ronne-Filchner ice shelves in Antarctica (see map on Fig. 24.2). The line where the ice begins to float is called the grounding line, and we will return to this point as it is important for the evolution of an ice sheet.

The thickness of the ice shelves can reach 2000 m at the grounding line, but they generally thin off rapidly as they reach the ocean and are about 200 m thick at the front. At the front, the ice shelves crack and break off, producing icebergs (calving phenomenon). Ice-shelves and sea ice should not be confused. The latter is made of seawater and, although some of the salt is expelled during the freezing process, they are still not composed of freshwater. Moreover, sea ice is only about a few meters thick.

Finally, the edge of a sheet (terrestrial or marine) is generally not uniform, but rather an alternation of fast-moving glaciers and relatively immobile zones. These fast-moving glaciers are called outlet glaciers. Their location depends on the topography of the bedrock (their flow is channeled into sub-glacial valleys) and on its geological properties with zones with relatively rapid flow (several hundred meters a year) being observed in the midst of relatively stagnant ice (a few meters per year). In Antarctica, some of these glaciers begin several hundred kilometers inside the ice sheet. When they are particularly wide (~ 40 km), they are also called ice streams.

Fig. 24.2 Satellite image of Antarctica (Blue Marble, NASA) with localization of the main ice shelves i.e. floating ice platforms. Siple coast is a region characterized by very large ice streams. Ice covers the whole of the emerged continent, with the exception of mountains arising above the ice, for example, in the peninsula and the trans-Antarctic Mountains

What Determines Sea Level?

It may seem easy to determine how sea level varied over time, but it is in fact a difficult characteristic to estimate. Distinct mechanisms intervene in these variations, involving important regional specificities. Since many observations are local, the representativeness of each measurement must be estimated according to the place, the time and the nature of the observation. Here, we identify the various processes that cause sea level to vary and assess how they impact on the interpretation of observations. Since we are interested in large time scales, we focus only on the level averaged over several years and so exclude the effect of tides and currents.

Sea level depends first of all on the mass of liquid water available, a quantity that changes over time due to its storage as ice on the continents during cold periods. This information is for the whole globe and it can be evaluated through reconstructions of the average isotopic composition of the ocean. Records of the isotopic composition of seawater are indirect and are essentially of two types:

(1) The isotopic composition of deep water during the Last Glacial Maximum (LGM) can be estimated by inversion of the isotopic composition diffusion profile of the interstitial water in the sedimentary column from the water-sediment interface (Schrag et al. 1996). In this way, in cores taken from the Pacific Ocean (the world's largest ocean basin), it was possible to establish that the average enrichment of the ocean during the LGM was 1.0 ± 0.1‰ compared to today.

(2) The changes in the isotopic composition of deep water can be reconstructed from isotopic analyses carried out on the shells of benthic foraminifera after subtracting the influence on these values of the variations in temperature of the deep waters as well as the local variations in the deep water isotopic composition (Waelbroeck et al. EPILOG 2002).

Variations in the average isotopic composition of the ocean can be translated into variations in the eustatic level of the seas. Current knowledge indicates that the relationship between these two quantities can be approximated initially using a constant multiplicative factor, so that an enrichment of 1‰ during the LGM corresponds to a decrease in the eustatic sea level of about 130 m (Fig. 24.3). Sea level changes have also been reconstructed from isotopic analyses

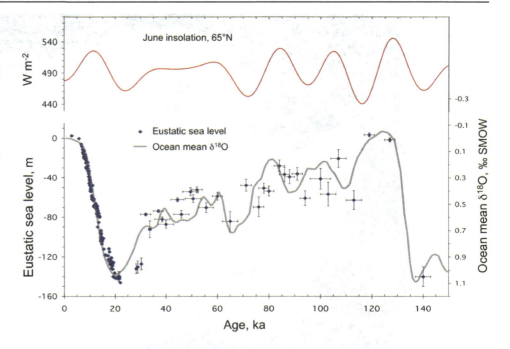

Fig. 24.3 Variations in summer sunshine levels at 65 °N and in the eustatic level of the sea during the last climate cycle. Eustatic levels calculated by Lambeck and Chappell (2001), based on measurements of the relative sea level. Reconstruction of $\delta^{18}O$ changes in the mean ocean from $\delta^{18}O$ of benthic foraminifera (Waelbroeck et al. EPILOG 2002)

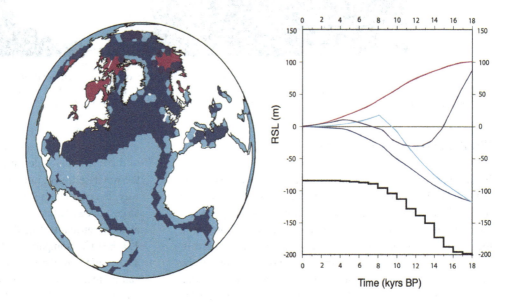

Fig. 24.4 Glacial isostasy

of the shells of planktonic foraminifera from the Red Sea, which record the large variations in salinity resulting from the closing and opening of the Bab el Mandab Straits caused by decreases and increases in sea level (Siddall et al. 2003).

At the glacial-interglacial scale, the changes in global ice volume constitutes the most important mechanism, and can cause variations in sea level of around one hundred meters (~130 m between the LGM and the present). These variations are called changes in the 'eustatic' level of the seas, and this corresponds roughly to the variations in the average level over the globe (see later for the concept of implicit ice). Alternatively, for the same mass of water, the volume changes with the temperature of the water because of thermal expansion (density depends on temperature). The impact of thermal expansion on the sea level is purely local, but its overall global average is sometimes added to the eustatic level of the seas.

In addition to the eustatic sea level, it is also important to take into account the isostatic variations which modify the shape of the ocean basins. Isostasy is the phenomenon of depression of the Earth's crust beneath the weight of the ice sheet (on land) and of water (under the oceans) (see Fig. 24.4). This means that both the changing quantity of liquid water and the changing shape of the basin need to be considered when attempting to measure the level of the water at the edges, which themselves can be affected by a vertical movement. Although it is now possible to observe the average level over the whole globe by satellite, for past periods, the vast majority of observations are taken at the coasts. Variations in relative sea level (RSL) is tracked at particular coastal points. It is important to emphasize that isostasy has a reaction time of several thousand years and that Scandinavia, for example, is still rising (at the rate of about one meter per century) following the disappearance of the ice sheet that covered northern Europe during the last glacial maximum. At each point, the relative sea level is therefore a combination of eustatic (global), isostatic, and possibly tectonic (local) variations. There are several ways to estimate this relative sea level, using traces of ancient beaches, coral analysis, speleothems etc.

Box 1. Glacial Isostasy

Post-glacial rebound: The relative changes in sea level are the result of tectonic processes and climate changes. These latter involve the accretion or melting of the ice sheets on the continents (more than 5 km thick). During ice ages, the pressure exerted by the accumulated weight of the ice sheets on the continents modified the shape of the surface creating large depressions in the lithosphere. These changes in shape brought about disturbances of the geoid (equipotential surface of the Earth's gravity field corresponding to the theoretical surface of the oceans) as well as vertical and horizontal displacements of the lithosphere and the mantle. When the shape of the Earth or the volume of the oceans changes, the geoid is thus modified as is the sea level. When the continental ice sheets melt, the lithosphere is released from the weight of the ice and the deflection is reabsorbed at a speed which depends on the viscosity of the mantle. This return is called the post-glacial rebound, post-glacial describing the period from −6000 (ka BP) to today. (6 ka BP marks the end of the melting of the ice caps of the northern hemisphere: the Laurentide and the Eurasian ice sheets having completely melted). For the Last Glacial Maximum (LGM), the relaxation time of the lithosphere is estimated at about 3000 years.

Sea level: Variations in sea level are caused by changes in the geoid surface, in the Earth's topography and in ocean mass. Sea level can be broken down into three contributing factors: the ice load (S^{ICE}), the oceanic charge (S^{OCE}) and the glacio-eustatic level (S^{EUS}).

$$S = S^{ICE} + S^{EUS} + S^{OCE} \quad (24.2.1)$$

Glacio-eustatic changes in sea level are controlled by the amount of ice stored on the continents. In glacial isostasy models, the glacio-eustatic sea level is defined for a totally rigid Earth without any disturbance of the gravity field. The glacial isostasy models allow us to estimate that the eustatic level during the Last Glacial Maximum was 130 ± 10 m below the current sea level (Lambeck, et al., EPILOG 2002). This estimate of the eustatic level is mainly based on the measurement of coral terraces in sites not affected by the postglacial rebound, such as the Seychelles, but also from calculations of the glacial isostasy models attempting to reproduce relative variations in sea level. The sea level at a given time and place in relation to the current sea level is defined as the relative sea level and can be described by the following equation:

$$\text{RSL}(\omega, t^{BP}) = S(\omega, t^{BP}) - S(\omega, t^P) \quad (24.2.2)$$

with ω being the geographical coordinates, t^{BP}, a time Before Present, et t^P, the present time. Equation (24.2.2) describes the relative level as the difference in sea level between the t^{BP} level and the current level. During the formation of ice sheets on the continents, the nearby regions and those further afield were affected differently depending on their distance from the ice masses and on the movements of the lithosphere and the mantle and therefore presented a sea level graph reflecting the trend of the regional ocean variations. These trends have been grouped into so-called 'Clark zones'. Figure 24.4 presents the result of modelled sea level changes due to postglacial rebound following the LGM according to the ICE-5G ice model (Peltier 2004). The graph shows changes in sea level for 18,000 years. The graphs are associated with their respective colors on the globe and each corresponds to a different regional trend of sea level variations. The pink and purple zones correspond to the regions that were either covered with ice during the LGM or were nearby, and so were very affected by the rebound of the lithosphere after the LGM. The dark blue zone represents regions showing subsidence of the lithosphere after the LGM. The light blue zone represents areas influenced by the late melting of ice from Antarctica after 9 ka BP (peak observed above the current sea level).

The level of the seas is an equipotential surface of the Earth's gravity field (the geoid) which is linked to the distribution of mass. This changes in the presence of an ice sheet or as a result of displacements of the asthenosphere (the ductile part of the Earth's mantle). This phenomenon must therefore also to be taken into account in the interpretation of relative sea levels. Although isostasy is a considerable complication for the interpretation of observations, this mechanism, on the other hand, it also provides

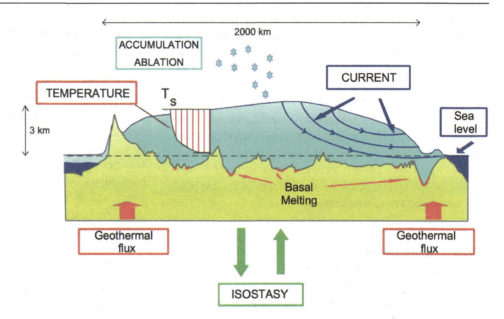

Fig. 24.5 Mechanisms involved in the evolution of polar ice sheets

information on the masses of the different ice sheets in the past. The extent of the ice sheet is generally estimated through geomorphological data. The inversion of relative sea level data using isostatic rebound models (including the geoid calculation) informs us about ice cap thicknesses in the past (Clark et al. 2002).

Finally, the ice sheets often lie on a bedrock that is below sea level. This is true for Western Antarctica currently. When the ice sheet melts, the sea replaces the volume of ice below the waterline and the impact on the sea level is thereby reduced. This volume of water is called 'implicit ice'.

Mechanisms Involved in the Evolution of an Ice Sheet

The evolution of an ice sheet obeys the law of conservation of mass. If the supply of snow is greater than the loss via melting or through the calving of icebergs, the sheet will grow or expand. In addition to this basic principle, there are many processes that interact with both external (atmosphere, ocean, solid Earth) and internal (for example, flow) interactions. In most cases, the geometry of the ice sheet is both a result of and a contributor to these feedbacks. The most important mechanisms are outlined in Fig. 24.5.

> **Box 2. Mechanisms Involved in the Evolution of Polar Ice Sheets**
> The geometry of an ice sheet depends on the supply of snow, the melting at the surface and the flow of ice to the edges where it may eventually drain into the ocean in the form of icebergs.

Surface mass balance is the difference between snowfall precipitation and ablation (fusion or sublimation). Redistribution by the wind can intervene as a positive or negative force, depending on the location. The accumulation zone is where the surface mass balance is positive (the central regions) and the ablation zone is where it is negative. We often refer to accumulation as the amount of snow accumulated before melting occurs. To take account of the density of snow, all these terms are generally expressed in water equivalent and the natural unit of time is the year (mass balance in m/year).

Flow: After snow is deposited, it becomes denser and transforms into firn and then into ice which flows due to the action of gravity depending on the slope of the surface. The displacement can therefore move across subglacial mountains or along with a rising bedrock. The flow can be broken down into two factors. One is the deformation of the ice which is linked to its mechanical properties. This means that it behaves like a highly viscous fluid, the viscosity of which depends on the temperature, the 'hotter' the ice, the more quickly it deforms. This deformation causes a variation in the velocity depending on depth and is generally concentrated in the underlying layers. The second is the ability of the ice to slide over the bedrock or, in the case of a sedimentary base, this base may deform. The two processes, the sliding and the sedimentary deformation, cause a horizontal velocity of the ice at the base. They can only act when the ice has reached melting point and their efficiency depends on the subglacial water pressure. We see that the ice flow depends on the field temperature, through its effect on the viscosity and on the melting point threshold at the base.

The temperature of the ice varies from the surface where it takes the annual average temperature (down to −60 °C in Antarctica) to the melting point at the many points of interface with the bedrock. The temperature thus increases overall with depth and this is mainly due to the geothermal flux which brings continuous heat to the base of the ice (of around 50 mW/m²). The flow affects temperature by transporting cold from the top to the bottom and from the center to the edges (advection). The heat produced by the deformation and by the sliding also has an impact and brings the base to melting point in active regions where speed of movement is high.

This interaction between velocity field and flow is called **thermo-mechanical coupling**.

Lastly, it is important to mention isostasy, which is the sinking of the bedrock under the weight of the ice (see Fig. 2.4).

Ice sheets have (and have had) different shapes and lifetimes, but they have followed a common pattern. An ice sheet begins to develop when, as a result of a cooling climate, snow becomes permanent in a region, i.e. when summer melting fails to eliminate all of the winter snow. It is usually in areas of high altitude where this process occurs first. In a mountain region, we see the emergence of glaciers in places where the slope is not too steep, that is to say, mainly in the high valleys. If the cooling continues, these glaciers will grow and descend until they fill up the lower valleys. When the region concerned is a high plateau, freezing is more abrupt with a threshold effect when the snow becomes perennial on the plateau. From this point on, the flow of ice plays a role by transporting ice to areas where snow is not permanent, whether by valley glaciers or by glaciers forming on the periphery of the plateau which bring ice to the surrounding plain.

From this moment on, two positive feedbacks supporting the ice expansion and cap formation are established. The first of these is the altitude-temperature feedback, which is based on the fact that atmospheric temperature drops with altitude. Areas where snow would normally not be permanent, freeze over due to the movement of ice, and this increases the altitude of their surface. As a result, the temperature at the surface of the ice drops and summer melting is reduced, so that eventually the snow becomes perennial. Over the years, this mechanism gradually causes the ice front to advance causing the accumulation zone (see Fig. 24.5) to extend further. The second feedback is the albedo. Because snow has a very high albedo (of around 0.9), the larger the accumulation zone, the greater the amount of solar energy reflected back to space instead of being used to heat the Earth's surface. The climate around an area of ice-cover will therefore change, facilitating the expansion of the permanent snow zone.

For a given climate, the expansion of an ice sheet will continue either until it reaches warm regions where melting prevails over the arriving flow of ice, or until it reaches the ocean and begins to float on the sea in the form of ice shelves, producing icebergs (see later the impact of geography).

It should be noted that we emphasize the impact of atmospheric temperature which governs the melting of snow and ice because this climate variable predominates. Precipitation, which governs the amount of snow accumulated, plays an important but lesser role and explains why, at the same temperature, certain regions become more easily covered in snow and ice than others. Among these regions are the glaciers in Norway today, but also the windward faces of reliefs (mountain or ice) where precipitation is increased by the orographic effect.

The same mechanisms apply to the retreat and disappearance of the ice caps (deglaciation) operating in the opposite direction within the context of global warming. The feedbacks mentioned are still active and they tend to amplify the retreat of the ice front. There are, however, two important differences. On the one hand, the growth rate of an ice sheet is limited by the supply of snow by precipitation, whereas the rate of decrease is limited by the amount of heat available to melt the ice. It is easy to understand that the latter limitation will not work well when the ice sheet reaches low latitudes and that deglaciation in this case can occur much faster than glaciation. This was the case for the two large ice sheets that existed during the Ice Age over America and Eurasia, and this had a strong impact on sea level. This explains the typical saw-tooth pattern of sea level rise and fall during a glacial period, with a slow drop followed by an abrupt rise (Fig. 24.3). On the other hand, the role of ice flow is not symmetrical between freezing and deglaciation. Although it helps the advance of the ice front as described, flow tends to thin the ice layer, so that its impact on the volume of ice is not uniform. In particular, it may happen that this flow 'runs away' and that the ice sheet then decreases abruptly (with the help of the altitude-temperature feedback), which can also lead to a rapid deglaciation phase.

Flow is a fundamental mechanism in the evolution of an ice sheet, and we explain below that it results from interaction between several processes which are themselves governed either by the major laws of physics or by laws (often empirical) describing the properties of the materials (ice of course but also the bedrock). We won't go into the detail of these equations which call upon continuum mechanics, but we shall nevertheless indicate what these laws are, and what implications they have on the flow of the ice caps. In the following, we shall only deal with the case of ice assumed to be incompressible. This hypothesis does not hold true for snow, whose density increases from the surface down to a maximum depth of 100 m, but in most cases, the mechanical effect of this layer of snow and firn is the same as a layer of ice of the same weight would be, which amounts to removing about twenty meters from the total thickness of ice at the chosen spot (the thickness is of the order of a few kilometers).

If we consider the balance of forces applied to an ice particle, the only body force is gravity. The other forces are surface forces arising from contact with other ice particles, or with the base or water (when it is floating). The ice flows sufficiently slowly so that we can ignore accelerations and inertial forces (Coriolis), and the balance of forces applied to ice sheets is often referred to as a quasi-static equilibrium. In terms of mechanical behavior and at the scale of the strain rate occurring in ice sheets, ice is considered as a viscous fluid, i.e. the strain rate (deformation per unit of time, directly expressed as function of the spatial derivative of the velocities) is connected to the stress. Water, for example, is also a viscous fluid, but its viscosity (lower than that of ice) does not depend on the stress (nor on the strain rate). This is called linear viscosity (also known as Newtonian). Ice, on the other hand, is characterized by non-linear viscosity which decreases with stress according to a power law (with an exponent of approximately 2): the more the ice deforms, the easier it is to deform. This type of behavioral law is not exceptional, it is also found for lava, mud and even chocolate. As with most viscous materials (maintaining the analogy with chocolate), the viscosity of ice decreases as its temperature increases (in an exponential relationship). Depending on the location of the ice sheet, the temperature can vary from −50 °C at the surface to melting point at the base, and this can influence the viscosity by a factor of up to 500.

Combining the quasi-static equilibrium, the law of viscous behavior, incompressibility and the various boundary conditions, we arrive at a system of equations which rigorously takes into account the mechanical equations. This system (called 'full stokes') can be solved numerically, but the cost in terms of computation time is such that it is not conceivable at this time to apply this method to the entire ice sheet, especially when the aim is to also study its temporal evolution. This approach is therefore limited to localized research. Fortunately, an approximation exists which takes a 'thin layer' approach making it possible to treat the ice sheet overall. Indeed, a characteristic of ice sheets is their very small aspect ratio, i.e. the ratio of thickness to expanse. For Antarctica, for example, the thickness is around 3 km and the expanse is 3000 km (ratio of 1/1 000). If it were made into a scale model 3 m wide, it would only be 3 mm thick and would look like a thin sheet of ice. Capitalizing on this aspect ratio, two separate approximations have been proposed: one for the part of the ice sheet resting on land (Shallow Ice Approximation, SIA), the other for the floating part (Shallow Shelf Approximation, see later). These approximations are used in ice sheet models (Ritz 2001). Moreover, they allow a qualitative understanding of the interaction between the geometry of an ice sheet and its flow. For the resting part for example, the SIA shows that the velocity of the ice (averaged over its thickness) is proportional to the thickness to the power of four and to the slope of the surface to the power of three. Although the thickness shows little variation over the entire ice sheet, the slope varies from 10^{-3} in the central regions to nearly 10^{-2} at the edges, thus implying a speed 1000 times greater (the size of variation observed in reality). This also explains why the thickness of an ice sheets is strongly related to their expanse with the amount of snow accumulation being of only minor importance. When an ice sheet grows, its slope at the surface increases and its drainage increases greatly, creating a negative feedback which limits thickening. Another result of the SIA, used in the interpretation of ice cores, concerns the fact that most of the deformation is concentrated in the layers near the bottom, and that higher up, horizontal speed changes little with depth and in a first approximation, the thinning of the ice layers decreases linearly with depth.

The ice flows through deformation but its velocity at the interface with the bedrock (basal velocity) also contributes to the flow. Two processes intervene, the sliding over the bedrock and the deformation of the underlying sediment. Based on Antarctic observations, it appears that it is the latter mechanism which is the most effective, leading to speeds of several hundred meters per year. In both cases, the basal velocity is negligible as long as the temperature at the interface is below melting point. On the other hand, at the melting point, not only is sliding possible but water is produced. This brings about another mechanism, the higher the water pressure, the greater the basal velocity, due to both a lubrication effect (less friction) and the fact that the water-saturated sediment is easier to deform.

The temperature field in the ice thus affects the flow in at least three ways: through the viscosity, through the threshold (melting point) from which basal movement is possible and through the subglacial water pressure. The temperature in the ice can be estimated reasonably well by solving the heat equation and taking into account any changes in surface temperature over time. In general, temperature increases with depth. It is very cold at the surface. At around 10 m depth, where seasonal variations are mitigated, it has the value of the mean annual temperature. In Antarctica, for example, the temperature at 10 m varies from about −20 °C at the coast to −60 °C in the center. At the base of the ice, the temperature is often close to melting point because geothermal energy (coming from the Earth) brings heat continuously to the base of the ice and the ice acts as an insulating material preventing this heat from escaping into the atmosphere. The thicker the ice, the higher the basal temperature. This temperature also depends on the amount of geothermal energy, which unfortunately is not well understood underneath the current ice sheets where it is difficult to measure directly. The variation in temperature with depth is not linear (this would be pure diffusion), because the flow transports cold from the top to the bottom and from upstream to downstream (advection). Moreover, the deformation of the ice produces heat (as it does with all materials). The same is true for the sliding on the bedrock and for the deformation of the sediment. Areas with rapid flow have a relatively warm base. If we ignore the phenomena related to flow (diffusive case), temperature would increase linearly with depth. Because of the flow, the ice is actually colder overall at the top and warmer at the base. Finally, if the base is at melting point, the additional heat is used for melting. It should be noted that the melting point drops with the ice pressure and therefore the depth. Below 3000 m of ice, the melting point is around -2.2 °C.

It can be seen that the temperature and velocity variables are connected by several terms. This interdependence is called 'thermomechanical coupling', and this coupling leads to a positive feedback which can have important effects on the evolution of the ice sheet. Suppose, for example, that an ice sheet is growing. Its basal temperature increases due to the insulating effect of the thickness of the ice, which reduces its viscosity and favors deformation. This produces heat which in turn increases the temperature. This positive feedback loop will quickly bring the base to the melting point where the water produced introduces a second feedback (the more water there is, the faster the basal velocity and the more heat is produced which in turn melts the ice). These feedbacks have been suggested as explanations for armadas of icebergs recorded as layers of ice rafted debris (IRD) in marine sediments during glacial periods (see Chap. 20). It is clear that this mechanism is involved in triggering a rapid flow. On the other hand, the most realistic numerical models have difficulties to simulate the return to a slow phase suggested by observations of rapid climate variability during glacial periods.

When the ice sheet ends with a marine edge, the ice streams enter the sea and form ice shelves. If the bay is sufficiently closed, these ice shelves limit the flow of the streams from which they originate or which flow into the same bay. This is called a buttressing effect. This effect occurs at all levels, from small bays of a few kilometers to large ice shelves such as the Ross and Ronne (see Fig. 24.2).

The geometry and dynamics of ice streams are balanced by this buttress but if an ice shelf disintegrates (for example, through the effect of oceanic heat melting it from below), the drainage of all the tributary ice streams can be greatly increased. This phenomenon occurred a few years ago in the Antarctic Peninsula after the disintegration of the Larsen ice shelf. The question arises of how much it might have contributed to glacial variations in the past and whether or not it offers an alternative explanation to the IRD layers found in marine cores and which testify to the arrival of icebergs.

To mechanically model the floating part, we can use the shallow shelf approximation to determine a relationship between the extension rate (variations in speed along the length of the flow) and the thickness of the ice shelf (at a specific power determined by the authors). If the ice-shelf is not confined (if it does not encounter resistance from the coasts or an island), then its speed increases from the grounding line to the front where it can exceed one kilometer per year. Above all, this approximation makes it possible to demonstrate that the grounding line cannot be stable if the base is below sea level and with a reverse slope (if it goes upwards as one goes from the center to the edge of the ice sheet, Schoof 2007). This instability can be explained qualitatively as follows: if the ice flow at the grounding line increases strongly as the ice thickens, any retreat of the grounding line, caused, for example, by the disintegration of the corresponding ice shelf, will intensify drainage, leading to a further retreat of the grounding line. This result is particularly important because it indicates that some ice sheets are inherently unstable because the location of the grounding line determines the extent of the cap and its volume. This is especially the case for Western Antarctica, but also for some parts of East Antarctica. For ice sheets in previous ice ages, it is possible that this instability could have contributed to rapid deglaciation, a process which is not yet fully understood.

Finally, subglacial isostasy (see Fig. 24.4) is a mechanism which plays an important role because it modulates many of the others. For example, subsidence of the Earth's crust under the weight of the ice does not directly change its thickness but it alters the altitude of the surface allowing more intense melting since the surface is lower. Moreover, the slope of the surface is also decreased, which slows down the flow and tends to make the ice sheet larger. Since isostasy occurs with a lag, this also leads to highly non-linear effects. The position of the grounding line is another example of an element that is very sensitive to isostasy because it is defined by floating (on the marine side), and this is determined by the relative sea level, itself affected by isostasy.

Reconstruction of Sea Levels and Ice Sheets in the Past

Over the past several millions of years, the Earth has seen a succession of glacial periods during which huge ice sheets covered North America, Eurasia and many mountain ranges around the world (including the Alps). These long glacial periods were interspersed with interglacials, as is the case in the present day, when a warm climate prevailed, confining freezing to the polar regions and to the highest mountains.

These evolutions of the cryosphere have been studied since the nineteenth century, but a global understanding of these phenomena has only become possible in the past few decades, thanks to direct observation of the current ice sheets and to the discovery of the paleoclimatic information contained in numerous sedimentary deposits.

Data that Enable the Reconstruction of the Geography of Ancient Ice Sheets

Reconstruction of sea levels using the $^{18}O/^{16}O$ ratio of the ocean provides a representation of how the amount of ice evolved over time (Fig. 24.3), which acts as a good indicator of the overall volume of all of the ice sheets. However, to know where these ice sheets were located (America, Eurasia, etc.) and how far they extended, we must rely on geomorphological information which we shall briefly review. Figure 24.1 (Last Glacial Maximum in the northern hemisphere) and 2.2 (Antarctic) show the different regions mentioned.

The expanse and dynamics of ancient ice sheets can be estimated directly from the deposits and traces that they left behind on the ground. In general, pre-glacial bedrocks are preserved in areas where the base of the ice remained cold; while in regions where the base of the ice reached melting point (temperate base), the flow of ice and of basal water reshaped the bedrock. We can then distinguish between formations due to glacial erosion and those due to sediment deposition. Glacial erosion abrades (polishes) the rock outcrops and creates incisions (streaks caused by the scraping of transported debris). The drop in pressure downstream of the obstacles causes freezing of the basal water and the fracturing and plucking of rocks. This passage creates 'roches moutonnées' (or 'sheepback rocks'). On a larger scale, glacial erosion hollows out valleys into troughs (U-shaped), especially on coasts as fjords are formed. At sea, it is possible, using multi-frequency sonars, to observe underwater channels that have obviously been dug out by the flow of ice, highlighting the location of 'paleo ice streams' (Anderson et al. EPILOG 2002). Observations of this type have been detected around the Antarctic and the Arctic Ocean. All these marks indicate the direction and route of the ice flow locally. In some cases, a change in direction over time has been observed and this information makes it possible to infer variations in the geometry of the ice sheet, in particular the displacement of the domes.

The deposition of transported material leads to the formation of a wide variety of moraines. We note the frontal moraines which mark the maximum extent of each advance of a glacier. Ground moraines are not very thick (a few meters on average) and may be flat or irregular. Drumlins, often grouped in fields, are elongated ovoidal hills. All these deposits are structured by the flow of ice, and thus indicate its direction. During periods of retreat, a sub-glacial hydrological network is formed if the base is temperate. The deposits associated with these phases are formed when the materials transported are abandoned, e.g. eskers at the site of former sub-glacial canals. An esker occurs in the form of an elongated ridge sometimes over hundreds of meters in length. These ridges are formed by materials being deposited in the tunnels of the subglacial rivers located at the base of the glaciers. Their often winding shape follows that of the tunnels that created them. All these traces left behind on the ground have been observed and compiled since the nineteenth century. For about twenty years, satellites have helped to give a large scale view of these lineations. Traces sometimes indicate a multitude of contradictory directions that reflect changes in the flow of the ice over time. The traces must therefore be classified chronologically and then interpreted in large coherent sets. Based on this synthesis, it is possible to recreate the geometry of the caps at different times. A major difficulty arises from the fact that an advancing glacier moves the deposits of previous glaciers and, through erosion, can erase previous traces. Therefore, we can only reconstruct the glaciers from previous glaciations if their traces have not been erased, so only if another more extensive glaciation has not occurred since. This is the case for Fennoscandia, where the moraines of the penultimate glaciation (Saalien or MIS 6, ~180–140 ka BP) are still visible much further south than those of the Last Glacial Maximum (see Fig. 24.6). Another difficulty with formulating an overview comes from the dating of geomorphological traces. For recent periods, radiocarbon dating allows very precise estimates for the past 30,000 years approximately, although it is essential to use organic matter. Prior to this period, which only covers from the end of the last glaciation to the present, or when there is no organic material, other techniques are necessary. Optically stimulated luminescence (OSL) indicates how long a rock has been exposed to solar radiation and has been used, for example, to date quartz crystals in moraines throughout the last glaciation. More accurate datings have radically changed our notions of the history of Fennoscandia (Svendsen et al. 2004).

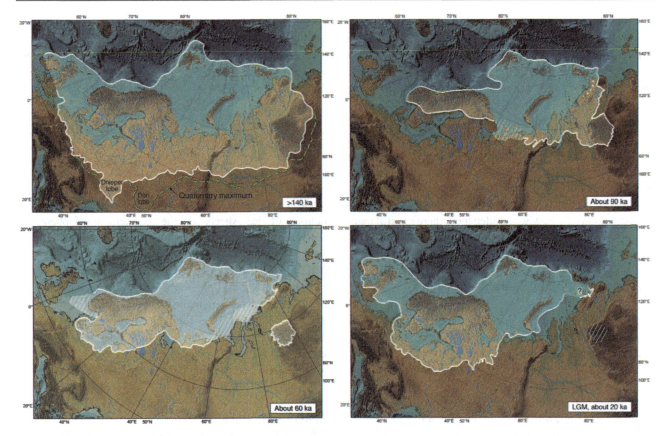

Fig. 24.6 The Eurasian ice sheet at various times in the past. Each map, centered on the Urals, represents the same region of northern Eurasia but at different times. The extent of the cap was determined from geomorphological data. During the Saalien (the penultimate glacial maximum, age >140 ka), the ice sheet extended particularly far towards the south reaching almost its maximum size during the Quaternary. The presence of lobes (the Dnieper during the Saalien and the Don for the maximum expansion) should be noted. These lobes may be the result of specific events called surges, during which the flow of glaciers or ice streams is particularly rapid. A tongue of ice then forms before disappearing, because its low altitude makes it prone to ablation. It can be seen that at the beginning of the Ice Age (90 ka), the ice sheet is prominent in the East but limited towards the West over Scandinavia. During the Last Glacial Maximum, the situation is reversed. Note that between each period represented here, there were periods when it was relatively clear of ice and snow, with the disappearance of the ice sheet located over the Sea of Barents and the separation into two sheets. Adapted from Svendsen et al. (2004)

The first major consolidated work to reconstruct the ice sheets at the end of the last glaciation was published by Denton and Hughes (1981). Their method of reconstruction involved the assumption that the profile of the ice sheets was parabolic (a hypothesis which is more or less valid for current ice sheets but which is based on the assumption of a stationary state) and the use of geomorphological data to constrain the extent. Since moraine dating was not as accurate at the time as it is now, in many areas it was assumed that the moraines were from the Last Glacial Maximum even though some had been formed during previous glaciations. Their reconstructions (even the 'minimal' one) are therefore considerably overestimated compared to the values acknowledged today. These reconstructions were also in contradiction with the global ice volume estimates from other sources (marine sediments, Mix and Ruddiman 1984).

Afterwards, understanding quickly evolved into a more dynamic view of ice sheets. For example, Boulton and Clark (1990) proposed a history of the Laurentide and in particular the migration and junction of the various domes (we will return to this in the section on the Laurentide). As for Fennoscandia, Lambeck et al. (2006) present its evolution throughout the last glaciation and Hughes et al. (2016) propose a chronology of the deglaciation. This type of overview often requires input from several groups of researchers and large-scale projects were established to estimate the extent and dynamics of ancient ice sheets (EPILOG 2002; QUEEN, Svendsen et al. 2004).

From the inversion of relative sea level data (RSL) and observations of the isostatic rebound which is still occurring, it is possible to estimate the ice load that produced these isostatic variations, and therefore the thickness of the ice sheets. This was the approach mainly used for the Last Glacial Maximum and for the last deglaciation (Peltier 2004; Clark et al. 2002; Lambeck et al. 2006).

Using all this information, it is possible to envisage a scenario representing the life of the ice sheets. This is the story outlined in the next section, with a focus on the last glacial-interglacial cycle for which there is more information available. It should be borne in mind, however, that although the main points are now well understood, there are still many regions and periods for which information is lacking.

The Last 50 Million Years

The Antarctic continent is located in the high latitudes since from the Late Cretaceous era ($\sim$240 Ma). However, it remained free of ice until the Eocene-Oligocene transition, 34 Ma ago (Chap. 6, Volume 2). Marine sediment recordings indicate an abrupt increase in benthic $\delta^{18}O$ at this date, suggesting that water depleted in ^{18}O was trapped in the form of ice. The appearance of the Antarctic ice sheet is attributed to the opening of an oceanic corridor between South America and the Antarctic Peninsula: the Drake Passage. The organization of the Antarctic circumpolar current would have led to the thermal isolation of Antarctica. However, more recently, based on modeling work, other authors believe that a reduction in CO_2 concentration was the cause of the cooling in Antarctica (De Conto and Pollard 2003).

In Chap. 6, a figure shows a new threshold around 15 Ma, to which the permanency of the East Antarctic ice sheet is attributed. There are indicators, however, that the ice may have retreated since 3 Ma, so that at 3 Ma the ice sheet could have been thicker than during the Last Glacial Maximum.

As for the northern hemisphere, scientists estimate that the first freeze-ups occurred 7 Ma ago on southern Greenland. Amplification of these glaciations then took place $\sim$3 Ma ago. Several hypotheses have been proposed to explain this amplification: (i) the closure of the Panama Strait occurred at this time and modified the oceanic circulation, causing warm water to travel up to the high latitudes of the North Atlantic, providing a considerable source of water vapor and therefore precipitation; (ii) the reduction in CO_2 concentration leading to cooling; (iii) an uplift of the Rocky Mountains modifying the planetary waves. Ice appeared between 2.75 and 2.55 Ma in Eurasia, then in Alaska and Canada, as evidenced by debris transported by icebergs which then appeared in North Atlantic sediments (Shackleton et al. 1984).

The Last Three Million Years

For the past three million years, the Earth's climate has oscillated between glacial and interglacial periods and most of the variations are due to the formation and subsequent melting of ice sheets in the northern hemisphere. In Antarctica they only fluctuated, with significant variations in Western Antarctica and in the peninsula (Anderson et al., EPILOG 2002; Ritz et al. 2001).

Until $\sim$1 Ma ago, the variations in the extent and volume of the ice sheets remained moderate, with variations of 60 m in sea level. The oscillations subsequently slowed down and gained amplitude. About 900 ka ago, the dominant periodicity of glaciations went from 40,000 to 100,000 years and the amplitude of the oscillations doubled, with variations of more than 100 m in the sea level. This event is known as the Mid-Pleistocene Revolution (MPR) (Raymo et al. 2006).

The freezing-up of the northern hemisphere has varied greatly over time: not all glaciations are alike, nor are all interglacials. The recent maximum expansion of the continental ice sheets is fairly well documented. However, many questions remain, in particular on the marine sides of the ice sheets. Below. we highlight some irregularities that are of interest.

The ice shelves in the Arctic Ocean

The Lomonosov Ridge is an underwater mountain range located at a depth of 1000 m. It stretches across the Arctic Ocean from Greenland to Siberia through the North Pole. Several striations aligned at the top of this ridge suggest the possibility of a gigantic ice-shelf that might have covered the center of the Arctic Ocean. These undated striations are unlikely to have been caused by icebergs which leave more random marks. They are also corroborated by marks in other parts of the Arctic Ocean (Jakobsson et al. 2008).

The Stage 11 Interglacial

During the Marine Isotope Stage 11, 400,000 years ago, the Earth experienced a long interglacial. Estimates (currently under discussion) of sea level vary between 20 m above and 8 m below the current level. If the high value is correct, it would require not only a massive deglaciation of Greenland and Western Antarctica but also a moderate loss of ice in East Antarctica. Because the orbital parameters were close to those of today, this interglacial is often considered to be one of the best analogues of the present time.

Stage 6: Saale glaciation

The penultimate ice age, called Saale glaciation in Eurasia, reached its maximum about 140,000 years ago. The Eurasian ice sheet covered a large part of Eastern Europe, Russia and Siberia, practically equivalent to the maximum expansion during the Quaternary in these regions (Svendsen et al. 2004). The cooling causing the development of this cap must have been much greater than during the LGM. Its expansion towards the south is particularly impressive because one of its lobes (the Diepner lobe, probably a transitory phenomenon) came within 500 km of the Black Sea. It appears in the southeast, that signs of maximum extension are found even further south than for the Saale glaciation, but it is difficult to know if these correspond to the same ice sheet and what its age might be (see Fig. 24.6).

Scenario of the Last Glacial-Interglacial Cycle

The temporal pattern of the last glacial-interglacial cycle is shown in Fig. 24.3. The external forcing associated with variations in the Earth's orbit is represented by the evolution of summer insolation at 65 °N latitude. The sea level indicates the overall volume of ice.

It is clear that the volume of ice reacts systematically to variations in summer insolation. As it is mainly the summer climate that is responsible for the ablation, an orbital configuration with strong summer insolation is unfavorable to the ice sheets. In addition, the slow freeze-ups and the rapid deglaciations are not symmetrical. The volume over the whole cycle is a sawtooth trend because the retreats are not complete and each advance (around 120, 90, 80, 60 and 40 ka BP) resumes from a still glaciated situation, even more so as the Ice Age progresses.

However, the geographical distribution reveals great diversity (the geographical aspects are summarized in Fig. 24.6 for Fennoscandia and 2.7 for Laurentide).

During the Eemian (125 ka BP), the eustatic sea level was up to 6 m higher than currently, testifying to a strong retreat of the ice. The southern part of Greenland and Western Antarctica were probably much less ice-covered than they are at present. However, ice core drilling indicates that there was an ice cap in the central part of Greenland at an altitude comparable to the current altitude (NGRIP Members 2004).

The Laurentide Ice Sheet

The Laurentide ice sheet over North America was the first to form, with ice appearing on the heights in the form of two separate caps, one over Keewatin (itself derived from the junction of the Keewatin and Baffin caps) and the other on Labrador (Fig. 24.7). Throughout the ice age, it underwent fluctuations, with the two caps joining during the colder periods followed by growth (both in volume and area) of the resulting ice sheet, generally with two distinct domes.

Conversely, during interstadials (warmer intervals during the glaciation) it shrank and in some cases separated into two (or even three) ice sheets. Another ice cap also developed on the Rocky Mountains, the Cordilleran ice sheet, but it took at least 60 ka BP before there was a junction between the Laurentide sheet and the still undeveloped Cordilleran sheet. The Cordilleran ice sheet only became heavily glaciated at a late stage (after 30 ka BP). This phenomenon was attributed to the influence of the Laurentide on atmospheric circulation. The Laurentide needed to be high enough to affect the jet stream, which had the effect of increasing the transport of moisture to the Cordillera.

It should be noted that despite fluctuations, a significant amount of ice persisted over North America throughout the ice age.

During the Last Glacial Maximum (21 ka BP), the area covered is shown in Fig. 24.1. During the deglaciation, the retreat of the Laurentide accelerated from 15 ka BP. A proglacial lake (Lake Agassiz) was formed south of the ice sheet. It emptied around 8200 years by breaking its ice dam and rapidly injected a large quantity of fresh water into the North Atlantic, with repercussions on the thermohaline circulation. The deglaciation in North America continued up to 6 ka BP.

The Fennoscandian Ice Sheet

Ice appeared later on Eurasia and initially in a limited way on the Arctic archipelagos (Svalbard, François Joseph) and on the mountains of Scandinavia. It was only towards 90 ka BP that a major glaciation appeared in Western Siberia. Then, during each cold period a large sheet developed with a junction between the Svalbard archipelago, northern Scandinavia and New Zemble (Barents and Kara seas). During each interstadial, the Barents and Kara seas were free of ice and often ice only remained in arctic archipelagos and on high points in Scandinavia (Svendsen et al. 2004). This behavior, which suggests a strong sensitivity to the climate, can be explained by several positive feedbacks:

– the terrestrial part (lying on a bedrock above sea level) was more limited in area and thickness than the Laurentide. As a result, it is more subject to the

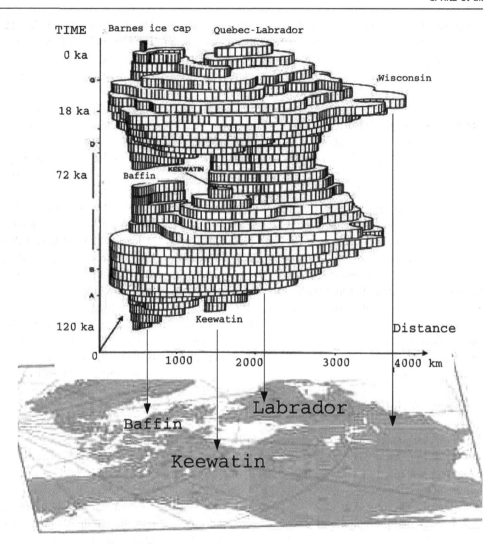

Fig. 24.7 Reconstruction of the Laurentide ice sheet during the glaciation. In general, there were three glacial centers: the first icecap developed on the Arctic archipelago (Baffin), followed quickly by another on Keewatin, and then these two caps grew together (then generally called the 'Keewatin cap'). At the same time, another cap developed on Labrador. Around 106 ka BP, the Labrador and Keewatin ice sheets were still independent, separated by the Hudson Bay. Later, they met to form a single mass of ice with two domes (Labrador and Keewatin/Baffin). This ice sheet again broke up into several small ones around 85–74 ka BP before finally reforming and growing until the LGM. Image adapted from Boulton and Clark (1990). According to Peyaud (2006)

altitude-surface temperature feedback. This instability is often referred to as the 'instability of small ice sheets';
– the Barents and Kara caps are ocean ice sheets whose base is below sea level (as is the case for West Antarctica) and this makes them sensitive to the instability of the marine ice sheets, which is linked to the dynamics of the ice flow. In addition, these ice sheets formed a glacial barrier which retained huge lakes. These lakes tempered the Siberian summers and reduced the melting of the glacial ice sheets, another positive feedback, but occurring during the cold period unlike the previous two.

Another interesting point concerns the location of the ice sheet during the glacial period. During the LGM, the Fennoscandian ice sheet was similar in size as it was at 90 ka BP, but located further to the west. This phenomenon is not explained but an interaction between the topography of the ice sheets and atmospheric circulation is suspected. For example, at the end of glaciation, the presence of the Scandinavian sheet prevented the transport of moisture towards the east.

The Fennoscandian ice sheet appears to have reached its maximum extent to the south between 20 and 18 ka BP. The deglaciation of the continental shelf began around 15 ka BP and from 13 ka BP onwards the ice retreated northward to the archipelagoes of the Arctic, while in the south, the Scandinavian cap receded to the Gulf of Bothnia and the Finnish border. After 10,000 years BP, the ice was confined to the Norwegian mountains and at the climate optimum of the Holocene, the ice cover was probably smaller than at present.

Antarctica

There is much less information available on the evolution of the Antarctic ice sheet partly because of the harsh conditions (logistics, climate), partly because the variations were smaller and partly because most traces of the Last Glacial Maximum are to be found at sea. Information comes from ice core drilling (mostly located on the Antarctic plateau (see Chap. 10, Volume 1), marine drilling at the edges, multi-beam sonar data at sea and marks of glaciations on the rocks and mountains beyond the current cap (trimlines or the limit between the glacial erosion model and the model of erosion by atmospheric processes). Digital modeling of the ice cores helps to connect the various data to explain the mechanisms involved.

For the past 3 million years, Antarctica has oscillated between two (or even three) states, with the East Antarctic and West Antarctic ice sheets behaving differently (oppositely in terms of volume). Four processes govern this evolution: (i) the surface temperature has little immediate influence. Indeed, since the maximum temperature on the coasts is about −10 °C, ablation is not a significant mechanism except on ice shelves. In the long term, climate temperature variations propagate through the ice and eventually reach the base of the glacier, where deformation and sliding are concentrated. The flow is then intensified but it takes about ten thousand years before this process is felt; (ii) precipitation has an immediate effect: the more it snows, the more the ice sheet grows. In general, precipitation is deemed to be related to atmospheric temperature. Therefore, it snows more during warmer periods because the air can then hold more moisture and this is borne out (at least on the Antarctic plateau) by ice core analysis; (iii) flow brings the ice towards the coast, is sensitive to basal conditions and to the thermomechanical coupling mentioned above, but the main feature of Antarctica is the presence of ice shelves that act as a buttress limiting the speed of the ice streams. If these ice shelves disappear, the upstream glaciers will accelerate leading to thinning of the ice sheet; (iv) the movements of the grounding line govern the size of the ice sheet, and the larger the sheet, the thicker it becomes. The grounding line is sensitive to sea level (if this lowers, ice shelves are able to ground) by a purely geometric effect. In addition, the flow of ice streams and the shift in the grounding line are linked, as a retreat occurs in association with an acceleration. For example, a disintegration of the ice shelves through an acceleration of the flow may force a retreat of the grounding line. Conversely, a retreat of the grounding line reduces the basal friction of the ice stream (because it begins to float) and allows it to accelerate. It should be noted that the precipitation effect acts in opposition to the others and tends to produce a smaller ice sheet during the glacial period.

In the interglacial period, conditions are roughly similar to the current ones. Compared with the present, the glacial period has lower temperatures, less precipitation, a lower sea level, and a colder ocean which help to maintain the ice shelves. The grounding line then advances, now limited only by the continental slope as shown by the grounding line map proposed by Bentley et al. (RAISED 2014) (Fig. 24.8). It is the amplitude of the variation in the grounding line that makes the difference between the East and West ice sheets.

Recent modeling results (Pollard and DeConto 2009) indicate that a third state, even more free of ice than the current one, may occur during particularly long and warm interglacials. In this third state, West Antarctica virtually disappeared and this could explain the high sea level at some times in the past (Eemien, MIS11).

Around East Antarctica, the continental slope is located a few tens of kilometers from the current coast (see Fig. 24.8).

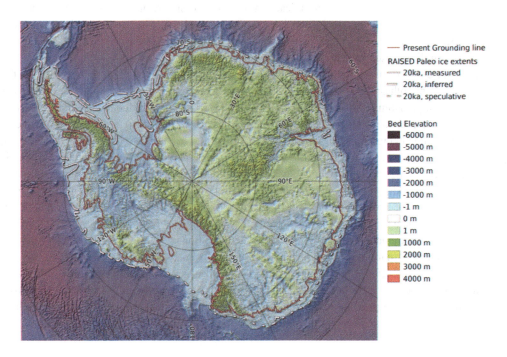

Fig. 24.8 Position of the grounding line in Antarctica. Current (red line) and during the last glacial (white lines). Note that during the glacial period this line is located near the continental slope and that the ice sheet has spread over most of the continental shelf (in light blue). Figure Quantarctica map, from Bentley et al. RAISED (2014)

This is the only margin of advance of the sheet. During glaciation, the East Antarctic ice sheet is therefore slightly thicker at the edges but this effect (confirmed by glaciation tracks on the mountains) remains confined to the edges. The central regions are thinner due to lower accumulation. It is estimated that the Antarctic shelf was about 100 m lower during the Last Glacial Maximum (Ritz et al. 2001). As for volume, the impact of the central regions is stronger and overall the volume of East Antarctica is less in glacial period than in interglacial.

West Antarctica has much more room to spread out (Fig. 24.8). Indeed, the two large embayments in which the ice shelves of Ross and Ronne-Filchner are located are shallow.

During glacial periods, these ice-shelves are grounded, advancing the grounding line by about one thousand kilometers. Here the effect of the grounding line far outweighs that of precipitation. This is why West Antarctica is more voluminous during glacial periods. However, there is still much debate about the exact volume. Estimates for the total contribution of the Antarctic to sea-level between the glacial maximum and now range from 7 m to over 20 m (sea level equivalent). These extremes correspond to two possible scenarios in West Antarctica during glaciation. For the same grounding line position, it is possible to have relatively stagnant ice streams and a very thick ice sheet at the edges, similar to the edges of East Antarctica currently. However, it could be that regions where there are ice shelves currently were very active, formed from huge ice streams comparable to the Sipple Coast in West Antarctica. These regions must have been relatively flat, which had repercussions as far as the center of West Antarctica. The first hypothesis was long supported by the trimlines which indicated altitudes at least 1000 m above the current level, but recent datings of these trimlines have in many places indicated that they are more than one million years old and therefore do not concern the Last Glacial Maximum (Bentley et al. 2010). Alternatively, the second hypothesis is supported by the ice core analysis from the Sipple Dome (the Ross ice-shelf slope) and Berkner Island (middle of the Ronne-Filchner ice-shelf).

For the time being, numerical modeling is the only tool to evaluate the evolution of Antarctic geometry during the recent glacial-interglacial cycles. Only a few polar cap models have tackled this problem because simulation of the movements of the grounding line remains a major difficulty. It should be noted that the mechanisms presented (which are included in the models) accurately reproduce the evolutions described above, that not all interglacials were similar (some were less frozen, like stage 11 or Eemian, others were intermediate between glacial and interglacial periods) and that the freezing-up process is slow, the volume increasing progressively throughout the glacial period. As for the last deglaciation, it appears to have occurred late (15 ka BP), but its speed depends on the model used (Huybrechts EPILOG 2002; Ritz et al. 2001), as well as on the volume of Antarctica ice during the glacial period. The speed of deglaciation depends on the model, but all agree that it has only just finished, which is in accordance with observations showing that the retreat of the grounding line in the Ross Sea continued until 3000 years BP (Conway et al. 1999).

Greenland

In terms of process, Greenland is an intermediate ice sheet. During the glacial periods, ablation was negligible there and its expanse was limited by the movement of the grounding line, in other words, it was limited by the continental slope. During interglacials, ablation plays an important role, as evidenced by the fact that its current edge is mostly terrestrial (there can be no coastline without ablation). This makes Greenland sensitive to warming climates and explains why it is assumed that this cap was significantly smaller during the Eemian. In terms of shape, Greenland changed from domed during interglacial (high and narrow) to flat and expanded during glacial periods.

Conclusions

The climate system is complex and the long-term component which includes sea level and the evolution of the polar ice sheets is no exception to this rule. In this chapter, we have seen that the mechanisms involved can be internal to the ice sheets such as thermomechanical coupling and feedbacks related to subglacial hydrology. However, most of the processes are related to interactions with other elements of the system:

- the atmosphere, due to the altitude-surface temperature link which causes the instability of small ice sheets. The influence of the ice sheets on the general circulation is also noted and we have seen that it could explain the interactions between the ice sheets;
- the ocean, since the ice sheets determine sea level, but also the oceanic general circulation. In the other direction, the movement of the grounding line, which is the key process of the Antarctic evolution, results from interaction between glacial dynamics and the local ocean;
- the solid Earth, through the mechanism of isostasy.

The data available for the past, in terms of sea level or through glacio-geomorphology reconstructions, indicate that all the mechanisms mentioned above are indeed active.

Moreover, numerical modeling makes it possible to simulate most of the recorded evolutions. However, the speed and amplitude of certain events remain hard to explain:

- During the deglaciation, there were periods of very rapid rise in sea level, of around 5 cm per year, for several centuries. This occurred in particular around 14,200 ka BP with an event called 'melt water pulse 1 A', the cause of which (Laurentide or Antarctica) is still controversial.
- The Heinrich events during which armadas of icebergs invaded the North Atlantic. Purely glacial mechanisms (thermo-mechanical coupling) have been suggested but this is difficult to reproduce correctly by the 3D ice sheet models. An interaction with the ocean that would melt an ice shelf at the mouth of the Hudson Strait has recently been suggested and would allow a better agreement with the ocean recordings than the previous hypothesis.

These two examples show that glacial dynamics may have played a more important role than was previously assumed, especially when the ice sheets are subjected to climate forcings (ocean or atmosphere), a hypothesis which is supported by the current observations of acceleration of outlet glaciers (in Greenland and Antarctica. For this reason, research in this area of glacial dynamics is being actively pursued in order to better assess the future behavior of the two remaining ice sheets, Greenland and Antarctica, in the context of climate change.

References

Bentley, M. J., Fogwill, C. J., Le Brocq, A. M., Hubbard, A. L., Sugden, D. E., Dunai, T. J., et al. (2010). Deglacial history of the West Antarctic ice sheet in the Weddell Sea embayment: Constraints on past ice volume change. *Geology, 38,* 411–414.

The RAISED Consortium, Bentley, M. J., et al. (2014). A community-based geological reconstruction of Antarctic ice sheet deglaciation since the Last Glacial Maximum. *Quaternary Science Reviews, 100,* 1–9.

Boulton, G. S., & Clark, C. D. (1990). A highly mobile Laurentide ice sheet revealed by satellite images of glacial lineations. *Nature, 346,* 813–817.

Clark, P. U., Mitrovica, J. X., Milne, G. A., & Tamisiea, M. E. (2002). Sea-level fingerprinting as a direct test for the source of global meltwater pulse I. *Science, 295,* 438–441.

Conway, H., Hall, B. L., Denton, G. H., Gades, A. M., & Waddington, E. D. (1999). Past and future grounding-line retreat of the West Antarctic ice sheet. *Science, 286,* 280–283.

De Conto, R. M., & Pollard, D. (2003). Rapid Cenozoic glaciation of Antarctica induced by a declining atmospheric CO_2. *Nature, 421,* 245–249.

Denton, G. H., & Hughes, T. J. (1981). *The last great ice sheets.* Wiley.

EPILOG. (2002). *Quaternary Science Reviews,* volume 21 (tous les articles dont Anderson, J. B. et al., pp. 49–70, Huybrechts, P., pp. 203–231, Lambeck, K. et al., pp. 343–360, Waelbroeck, C. et al., pp. 295–305).

Hughes, A. L. C., Gyllencreutz, R., Lohne, S., Mangerud, J., & Svendsen, J. I. (2016). The last Eurasian ice sheets—A chronological database and time-slice reconstruction, DATED-1. *Boreas, 45,* 1–45. https://doi.org/10.1111/bor.12142.

Jakobsson, M., Polyak, L., Edwards, M., Kleman, J., & Coakley, B. (2008). Glacial geomorphology of the Central Arctic Ocean: The chukchi borderland and the Lomorosov Ridge. *Earth Surface Processes and Landforms, 33,* 526–545.

Lambeck, K., & Chappell, J. (2001). Sea level change through the last glacial cycle. *Science, 292,* 679–686.

Lambeck, K., Purcell, A., Funder, S., Kjaer, K. H., Larsen, E., & Möller, P. (2006). Constraints on the Late Saalian to early Middle Weichselian ice sheet of Eurasia from field data and rebound modelling, *Boreas,* 35. https://doi.org/10.1080/03009480600781875.

Mix, A. C., & Ruddiman, W. F. (1984). Oxygen-Isotope Analyses and Pleistocene Ice Volume. *Quaternary Research, 21,* 1–20.

NGRIP Members. (2004). High resolution record of Northern Hemisphere climate extending into last interglacial period. *Nature, 431,* 147–151.

Pollard, P., & DeConto, R. M. (2009). Modelling West Antarctic ice sheet growth and collapse through the past five million years. *Nature, 458,* 329–332. https://doi.org/10.1038/nature07809.

Raymo, M. E., Lisiecki, L. E., & Nisancioglu, K. H. (2006). Plio-Pleistocene ice volume, Antarctic climate, and the global delta ^{18}O record. *Science, 313*(786), 492–495. https://doi.org/10.1126/science.1123296.

Peltier, W. (2004). Global glacial isostasy and the surface of the ice-age Earth: The ICE-5G(VM2) model and GRAC. *Annual Review of Earth and Planetary Sciences, 32,* 111–149.

Peyaud, V. (2006). *Rôle de la dynamique des calottes glaciaires dans les grands changements climatiques des périodes glaciaires-interglaciaires.* Thèse de doctorat, université Joseph Fourier, Grenoble 1.

Ritz, C., Rommeleare, V., & Dumas, C. (2001). Modeling the evolution of Antarctic ice sheet over the last 420,000 years: Implications for altitude changes in the Vostok Region. *Journal of Geophysical Research, 106,* 31, 943–31, 964.

Schoof, C. (2007). Ice sheet grounding line dynamics: Steady states, stability, and hysteresis. *Journal of Geophysical Research, 112,* F03S28, https://doi.org/10.1029/2006jf000664.

Schrag, D. P., Hampt, G., & Murray, D. W. (1996). Pore fluid constraints on the temperature and oxygen isotopic composition of the glacial ocean. *Science, 272,* 1930–1932.

Siddall, M., Rohling, E. J., Almogi-Labin, A., Hemleben, C., Meischner, D., Schmelzer, I., et al. (2003). Sea-level fluctuations during the last glacial cycle. *Nature, 423,* 853–858.

Shackleton, N. J., et al. (1984). Oxygen isotope calibration of the onset of ice-rafting and history of glaciation in the North Atlantic Region. *Nature, 307,* 620–623.

Svendsen, J. I., et al. (2004). Late quaternary ice sheet history of Northern Eurasia. *Quaternary Science Reviews, 23,* 1229–1271.

Modeling and Paleoclimatology

Masa Kageyama and Didier Paillard

Why Develop Paleoclimate Models?

Today, climate models are widely discussed because of the climate predictions they produce for the next century (see Chap. 31), particularly when the IPCC (Intergovernmental Panel on Climate Change) assessment reports are published. The models used for climate prediction are developed first and foremost based on the data from observations of recent decades. Before being applied to forecasting, these models are evaluated to assess their ability to reproduce the present climate and its recent variations over the last decades. They represent our understanding of the current climate and of the mechanisms that play important roles in recent variations, but also our ability to translate this understanding into digital codes which is necessarily limited by computing capabilities. Thus, even though the most powerful computers are currently used to produce simulations with the most complex climate models, they are limited by computing power in terms of resolution and system complexity. We will return later to this crucial question of finding the best compromise between the desired timespan for a climate simulation and the complexity of the climate model used, a question that is even more critical in paleoclimatology, where the time scales are much longer than for the IPCC predictions (which last a few centuries at most).

It is easy to understand why models are used for the forecasting of future climates: this approach, based on our understanding of the physics of the climate system, provides the only means of obtaining these forecasts. But modeling can also contribute a lot to our understanding of current and past climates which are characterized through observations and reconstructions. A first justification is the evaluation of the models used for future forecasts. During the development phase of a model, the first step in evaluation is to compare it to the available observations for the current climate and its short-term variability over time scales ranging from a few years to a few decades. In recent decades, anthropogenic disruption has created previously unknown levels of atmospheric greenhouse gas concentrations, so models need to be tested in climate configurations different to the current climate. There is no perfect paleoclimate equivalent for the forcing of anthropogenic disturbance, but paleoclimates, even if they are not recorded with the same precision as the present climate, offer examples of climates very different from the current one and of transitions of varying rapidity between states. If we have correctly understood the climate system and if we want to use this understanding to predict climate in the future, we must be able to reproduce the variations in past climates.

An example of an evaluation based on paleoclimates is one dealing with entry into the last glacial period, also called last glacial inception. This occurred about 115,000 years ago, and the major cause of this disturbance, external to the climate system, is the difference in insolation received by the Earth. In principle, a climate model, even if it does not include an ice sheet model, should be able to simulate perennial snow cover at the formation sites of the first ice caps located north of present-day Canada. Following various trials, it was found that very few atmospheric general circulation models were capable of simulating this perennial cover. Furthermore, those that did manage to do so were often ones that simulated a current climate that was much colder than observations. These trials were therefore far from satisfactory. However, as components (ocean, vegetation) were added one by one to the climate models, it was found that these components could play an important role in amplifying the initial insolation signal. Thus, over a longer period, the disappearance of forests or the appearance of sea ice, initiated by changes in insolation, support the development of perennial snow cover by modifying the albedo. It can therefore be concluded that it would be difficult to

M. Kageyama (✉) · D. Paillard
Laboratoire des Sciences du Climat et de l'Environnement, LSCE/IPSL, CEA-CNRS-UVSQ, Université Paris-Saclay, 91190 Gif-Sur-Yvette, France
e-mail: masa.kageyama@lsce.ipsl.fr

© Springer Nature Switzerland AG 2021
G. Ramstein et al. (eds.), *Paleoclimatology*, Frontiers in Earth Sciences,
https://doi.org/10.1007/978-3-030-24982-3_25

simulate permanent snow cover using only a simple atmospheric model because of the absence of positive feedback mechanisms. It can be seen here that while a simple evaluation can lead to a better understanding of the system, it is also only the start of an extensive exploration of increasingly complex models to find the one that best fits with the data.

Paleoclimate simulation is not only used to evaluate the models used for climate predictions over the next century. This would be an extremely narrow application, especially given the time scales that can be handled by these models. The starting point for the development of paleoclimate modeling is the assumptions applied when interpreting the data. A paleoclimate model seeks to formalize the assumptions based on the physical principles of the climate system and to test whether these principles can explain the observed climate variations. It is clear then that the models can be extremely varied depending on the data they are trying to interpret. Indeed, even though the components of the climate system are all interdependent, which is a defining feature of this system, it is not always necessary to represent them all in detail in order to reproduce an observed phenomenon. In fact, it is more interesting to isolate the processes or key components responsible for a phenomenon. This is one of the approaches to paleoclimate modeling which tries to build a 'minimal' model to explain a phenomenon. This is very different from the models used to predict the climate of the next century, but these approaches are important to provide a better understanding of the climate system and its evolution.

Modeling can also highlight the importance of a particular forcing or process. By comparing experiments which include a certain process or forcing with experiments which exclude it, it is possible to study its impact and identify which mechanisms explain this impact. These 'sensitivity experiments' are not necessarily very realistic but they complement the more realistic simulations of paleoclimates, by helping to better understand them. One example, in Section "General Circulation Models, Complex Models of the Earth System", an attempt is made to understand the impact of ice caps versus the impact of a lower atmospheric CO_2 concentration on the climate of the Last Glacial Maximum (LGM). In order to understand this, simulations are created where the ice caps of the LGM are placed in the context of the current CO_2 concentration, and also where the CO_2 concentration of the LGM is positioned with the current ice caps. Although these simulations do not correspond to real situations, they provide a better understanding of the simulated glacial climate by imposing both ice cap and CO_2 concentration forcings from the LGM.

This chapter starts by presenting the basic concepts of modeling, definitions essential to our understanding of models and the digital experiments used in climatology and paleoclimatology. We then focus on the three main families of paleoclimate models: the most complex general circulation models, climate models of intermediate complexity, and conceptual models. For each of these families of models, we give examples of their use in paleoclimatology.

Some Basic Modeling Concepts

Vocabulary

Before showing how climate modeling contributes to the study of past climates, it is useful to define the concept of a model. Indeed, this word has quite different meanings in the various scientific disciplines. In general, a 'model' is a representation of a set of scientific ideas, formulated within as rigorous a framework as possible, which explains a complete set of phenomena. The model is judged to be even more effective when it is simple and concise, and when it offers a maximum number of solutions. It becomes quantitative when it is based on mathematical relationships. When we talk about climate modeling, we mean 'physical' models of the climate system incorporating a set of mathematical equations that trace the evolution of the system from a starting position within boundary conditions. It is therefore a system of first-order differential equations, which can generally be written in the following form:

$$\frac{dX(t)}{dt} = f(X(t), t) \quad (25.1)$$

where $X(t)$ is a vector dimension N which provides an overall description of the state of the model at each instant t, and $f(X, t)$ is a function of X and of time t which describes the evolution of the system. The dimension N of the vector $X(t)$ thus represents the 'size' of the model, which is sometimes called the 'number of degrees of freedom' of the system, and the space of dimension N of all the vectors X is called 'the phase space'. If the state of the system is known at a given instant t_0, denoted by $X_0 = X(t_0)$ and called the 'initial condition', then Eq. (25.1) makes it possible to know the state of the system at all times.

When referring to a climate model, we may imagine a very 'complex' system, with a large number of degrees of freedom. This is often the case, but not always. Indeed, it is important to highlight two contradictory aspects of climate modeling. On the one hand, modeling aims to improve our understanding of how the system functions, and on the other hand, it is trying to provide the best possible representation of it. In order to explore and understand what is happening within a system of equations, it is preferable that the number of degrees of freedom N be small. Conversely, to achieve a good representation of a system as complex as the climate, the number of degrees of freedom N needs to be large and will be limited only by computing power. Although both of these qualify as modeling, the second case is more accurately referred to as 'climate simulation' where the primary objective is to achieve a maximum of realism, at the expense of an in-depth understanding of how the system operates. In the most sophisticated climate models, there are several million degrees of freedom thus making it difficult to understand and analyze the detail of the sequence of processes involved in the simulations carried out by these models. A typical strategy is therefore to multiply the number of simulations, as described later, by carrying out sensitivity experiments. Conversely, much simpler models, which may produce less realistic results, can provide insight into the root causes of certain mechanisms that underlie the phenomena being represented. If the aim of modeling is summarized in the maxim 'understand so as to better simulate', it is obvious that a whole spectrum of models of varying complexities is necessary in order to tackle the different aspects of a problem.

Before further describing climate modeling in general and the problems encountered in paleoclimatology in particular, we will revisit Eq. (25.1) in more detail in order to explain some concepts that are widely used either implicitly or explicitly. The vector $X(t)$ which describes the whole system is also called the **prognostic variable set** of the model. This refers to all quantities $X_i(t)$ in the system (25.1) possessing an equation of evolution. Moreover, it is often useful to include additional variables to represent the physical quantities used in the equations, quantities which depend directly on the prognostic variables $X_i(t)$ without recourse to an associated evolution equation. For example, the quantity $y(t) = X_1(t) + X_2(t)$ is deduced from the quantities $X_i(t)$ and so the evolution equation for the derivative $dy(t)/dt$ is redundant in the system of Eq. (25.1). These additional variables are called the **diagnostic variables** of the model, because they are mainly used to provide a better understanding of the model in terms of the customary physical values. Thus, typically, in an atmospheric circulation model, the only prognostic variables at each point of the grid of the model are temperature, humidity, and wind velocity on the horizontal plane, with evolution equations representing the conservation of energy and water (transport equations) and the conservation of momentum (i.e. the Navier-Stokes equation) on the horizontal plane. All other values (vertical velocities, energy fluxes, precipitation, clouds etc.) are deduced more or less directly. These are merely diagnostic or secondary variables, but they are nevertheless very useful at all the stages of modeling, from the design of the model to the analysis of the results. These diagnostic variables, often more numerous than the prognostic variables, do not mean additional degrees of freedom.

Moreover, the notation of the system of Eq. (25.1) always involves values which are established at the outset, deeming these to be either physical values external to the model under consideration, or more or less well defined constants. These values are the **model parameters**. When these parameters are spatialized, i.e., dependent on their geographical location, they are then considered to be **boundary conditions**. When the parameters are time dependent, they may be referred to as model **forcings**. For example, for an atmospheric model, the surface temperature of the oceans is a boundary condition (and also a forcing, if it depends on time), and the atmospheric concentration of CO_2 is a parameter (and also a forcing, if it depends on time). For a coupled ocean-atmosphere model, this same sea surface temperature is a prognostic variable while pCO_2 remains a parameter. For a climate-carbon coupled model, pCO_2 is explicitly calculated and thus becomes a model variable as well. It is often interesting to explore how the model outcomes change when the values of certain parameters change. These are called **sensitivity experiments**, because the objective is not to perform realistic climate simulations, but to see how sensitive the model is to certain parameters (examples of experiments of this type are shown in Sections "General Circulation Models, Complex Models of the Earth System" and "Examples of Long-Term Simulations and Studies of Sensitivity to Forcings"). When this type of study systematically includes many parameter values and many parameters, this is called the **exploration of the parameter space** of the model, and is sometimes imprecisely referred to as the exploration of 'the phase space' (although, strictly speaking, it is the space of the prognostic variables and not of the parameters).

Dynamic Systems

It is also important to briefly outline the general results that can be obtained from an equation system such as system 25.1. First, the choice of functions $f(X, t)$ must be restricted to cases likely to have a physical meaning. Instead of starting from a single initial condition X_0, we start with a set of proximate initial conditions, which fill an initial volume V_0 in the phase space. For the 'physical' cases, the second principle of thermodynamics implies that, at time t, the corresponding states $X(t)$ fill a volume $V(t)$ which decreases with time (in the case of dissipative systems) or remains constant (in the case of conservative systems). While the conservative systems retain the memory of the initial condition, since the volume $V(t)$ remains constant, this information is gradually lost in dissipative systems. Indeed, in general, this volume tends towards zero as time t approaches infinity. Climate (like many other physical systems) is a dissipative system. Figure 25.1 gives examples of typical behaviors of a system for two different initial conditions.

As dissipative systems gradually forget their initial condition, this may turn out to be positive: because this initial information is in any case lost after a certain time, this information is not relevant to the long-term behavior of the

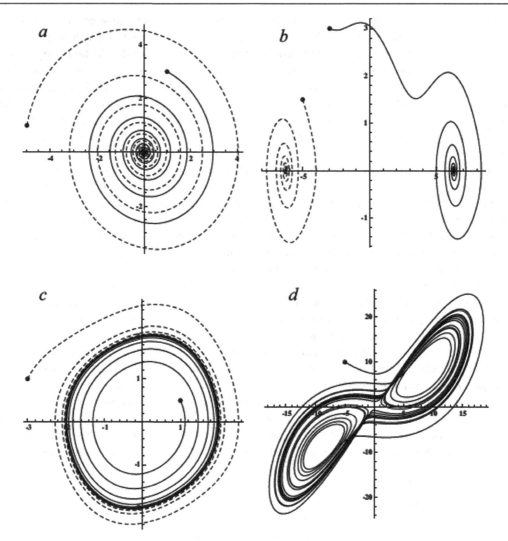

Fig. 25.1 Examples of behaviors of simple dissipative dynamic systems. The departure points are represented by the black dots. **a** Convergence of trajectories towards a single equilibrium point. **b** Convergence to a point of equilibrium dependent on the initial position. A 'catchment area' can thus be defined for each point of equilibrium. **c** Convergence towards a limit cycle. After a transitional phase, the system has a periodic oscillation. **d** Here, the trajectories converge towards a more complex object than a simple point or cycle. This object is called a 'strange attractor' (a well-known example of this case is the Lorenz system, 1963)

system. We can therefore focus on the asymptotic behavior alone in the model, that is to say its behavior from moment t_1 onwards, after this initial condition is forgotten. Conversely, the transitional phase between the beginning of the simulation at t_0 and time t_1, is highly dependent on the initial condition selected, and the results will only be relevant if this condition is correctly understood. Knowing that the duration $t_1 - t_0$ is a few weeks for atmospheric dynamics, we see here an essential difference between a climate model and a meteorological model. For the latter, this transitory phase is the most interesting. A major difficulty in weather forecasting is to provide, in real time, an initial condition that represents the state of the atmosphere 'now' in order to be able to anticipate its state in the hours and days to come. For the climatologist, the initial state of the atmosphere is of little importance, since it will be quickly forgotten. It should be noted, however, that this may not be the case for the initial state of other physical components of the system, such as the ocean, which have much longer time constants.

Nevertheless, this initial state of the atmosphere is not completely irrelevant. The discovery of chaotic systems in the 1970s demonstrated that the 'convergence' of the trajectories in a dissipative system does not necessarily mean that the model 'converges' towards a point of equilibrium (Fig. 25.1a), or even towards a simple trajectory such as a limit cycle (e.g. periodic oscillation, Fig. 25.1c). In fact, if the volume $V(t)$ tends to zero, this does not imply that it is a limit point, nor even a simple line (such as a closed curve, for an oscillator). The system can eventually 'converge' towards much more complex objects, known as 'strange attractors'.

A classic example is the Lorenz attractor, in the shape of butterfly wings, which has a fractal structure (Fig. 25.1d). This is the attractor of a very simple model, with only three degrees of freedom, which was formulated by Lorenz (1963), to illustrate the chaotic behavior of the atmosphere.

The result is that although the initial condition is effectively forgotten, it is nevertheless critical to the determination of the 'true trajectory' of the system. It is so critical that it makes it absurd for 'long-term' simulations, that it, beyond time t_1, to focus on a 'single real trajectory' for the model originating from a given initial state. Chaotic systems, like the atmosphere, are in fact characterized by a high dependence on the initial conditions: a tiny difference between two different starting points becomes exponentially greater. It can be shown that they are also highly dependent on the parameters used. For these systems, long-term deterministic predictions simply do not make sense. It is therefore necessary to focus on a set of trajectories, not to define a single result, but instead to assemble a set of possible results. Indeed, only the 'average' trajectory is significant as it represents a particular statistic of the attractor of the system.

In practice, it is therefore meaningless to try to calculate what the weather will be like on a given day at a given location, beyond a few weeks into the future. Only climate magnitudes (averages, differences, etc.) have some meaning. To calculate these climate averages, two solutions are possible. If we consider that the system is stationary, that is to say that its statistical features do not change over time (this would be the case, for example, in the absence of forcings), then it would simply be a matter of averaging the model results over several decades, as is done by geographers who use an average of weather variables over thirty years to define a 'climate'. We consider that the trajectory followed by the model represents the system that can be interpreted statistically, as is the case for the real climate. However, when the system is subjected to a forcing, such as the current anthropogenic disturbance, the system can no longer be considered 'stationary', and its statistical characteristics (i.e. climate values) will evolve over time. A temporal average is therefore no longer relevant and several simulations need to be carried out, differing only in their initial condition. This ensemble simulation establishes the range of different trajectories covering the range of possibilities.

Climate and Determinism

Before further describing (paleo)climate models, certain paradoxes surrounding the idea of climate should be highlighted. As we have seen above, climatology conveys a statistical approach, as opposed to meteorology, which has, above all, a deterministic perspective. The fundamental reason for this distinction stems mainly from the chaotic nature of the atmosphere, which becomes inherently unpredictable in a short period of time. If all the boundary conditions of the atmosphere (and all parameters) are considered to be either constant, or having a simple annual cyclical forcing, then the system will be stationary, in other words the statistical variables that define the climate will be stable over time. Although the terms of the system of Eq. (25.1) are never zero and the atmosphere changes endlessly, we refer to the climate model as being **in equilibrium**. Conversely, the climate will change only when the boundary conditions of the atmosphere, or some of its parameters, change over a time period of a decade or more, in other words, a timeframe compatible with the concept of climate. In a way, although a climate model aims to represent primarily atmospheric variables, it is only the slower physical components, other than the atmosphere, that cause the 'climate' of the model to evolve, i.e. to change the statistical distribution of the results. This is the case, for example, when there is a change in the ocean, carbon cycle, ice caps. The evolution of the climate is therefore only predictable if the 'non-atmospheric' components are predictable. The chaotic nature of the atmosphere does not imply that climate is unpredictable.

The Framework of a Climate Model

Selecting Components of the Climate System: Model and Boundary Conditions

In practice, it is not possible (or even desirable) to have a mathematical model that simulates all the phenomena that can interact with and modify the climate, including not only the atmosphere and the oceans, but also the terrestrial and marine biosphere, biogeochemical cycles, ice cap dynamics, hydrology and continental erosion. The first difficulty is to choose a relevant subsystem to define the variables of the model, and that can be the object of a temporal evolution given by a system of evolution equations, as above in (25.1). The other factors have to be imposed, in other words fixed as boundary conditions or forcings. For example, when coupled ocean-atmosphere models are used to simulate the past, the following will typically be imposed: (1) changes in coastlines, everything related to continental surfaces, particularly changes in topography, ice sheet extent and height, changes in ocean bathymetry; (2) everything related to atmospheric concentrations, in particular greenhouse gases, but also, in some cases, dust and other aerosols; (3) changes in the orbital parameters of the Earth. We then examine the response of the ocean-atmosphere system to these boundary conditions and forcings, either taken together in order to obtain a realistic simulation of the climate or taken separately to study the role of each of them individually. This is referred to as a sensitivity experiment.

Coupling of Several Components

In general, depending on the scientific question posed and the computing power available, the first step is to identify the part of the system which will be explicitly represented by the model. If, for example, the intention is to represent the evolution of the ice caps under the influence of the orbital parameters of the Earth, it is essential to concentrate on a model which explicitly represents the dynamic of the ice caps. However, when the focus is on fast-reacting components (ocean, atmosphere), it is acceptable to fix the slow-moving components by establishing these as boundary limits. The converse is not true. Thus, the evolution of the ice caps only makes sense when its interaction with a changing climate system is considered. Herein lies one of the main difficulties with modeling the climate over long periods of time (both past and future). How can the interactions (exchanges of energy and matter) between physical objects with very different time constants be calculated in a meaningful way? Several strategies are possible.

A—If only a state of equilibrium in the system is of interest, then the equilibrium of one of the systems (for example, the ice caps) can be found by fixing another (e.g. giving the atmosphere fixed boundary conditions), then by performing the inverse operation (the atmosphere is calculated while the ice caps are fixed), then reiterating the process until the results converge. This is called asynchronous coupling. In essence, the two physical components are coupled, but in a way that does not reflect (or very inexactly) the real flow of time. This makes it possible to achieve equilibrium of the coupled system over a relatively short period of time (for example, a few decades for the atmosphere), since the fast-reacting component (the atmosphere is the factor which demands the most computing time due to the fact that it has inherent small-scale variations) is always calculated by assuming an equilibrium with the slow component.

B—Although asynchronous coupling can be a good method of calculating climate equilibria, it is not a very rigorous way to calculate a meaningful evolution of the system. This may provide a useful approximation if the time constants of the physical systems are very different (e.g., atmosphere and ice caps), in which case it amounts to considering that the slow component is the sole driver of the evolution of the system with the fast component merely adapting to the slow component. If physical components with 'intermediate' times are included (such as the ocean, which reacts more slowly than the atmosphere but faster than the ice caps), this strategy may fail. Therefore, it is sometimes desirable to simplify the physics of fast-reacting components (especially the atmosphere) in an attempt to calculate only the long-term variations that can be used directly by the slower components. This has led to the development of models known as 'intermediate complexity' models (see Section "Earth system models of intermediate complexity (EMICS)").

Comparison with Paleoclimate Data

If our objective is to explain climate variations as reconstructed from paleoclimate records via modeling based as much as possible on the physics, it is important to set up the model and experimental design so that this comparison is easiest. This gives rise to a second major difficulty in paleoclimatology simulations: paleoclimate indicators (proxies) are never clear-cut in terms of the physical variables simulated by models. These indicators are dependent on particular climate parameters, but are these the ones being simulated? It is therefore risky to rely only on the same physical models as those specially developed for comparison with current observations, whose important parameters can be measured by oceanographers, glaciologists and meteorologists. The ideal situation is to be able to quantitatively compare paleoclimate simulations with paleoclimate indicators. The most promising strategy is to explicitly simulate these indicators in the models so that the multiple factors likely to influence them are taken into account. For example, it is useful to explicitly simulate the water isotopes ($\delta^{18}O$ and δD) in atmospheric and ocean models, and the carbon isotopes ($\delta^{13}C$) in biogeochemical models, in order to have a more direct comparison between the output and the measurements. An example of the application of one of these models is given in Chap. 29.

Another important difficulty concerns the chronology of events. Climate models need boundary conditions and forcings (e.g. variations of insolation, concentrations of atmospheric CO_2, sea level) in order to produce results such as temperatures or precipitation. It is difficult to enter all of the forcing parameters at the same chronological scale into the model. It is also often problematic to compare the results of these simulations with paleoclimate data whose time scale is not clearly defined. This is especially true when using short-term simulations, such as the results of general circulation models (atmospheric or coupled ocean-atmosphere models). For example, to better understand how a glaciation starts, simulations of the atmosphere, or of the ocean-atmosphere system, are performed. But when is the start of a glaciation? If marine isotopes can be relied upon, the ice caps began to grow at around 120 or 122 ka BP. This moment could then be simulated by imposing to the model the insolation, greenhouse gas forcings etc. which best fit this time period. In the case of an 'equilibrium' simulation that numerically integrates the atmosphere over a few decades, or the ocean-atmosphere system over 1000–2000 years, with constant forcings, this strategy is likely to be disappointing, since the forcing is 'barely' adequate to simulate the desired objective i.e. an accumulation of perennial snow on continents at high northern latitudes. The modeling strategy is then to start at 115 ka BP, when the astronomical forcing seems most favorable, to maximize the response of the system. Since it is difficult to modify all the boundary conditions of the model in a coherent way so that it aligns with the situation that existed 115 ka BP ago, the best approach is often to maintain conditions close to the control situation, the 'current' state, or rather the 'pre-industrial' state, such as sea level and greenhouse gases. Often, when modelers talk about experimenting with entry into glaciation using OAGCMs, this is in fact a simulation with the same boundary conditions and forcings as for the pre-industrial period except for the insolation, which is changed to correspond to the astronomical forcing of 115 ka BP ago. The objective is not to make an ice cap 'grow' (which would require a comprehensive ice cap model and thousands of years of integration) but simply to check that when insolation is modified, snow may accumulate in certain locations. There is only a very distant connection between this and the available paleoclimate observations, and therefore comparison with the data is not easy since the conditions imposed on the numerical experiment are idealized.

In general, models can only represent a small part of the global climate system with the other parts being imposed or ideally represented. Although their aim is to describe certain aspects of the problem in the most realistic way possible, they cannot claim to be exhaustive. Before embarking on any modeling exercise, it is essential to formulate a precise hypothesis corresponding to the selected configuration. Taking the example above of entry into glaciation, it is not possible to 'simulate the start of an ice age' in all of its aspects. However, some questions can be formulated and an attempt made to answer them. For example, for a general atmospheric circulation model: 'Taking a control situation i.e. a pre-industrial climate as a starting point, does simply changing the radiative forcing at the top of the atmosphere without changing ocean surface temperatures (which would require an ocean model), or vegetation (requiring a vegetation model), or the expansion of the ice sheet (requiring an ice sheet model), or anything else, bring about persistent snow cover in some northern regions?' Formulated in this way, it is easier to understand the gap between this and a broad-ranging simulation of a glacial inception. More comprehensive models can answer more general questions, but there is no 'all-encompassing' model. It is therefore important to correctly define the hypothesis to be tested, and to choose the relevant model configurations to do this.

The following sections describe the major families of climate models, from the most complex to the conceptual. Each section gives examples of the application of these models to paleoclimate questions.

General Circulation Models, Complex Models of the Earth System

Equations, Discretization and Parametrization: Example of Atmospheric General Circulation Models

A natural approach to simulate the characteristics of the Earth's climate is to look at the basic equations describing the behavior of the atmosphere and the ocean. First, we will describe how atmospheric general circulation models (AGCM) are constructed in order to represent the evolution of atmospheric characteristics (temperatures, winds, precipitation, etc.) on a global scale. We know (see Chap. 1) that at this scale, the atmospheric circulation is driven by the differential in insolation between the equator and the poles. The fundamental equations are therefore energy conservation, supplemented by mass conservation (of air and water), momentum conservation and the law of perfect gases.

Conservation of energy:

$$DI/Dt = -p(D\rho^{-1}/Dt) + Q \qquad (25.2)$$

where

I is the internal energy per unit of mass ($I = c_p T$, c_p being the specific heat of air at constant pressure), p is the pressure,
ρ is the density of the atmosphere,
Q is the heating rate of the atmosphere per unit of mass,
D/Dt is the Lagrangian (Material) derivative: $D/Dt = \frac{\partial}{\partial t} + u\frac{\partial}{\partial x} + v\frac{\partial}{\partial y} + w\frac{\partial}{\partial z}$
u, v, w being the wind components with the dimensions x (longitude), y (latitude) and z (altitude).

Conservation of momentum:

$$\frac{D\mathbf{v}}{Dt} = -2\Omega \times \mathbf{v} - \frac{1}{\rho}\text{grad}(p) + g + F \qquad (25.3)$$

where

$\mathbf{v} = (u, v, w)$ is the velocity of the wind relative to the surface of the Earth,
Ω is the rotational angular velocity of the Earth,
p is the atmospheric pressure,
g is the acceleration due to gravity,
F is the force exerted per unit of mass.

Conservation of mass (of air and water):

$$\frac{D\rho}{Dt} = \rho - \text{div}(\mathbf{v}) + C - E \qquad (25.4)$$

where C is the creation rate of the species under consideration, and E is its destruction rate.

The law of perfect gases:

$$p = \rho r t \qquad (25.5)$$

As described above, these equations are very general and are valid for both small and global spatial scales. The whole art of the modeler involves simplifying these equations for a given problem and expressing them in a form so that they can be solved numerically for this problem. It is the choice of the simplifications and of the expression of the equations that makes the differences between the models. These are always based on a set of assumptions deemed important for the problem being studied. Numerical simulations are then a test of our understanding of the system, expressed as a set of equations which define the numerical model.

A first simplification of Eqs. (25.2)–(25.5) is often done in current atmospheric general circulation models: the hydrostatic approximation. The objective of these models is to represent the characteristics of the troposphere, the lowest layer of the atmosphere which determines the climate on the Earth's surface. This layer, which reaches altitudes from 10 km (at the poles) to 20 km (at the equator), is extremely thin compared to the radius of the Earth (~ 6400 km) and is a fine layer in which particles of air travel much further and faster in a horizontal direction than a vertical one. From these considerations of scale, it can be deduced that when we consider atmospheric circulations with a horizontal scale much greater than the thickness of the troposphere, the atmosphere is close to the hydrostatic equilibrium, as described by the equation:

$$\Delta p_{atm} = -\rho g \, \Delta z \qquad (25.6)$$

where Δp_{atm} is the difference in atmospheric pressure between two levels separated by altitude Δz, ρ is the density of air, g is the acceleration due to gravity.

This direct relationship between pressure and altitude leads atmospheric specialists to often present variations in vertical atmospheric properties as a function of pressure: for example, a pressure of 1000 hPa indicates a level close to the surface, a pressure of 500 hPa indicates the mid-troposphere and a pressure of 200 hPa indicates the altitude where the subtropical jet streams are most intense. It is just above this level of pressure that the transition between troposphere and stratosphere is found. The hydrostatic approximation considerably simplifies the solution of the system of Eqs. (25.2)–(25.5), because by judiciously choosing the vertical coordinate, the vertical speed is diagnostically deduced from the horizontal components of the wind (thanks to the continuity equation). The prognostic variables of the system of equations are therefore temperature and humidity, and the two components of the horizontal wind. All other characteristics of the atmosphere can be deduced from these four variables. It is therefore the evolution of these four variables that have to be calculated, using the fundamental equations, simplified by the hydrostatic approximation.

These equations are solved for the boundary conditions and forcings chosen by the modeler to answer the posited questions. For the atmosphere, these are greenhouse gas concentrations, insolation (the amount of energy entering the atmosphere at its summit) and surface conditions: distribution of the different surface types (oceans, land, ice caps, different types of vegetation), orography, ocean surface conditions (temperature and sea ice coverage). An initial state for all the prognostic variables of the model is also chosen. From this initial state, the evolution of the atmosphere is calculated over the time necessary for its characteristics (temperature, precipitation, wind etc.) to be in equilibrium with the imposed boundary conditions. It should be noted here that many climate models of the 'general circulation model' type have been developed from meteorological forecasting models, at least for their atmospheric part. However, this does not mean that these models can predict the weather (meteorology) on a specific date in the past or the future. Given the chaotic nature of the atmosphere, it is impossible to make weather predictions further out than ten days. What we are trying to establish is a statistical equilibrium for boundary conditions and for specific forcings, not the weather on a particular date.

How is this done in practice? It is not possible to solve the equations analytically, that is to say, it is not possible to obtain general formulae describing the temporal evolution of the prognostic variables of the system for a particular point of the troposphere. The equations are solved using numerical methods which involve discretizing them. The state of the atmosphere is described using a finite number of values which is nevertheless large for general circulation models (around 10^5–10^6). There are many methods of discretization and we will come back to this. One of the simplest ways is to describe the state of the atmosphere using the prognostic variables of the equations on a three-dimensional grid covering the globe. Let $X(t)$ be the set of these values describing the state of the atmosphere at time t. The basic unit of temporal discretization is called the '**time step**'. Starting from the initial state, describing the state of the atmosphere X_0 at time $t = t_0$, the equations enable the state of the atmosphere X to be calculated with the following $t = t_0 + \Delta t$ time step:

$$X(t_0 + \Delta t) = X_0 + \Delta X$$

ΔX can be obtained directly through the differential equations chosen to describe the evolution of the atmosphere, which allow us to calculate $\Delta X/\Delta t$ and then, once Δt is fixed, ΔX and $X(t_0 + \Delta t)$. Thus, progressing time step by time step, the evolution of the atmosphere can be calculated over a period long enough to obtain robust statistics, allowing a simulated climate to be defined based on the results of the model.

The time step Δt cannot be freely chosen. Obviously, the smaller the Δt, the longer it will take to get to the result. However, there is a maximum time step, equal to $c \times \Delta x$, where Δx is the chosen spatial resolution and c is the characteristic speed of propagation of the information from one point to another. This is called the Courant, Friedrichs and Lewy criterion (or CFL criterion), named after the mathematicians who formulated it. Thus, obtaining a simulation with a fine spatial resolution takes a long time because it requires not only calculations to be made on more points, but also a smaller time step. A compromise must therefore be made between spatial resolution and time taken to produce the simulation. Paleoclimate studies which require long simulations over several hundreds or even thousands of years often used models of coarser resolution rather than those used for climate forecasts into the next century. However, nowadays, there are specific projects in which the same models are used to compare the mechanisms of past and future climate changes.

Within the atmospheric general circulation models, two types of processes are often differentiated: **dynamic processes** and **physical processes**. The first type deals with the evolution of the circulation and can only be calculated from the three-dimensional spatial distribution of other variables, such as temperature. It is through the use of the dynamic laws [Eqs. (25.2)–(25.5)] that we can run the simulation forward, time step by time step. The second type are calculated for each vertical column separately for a given time step. These are mainly radiation, clouds, precipitation and surface exchanges. The distribution of the three-dimensional variables used at the dynamic stage of integration is obviously closely dependent on the evaluation of the physical processes for each vertical column. The dynamic and physical calculations are therefore carried out alternately, sometimes using different time steps. Taking the example of the atmospheric model included in the IPSL_CM6 model used in the Sixth IPCC Assessment Report (publication planned for 2021) the time steps are from 430 s (high-resolution version with 50 km and 79 vertical levels) to 2 min (low-resolution version with 300 km and 39 vertical levels) for the dynamic processes and 15 min for the physical ones.

In a general circulation model, we try to achieve the best representation of both types of processes. Circulation is calculated based on the basic laws of fluid mechanics, expressed for the particular case of a thin atmospheric layer surrounding a rotating planet. We have seen how these equations can be simplified for this specific context, based on the characteristic scales of global atmospheric circulation. However, there is a second type of simplification inherent to the construction of a model, and this is related to the physical processes defined above. The fine details of these processes are not always well understood. Moreover, their characteristic spatial scale is often much too small for them to be explicitly represented in current models, whose spatial resolution is of the order of a hundred kilometers. Therefore, the modeler will not attempt to represent the process in detail, but rather to represent its impact on the atmospheric characteristics at the resolution of the model. Thus, for example, each cloud is not represented individually; rather the impact of clouds on the radiative balance and on precipitations is formulated. This is called **parametrization** of a **subgrid process**. These parameterizations represent simplifications of reality in the sense that we have an incomplete knowledge of it and the process itself is not represented but rather its impact at the relevant scale. The parameterizations, as well as the methods of discretization of the equations used, along with the spatial and temporal resolutions, constitute the main characteristics of a model.

Returning to the methods of discretization of the equations governing the evolution of the state of the atmosphere, two approaches can be identified. The first is a description of the atmosphere in a finite number of points, generally organized into a three-dimensional grid. These are called **'grid-point models'** or 'grid-box models', with the 'box' referring to the smallest unit volumes of the grid. The resolution of the model is defined by the size of this box, or by the number of points used to describe the longitudes, latitudes, and the number of vertical levels. There are many examples of grids, among which grids whose points are regularly spaced in terms of longitude and latitude, and grids whose points are regularly spaced in terms of longitude and the cosine of latitude. In general, the vertical levels are not evenly distributed. In particular, they need to be closer together in the boundary layer of the atmosphere, the layer closest to the surface.

A second type of approach involves using spherical harmonics to describe the variations in the atmosphere on the horizontal plane. The grid point method is retained for the vertical dimension. These **'spectral' methods** are particularly suited to the atmosphere, which forms a continuum on the surface of a sphere. The calculations for this method are faster, in particular due to the fact that the first and second derivatives on the horizontal can be easily expressed for this type of decomposition. The spectral models are well suited for the representation of waves in the atmosphere with a smaller number of degrees of freedom than in the grid point models. The advantages of the spectral models are, however, less significant for fine resolutions, as there are many calculations, especially for physical processes, which still have to be carried out on a grid model. In general, the number of points in the grid exceeds the number of degrees of freedom in the spectral method so as to avoid problems with aliasing. These grids therefore give the impression of a finer resolution than the real number of degrees of freedom of the model. This is why the description of the resolution of these models refers to the number and type of harmonics chosen.

The oceanic general circulation models are also constructed based on fluid dynamics equations with the additional constraint of salt conservation. The discretization used is in grid point because of the geometry of the basins. The specificities of oceanic general circulation models are not detailed further here.

Towards an 'Integrated' Model of the Earth System

Historically, the first climate simulations carried out with general circulation models employed 'only' atmospheric models. Interactions with the surface, especially with the ocean, were very limited because the majority of the surface characteristics were imposed (surface temperature of the oceans, presence of sea ice, surface albedo, roughness of terrain etc.). As a result, the atmospheric circulation obtained was in equilibrium with these surface conditions and other forcings. In particular, it was then possible to evaluate the response of the atmosphere to changes in the ocean surface. It is clear, however, that the ocean, like vegetation and land surfaces, does not remain unaffected when faced with climate change. Modelers therefore quickly sought to estimate the impact of feedbacks from the other components of the climate system on the atmosphere, which in turn defines the climate at the surface. Figure 25.2 shows the evolution of climate models since their inception. It shows the coupling first with ocean surface models, then with complete ocean circulation models. In parallel, land surface models have progressed from simple hydrological models, with fixed albedo and surface terrain, to models including interactive vegetation, allowing the surface characteristics to be calculated according to changes in vegetation caused by changes in climate or by man. Finally, models increasingly include a representation of atmospheric chemistry and aerosols, which have a significant influence on radiation, as well as the biogeochemical cycles such as the carbon cycle. In this type of model, the atmospheric concentration of CO_2 is no longer imposed and is instead calculated from emissions.

Climate models developed in this way require enormous computing power. Simulations are generally carried out on supercomputers adapted to this type of coding. These computers are scarce, which explains the limited number of general circulation models in the world. The models, their resolution, as well as the components of the climate system and the processes to be included are chosen at the outset according to the issue to be addressed, but also in keeping with the current limits in computing power. The performance of the models will vary depending on the model and the computer used, but for example, the approximate computation times for the IPSL model are: run at very low resolution (IPSL-CM5A2: atmosphere 96 × 95 × 39, ocean 2°), about 70 years per day; at low resolution (IPSL-CM6-LR, atmosphere 144 × 133 × 79, ocean 1°) 16 years per day, and at medium resolution (IPSL-CM6-MR, atmosphere 280 × 280 × 79, ocean 1°) about 6 years per day.

Thanks to the improvements in supercomputers over recent decades, the development of coupled atmosphere-ocean models, followed by atmosphere-ocean-vegetation models has become possible. These models require a longer computing time, not because there are many additional calculations to be performed for a given duration, but because vegetation, and even more, the ocean, are components of the climate system whose response time is far greater that of the atmosphere. While we consider that simulations using an atmospheric model alone, forced by boundary conditions which repeat each year, must be integrated over a period of 20 to 50 years to obtain a response from the atmosphere in equilibrium with these boundary conditions, a coupled atmosphere-ocean model, in principle, needs to be integrated over one or even several thousand years. The biggest challenge then is to close-off the water and energy balances in the model to avoid a gradual drift related not to the imposed forcing but to the model itself.

'Realistic' Modelling of Paleoclimates

Boundary Conditions and Initial Conditions

Many paleoclimate simulations aim to 'recreate' past climates as accurately as possible. The models can then be evaluated in the context of climates documented by paleoclimate indicators which are different from the current climate. It also provides a better understanding of the possible connections between differences in climate between distant regions and supports reconstructions by providing better spatial and temporal coverage or by including a regional phenomenon not covered by the reconstructions, thereby improving our understanding of them. We will return to the comparison between models and data and the value of this exercise at the end of the section.

How can a 'realistic' simulation of a paleoclimate be achieved? First, depending on the question at stake, the part of the climate system being assessed needs to be defined. For example, if we want to study the continental climate in the context of given ocean conditions, it is best to use an atmospheric model, possibly coupled with a dynamic land surface model or vegetation model. Once the subsystem is selected, a paleoclimate simulation is carried out by imposing the most realistic forcings and boundary conditions possible on this subsystem for the simulation period. Thus, the more the subsystem is constrained, the more conditions there are to be imposed, conditions that need to be known for the study period. Continuing with the example of the simulation using only a general atmospheric

Fig. 25.2 History of climate models used in successive IPCC reports. From the 4th IPCC report (2007), Solomon et al. (Ed.), Cambridge University Press, http://www.ipcc.ch/publications_and_data/ar4/wg1/en/contents.html

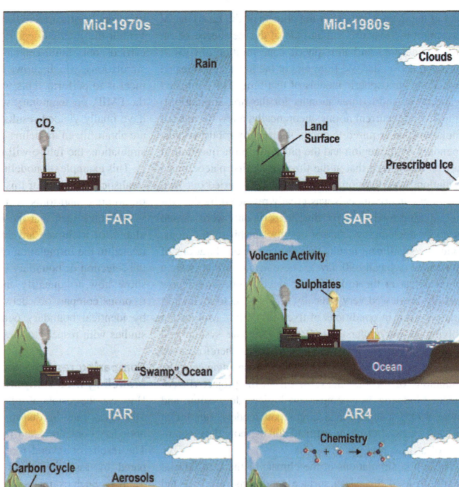

circulation model, the ocean surface temperatures and the extent of sea ice have to be defined for all ocean-type points in the atmospheric model. In practice, this is extremely restrictive since at the very least, reconstructions of ocean surface conditions with a spatial resolution similar to the model for a typical seasonal cycle would be required. These numerical experiments are almost always based on strong assumptions as to the ocean surface conditions as well as the land-ocean distribution, topography, and extent and altitude of the ice sheets. The other forcings are better known, at least for the relatively recent periods of the Quaternary: insolation (Berger 1978; Laskar 2004) and atmospheric concentration of greenhouse gases, with measurements from Antarctic ice cores now dating as far back as 800,000 years.

Thus, a 'realistic' simulation of a paleoclimate is based on several series of assumptions: those related to the design of the model itself and those related to the fixing of boundary conditions for a specific experiment. So, this simulation summarizes both our knowledge of climate characterization (incorporated into the climate model), and our understanding of the forcings of this climate (expressed in the forcing and boundary conditions). It is also constrained by the limitations of computers and technology of its time.

Take the example of the climate simulations for the Last Glacial Maximum (LGM, about 21,000 years ago). The first simulations of the climate of this period (Gates 1976) were carried out shortly after the first reconstructions of the ocean surfaces (sea surface temperature, extent of sea ice) and of land (ice caps) were produced (CLIMAP 1976). At this time, the LGM atmospheric concentration of CO_2 was not known. It was only in the 1980s (Manabe and Broccoli 1985) that the first numerical simulation was carried out taking into account the main forcings for the climate of the Last Glacial Maximum: the expanse and altitude of the ice sheets, ocean surface conditions and atmospheric concentration of CO_2. It was a simulation derived from one of the most sophisticated stand-alone atmospheric models of that time. The simulated duration was short (three months for the first simulation!) compared with current norms: theoretically, a few decades of simulations are required to obtain statistically robust results, depending on the region and the phenomenon in question. It must be recognized that at the time of the first paleoclimate simulations, the duration of the simulations carried out with general circulation models was greatly restricted by the cost of computing time. To save time, experiments were carried out under unchanging January or July sunshine conditions, producing significant results for relatively short durations of simulations (typically 90 days). These results, which were a technical feat at the time of their publication, would now probably be viewed very critically, mainly due to their short duration, even in conditions of fixed insolation. Models are evolving as our understanding of the climate system and computer capabilities improves. Simulations therefore need to be revised periodically in the light of these advances. The simulations of the first phase of the PMIP project (Paleoclimate Modeling Intercomparison Project, Joussaume and Taylor 1995) using general circulation models, ran for at least ten years after an adjustment for the boundary conditions of at least one year.

These first simulations of the climate of the Last Glacial Maximum used the ocean surface conditions reconstructed by the CLIMAP project (1976, 1981). These reconstructions were the result of a major work of data synthesis, but problems were quickly identified, particularly for the sub-tropical regions, where a higher temperature than is the case currently was reconstructed, and for the North Atlantic, where the winter sea ice cover was overestimated. These problems were partially amplified by the methods used to extrapolate the reconstructions for each site to cover the globe with an even grid. The CLIMAP project, at the specific request of modelers, provided reconstructions for the months of February and August. However, it is entirely possible that certain species used for reconstructing SSTs are not particularly sensitive to these specific months, but to other factors. Thus, manipulating data to construct boundary conditions for models, especially atmospheric ones, can prove to be extremely restrictive for the interpretation of data records. Furthermore, since they are used to establish the boundary conditions, they cannot also be used to validate the model. Therefore, as soon as they became available, it was very useful to use coupled ocean-atmosphere models to simulate paleoclimates. It is worth highlighting again the challenge represented by the first coupled simulations of the climate of the LGM. Again, in this case, the first published simulation was only about thirty years long, a very short time frame compared to the response time of the deep ocean! Within the international PMIP2 project, eight groups have carried out multi-centennial coupled ocean-atmosphere simulations, which shows how difficult this type of experiment is to perform. This was confirmed in the 3rd phase of the PMIP, contemporary with CMIP5, for which 9 models have finally yielded results for the LGM. For PMIP4, there are about fifteen modeling groups planning to undertake this simulation, the future will tell us how many succeed.

This example of modeling of the LGM climate shows that 'realistic' modeling of this climate has evolved in line with the forcings and tools available.

We have seen that the uncertainties in a 'realistic' simulation of a climate are due to two types of factors: those related to the formulation of the model and those inherent in the selection of boundary conditions. The next two sections show how to quantify these uncertainties, both through rigorous comparisons between the results of models forced by identical boundary conditions, and through sensitivity studies with respect to these boundary conditions.

Comparing results from different models: Modeling Intercomparison Projects

How can the results of different models be compared? These differences may be due to the models themselves, or to the boundary conditions and forcings imposed on these models. The results of several models can only be rigorously compared by assigning them the same boundary conditions/forcings. Such exercises have been proposed for the modeling of current climates using atmospheric general circulation models (AMIP project, Atmospheric Model Intercomparison Project, http://www-pcmdi.llnl.gov/projects/amip/), followed by coupled models, both for current and future climates (CMIP project, Coupled Model Intercomparison Project, http://www-pcmdi.llnl.gov/projects/cmip/). The CMIP5 exercise corresponds to the results produced for the 5th IPCC report and CMIP6, currently underway, will provide its first results for the 6th IPCC report, which will be published in 2022. In the same vein, PMIP (Paleoclimate Modeling Intercomparison Project), the project to compare paleoclimate models came into being in the 1990s (http://pmip.lsce.ipsl.fr). At first, this project involved atmospheric general circulation models (PMIP1 project, http://pmip1.lsce.ipsl.fr/) for the Middle Holocene (6000 years ago) and the Last Glacial Maximum (21,000 years ago). It was then extended to coupled atmosphere-ocean and atmosphere-ocean-vegetation models (PMIP2 project, Braconnot et al. (2007a, b), http://pmip2.lsce.ipsl.fr/). A new feature of PMIP3 was to use climate models strictly identical to those used for CMIP5. PMIP4 coordinates both CMIP6 simulations, which will therefore use the same models as those used for climate projections, and simulations based on other models, usually longer ones or for older climates. The PMIP4-CMIP6 simulations concern the following climates: the last millennium, the Middle Holocene, the Last Glacial Maximum, the last interglacial and the Middle Pliocene (Kageyama et al. 2018).

The PMIP project first focused on defining precise boundary conditions and forcings for the Middle Holocene and the Last Glacial Maximum. This made it possible to rigorously compare the results of the models participating in the project with the paleoclimate reconstructions. Below is an example of a comparison of PMIP2 model results for Europe during the Last Glacial Maximum. Figure 25.3 shows the temperature of the coldest month in an average seasonal cycle simulated for this period by the eight models whose results were available in the database in November 2009. The differences with the current climate are shown. The color of the diamonds indicates the average temperature reconstructed from pollen data by Wu et al, (2007), on the same color scale as for the one used for the model results. In this figure, it can be seen that the most noticeable cooling of at least 12 °C is simulated by the models in the northern part of the area under study, on the Fennoscandinavian cap and on the sea ice off the coast of Scandinavia. This cooling lessens towards the south, where it is about 3 °C. Even if the same boundary condition forcings are applied to the models, the climates obtained differ from one model to another. For example, the cooling simulated over the ice cap is between 12 and 18 °C for the ECBILTCLIO model, whereas it is greater than 30 °C in the HadCM3M2 model. Around the Mediterranean, the CCSM3 model simulates practically no temperature change for the coldest month, whereas the ECHAM5.3-MPIOM-127-LPJ model simulates a drop in temperature of between 3 and 6 °C.

This shows how models developed to first represent the current climate can diverge in their representation of climates different from the current one. This discrepancy is found in the forecasts of future climates, but only paleoclimate simulations allow climate simulations different from today's climate to be compared with the data.

Comparisons Against Paleoclimate Reconstructions

In Fig. 25.3, reconstructions of the temperature of the coldest month by Wu et al. (2007) are indicated by diamond shapes with the same color code as the output of the models. It should be noted that for Western Europe, all of the models simulated temperatures that are warmer than the reconstructed temperatures. However, it is important to take into account both the dispersion of the results of the models, which is done to a certain degree in Fig. 25.3 by including the results of all the models as well as the level of uncertainty of the reconstructions, which cannot be shown on the maps. Figure 25.4 compares the same model and reconstruction results from another perspective. Here, we have chosen to show the average temperature by longitude for Western (10° W–15° E) and Central Europe (15–50° E) and the reconstructions with their uncertainty range. This time we see that the temperatures simulated by the models are compatible with the reconstructions, if we take into account the uncertainty characterizing the reconstructions, including for Western Europe.

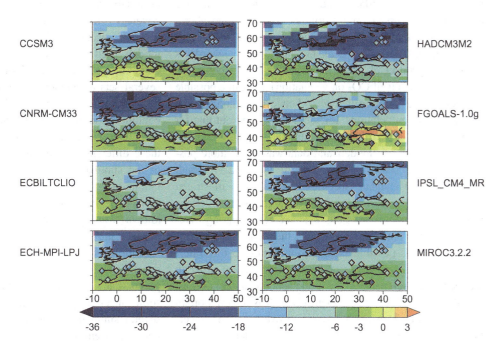

Fig. 25.3 Maps: temperature of the coldest month in an average seasonal cycle, as simulated by the coupled ocean-atmosphere models participating in the PMIP2 project (November 2009 PMIP2 database); Diamonds: same variable, as reconstructed by Wu et al. (2007)

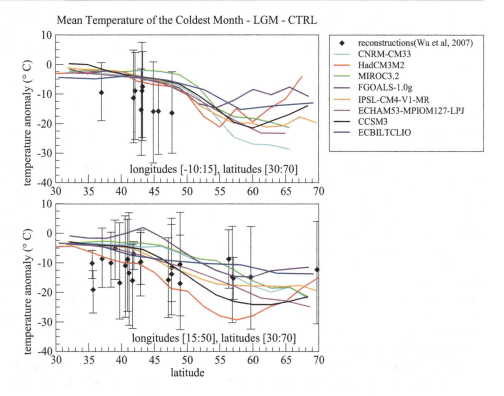

Fig. 25.4 Comparison between the simulated temperatures of the coldest month (during an average seasonal cycle) for the LGM in Europe (continuous lines, given for different models of the PMIP2 database), and reconstructed temperatures by Wu et al. (2007) (diamond shapes and the uncertainty bars associated with them) for the Atlantic (top) and Eastern (bottom) regions. The values shown are differences from the reference (pre-industrial) climate

Ramstein et al. (2007) summarized the efforts of the years 2000's to reach a congruent comparison between reconstructions and simulations for winter temperatures during the Last Glacial Maximum. The simulations of the PMIP1 project, using standalone atmospheric models forced by the CLIMAP (1981) ocean surface conditions, resulted in overly high winter temperatures compared to the initial pollen-based reconstructions of Peyron et al. (1998). By working on the boundary conditions (expansion of the ice caps), on the models (transition to coupled atmosphere-ocean models) and on reconstructions (new reconstructions by Wu et al. (2007), based on the same pollen records as used by Peyron et al. (1998), but taking into account the effect of low levels of CO_2 on vegetation) it became possible to reduce the large differences between simulated and reconstructed temperatures. It should be pointed out that for other variables (in particular the summer temperatures), the comparison between models and data was much more positive from the start. The example of the coldest month temperatures was taken specifically because it illustrated what can be learned from the models and the reconstructions through the sometimes tedious exercise of comparing models with data. This example shows that it is important to consider all the possible factors contributing to the differences between simulations and reconstructions in order to reduce these discrepancies: the models, boundary conditions, but also the reconstructions themselves. It is also important to have results from many models to overcome the uncertainty associated with the use of a single model.

Sensitivity Experiments

We have seen that the uncertainties in the results of numerical models stem from the formulation of the models themselves, since these models are built on assumptions considered relevant to the given problem, and on the conditions imposed on the model, which are themselves based on assumptions because we lack the necessary level of precision and spatial and temporal coverage. How can these uncertainties be calculated? One method, discussed above, is to increase the number of models used. Similarly, if we are unsure of the boundary conditions to be imposed on the model, or if there are several sets of boundary conditions possible, we can carry out several experiments with different sets of boundary conditions so that the climate responses to these conditions can be compared and we can determine whether these differences in boundary conditions cause differences in the simulated climate. For example, for climate simulations using an atmospheric general circulation model, the surface temperatures of the oceans are generally not known with great certainty for all months of a given period and for all the grid points of the model. Assumptions are then made so as to reconstruct the seasonal cycle of ocean surface temperatures based on the points that are available and about which there is also some uncertainty. It is possible for the model to perform several simulations based on different ocean surface temperature scenarios. We can then analyze the one that produces the land temperatures/precipitation values most compatible with the reconstructed data. These are known as sensitivity experiments to sea surface temperatures.

In the case discussed above, we examined several scenarios with different ocean surface temperatures to find out which was most realistic for the period in question. More generally, it is also instructive to analyze the signature or influence of each forcing among several forcings and changes in boundary conditions. To continue with the example of the climate of the Last Glacial Maximum, if a coupled atmosphere-ocean model is used, the forcings applied to the model to obtain a simulation of the climate of this period are:

1. the insolation;
2. the greenhouse gas concentration (CO_2, CH_4, N_2O);
3. the altitude and extent of the ice caps and the change in land-ocean distribution caused by changes in sea level.

To better understand why the glacial period climate is different from the current climate, simulations can be carried out in which only one or two of these boundary conditions are imposed and the simulated climate is compared with a more 'realistic' simulation where all of the boundary conditions are applied. By carrying out these simulations where one or two boundary conditions or forcings are imposed with glaciation values, the aim is not to try to achieve a realistic simulation. Instead, the purpose is to gain a better understanding of the response to each type of forcing. These simulations are also called sensitivity experiments with 1, 2 or 3 forcings.

This approach is illustrated in Fig. 25.5 where the response of the IPSL climate model to LGM conditions is

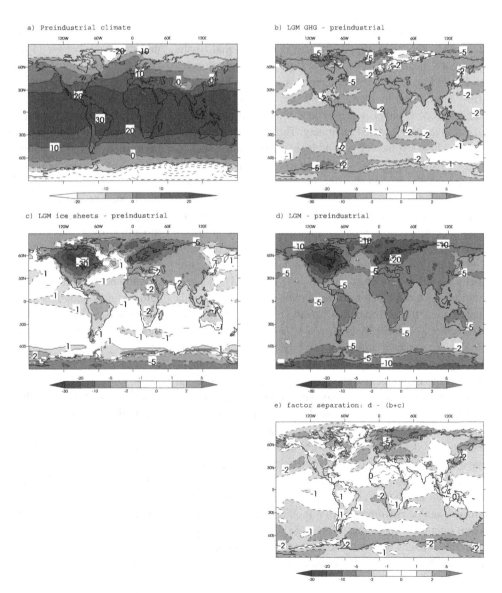

Fig. 25.5 Average annual air temperature at 2 m simulated by the coupled ocean-atmosphere IPSL model: **a** for pre-industrial conditions; **b**, **c** and **d** anomalies compared to this pre-industrial climate: **b** for a pre-industrial simulation, where the atmospheric greenhouse gas concentration has been replaced by ice age values; **c** same for ice caps, **d** simulation with all LGM conditions; **e** difference d − (b + c), allowing the quantification of the proportion of the difference between pre-industrial and the LGM related to the interactions between the impacts of the ice caps and the reduction in greenhouse gas concentrations. For figure (**a**), isolines every 10 °C, dashed lines for negative values, long dashes for 0 °C, and continuous lines for positive values

analyzed by separating the impact of the ice caps from the impact of the reduction of atmospheric greenhouse gas concentration. This figure refers to the annual average temperature of air 2 m above the surface. The values for this temperature in the pre-industrial climate simulation are shown in Fig. 25.5a and the differences between the simulated LGM and pre-industrial climates are shown in Fig. 25.5d. This map shows an overall global cooling, moderate (by a few degrees approximately) over the oceans but very strong (more than 30 °C) over the ice sheets of the northern hemisphere. Figure 25.5b shows that the contribution of greenhouse gases to this response is less extreme over the continents and a little weaker over the oceans than the response to all of the LGM conditions. Figure 25.5c shows the impact of the ice caps alone. It shows that these are responsible for a significant cooling over the continents of the northern hemisphere, but also for a cooling of between 0 and 2 °C over most of the oceans, with the notable exception of the Southern Ocean. It is worth noting that the sum of the anomalies shown in Fig. 25.5b and d is not equal to the difference between the LGM and pre-industrial climates shown in Fig. 25.5d. This difference is shown in Fig. 25.5e. This shows that in many regions the impact of the two factors taken together is greater than the sum of the impacts of each factor considered separately. This is referred to as the synergy between the various forcings. In other areas, such as north of the Nordic Seas, the impact of the two forcings together is lower than the sum of the impacts of the individual forcings. This shows that to quantify the impact of a specific forcing within a group of two forcings (as in this case, the impact of the ice caps and the reduction in greenhouse gases), four simulations must be carried out: a control simulation (in this case, the pre-industrial climate), one including all the forcings (in this case, the LGM) and simulations with each factor taken individually. This method is called the 'factor separation' method developed for the atmospheric sciences by Stein and Alpert (1993).

So far, we have studied sensitivity experiments on the forcings and boundary conditions imposed on models. Sensitivity experiments may also be applied to internal processes of the climate system. To examine the importance of this processes in the response by the climate system to a disturbance, its formulation in a model can be modified. For example, if deep convection at the equator is suspected to have an important influence on some aspect of the climate in the mid-latitudes, the formulation for deep convection in the model can be altered and a sensitivity experiment for this process can be performed with a modified model, under the same boundary conditions. This type of sensitivity experiment also makes it possible to evaluate the importance of a feedback by excluding or activating it.

Outlook

In this section, we mainly describe atmospheric general circulation models, including some coupled with oceanic general circulation models. These models, now complemented by vegetation models, carbon cycle models and atmospheric chemistry models, are becoming increasingly complex, with more components of the climate system, more processes and more associated feedbacks being included. The complex models of the Earth system have sometimes been described as the 'biggest' models that can run on the 'biggest' computers, a sort of 'maximum' model. These limit the number of numerical experiments that can be run for each given problem and the number of sensitivity experiments that can be conducted to better understand the influence of a particular process or mechanism. This situation is changing, as advances in computing now allow modelers to carry out more experiments for a given period. These experiments are essential to improve our understanding of the importance of forcings or processes within a change in climate. New developments in complex climate system models, the improvement in their resolution, the inclusion of new processes or new components must consider the necessary compromise between the computational time required for a simulation and the number of simulations that can be performed with the computer available. With increasing computing capabilities, it is possible to improve the resolution of the models and increase the number of processes included. The models used for the IPCC assessment exercises provide a good indication of the progress of climate modeling over the past two decades (Fig. 25.2). In the future, we will see new components of the climate system being integrated as for instance ice caps. The aim of these developments is mainly to provide a better prediction of the future climate, but they also contribute greatly to the study of paleoclimates, which in turn makes it possible to evaluate these models under climate conditions different to current ones.

Earth System Models of Intermediate Complexity (EMICS)

Basic Principles and History

We have seen that climate models developed from general circulation models were intended to be as comprehensive as possible in their representation of the climate system. This requires considerable computing power and calculation times, which in practice forces the modeler to limit the number of simulations performed. These simulations are also quite short compared to the typical timeframe in climate evolution. Thus, in parallel with the development of general circulation models, simpler models, more adapted to the study of paleoclimates, have been developed. The aim was to represent from the outset the slow-moving components of the climate system, the ocean and the ice caps, in order to study long-term climate change (i.e. for time scales in excess of a thousand years). To develop these models, representation of the rapid components of the climate system, particularly of the atmosphere, has to be simplified. In fact, the term 'simple model' is misleading because it refers above all to models which are more efficient in their use of computing time. Developing a model of this type is not necessarily 'simple' because one cannot simply retain the basic equations of atmospheric dynamics. The saving of computation time is generally achieved by establishing parameters for the transport of heat and moisture by the stationary and/or transient waves (such as, for example, by the depressions of the mid-latitudes), in other words, by trying to represent the effect of these phenomena without explicitly calculating them. This makes it possible to extend the duration of a time step and to use a coarser spatial resolution. The number of degrees of freedom of these models lies between conceptual models (around ten) and general circulation models ($\sim 10^5$–10^6) and they are called 'EMICs', 'Earth system Models of Intermediate Complexity' (Claussen et al. 2002). In fact, the terminology was created long after the development of the first models in this category. It emerged at a time when the modelers who specialized in these models decided to join together to define the specificity of their models compared with others. These models are characterized by a more complete representation of the climate system than the 'simple' ocean-atmosphere models and by a relatively short computing time compared to the general circulation models, characteristics which allow the evolution of the climate system to be studied over long time scales and many different scenarios to be explored. There are many EMICs, corresponding to the many different ways the representation of the climate system can be 'simplified'. It should be noted that some models have been developed by 'downgrading' a general circulation model, i.e. by reducing its vertical and horizontal resolutions. These are the most complex models in the EMIC category.

As with general circulation models, climate models of intermediate complexity can be used to obtain realistic climate simulations or to study the sensitivity to certain forcings, processes or feedbacks of the represented system. In the following sections, we give examples of the use of EMICs, both for long-term simulations and for studies requiring numerous experiments. Experiments of this type could not have been carried out using general circulation models given current capacity of computing power. This shows the complementarity of the two types of models, one type being useful for its 'efficiency', the other for the spatial and temporal detail in its representation.

Examples of Long-Term Simulations and Studies of Sensitivity to Forcings

One of the first models of intermediate complexity of the climate system is the one developed by the Catholic University of Louvain-la-Neuve. This model includes simplified representations (by latitude and vertically, for the northern hemisphere) of the atmosphere, the ocean, sea ice and the polar ice sheet. It was developed specifically to study the glacial-interglacial cycles, as demonstrated by the first simulations of Gallée et al. (1992). Since then, Berger et al. (1998) and Loutre and Berger (2000) have taken up this model and carried out sensitivity experiments to identify the respective roles of orbital variations and greenhouse gases in the last glacial-interglacial cycle.

Figure 25.6 shows a selection of the results of these two articles in terms of volume of ice (top) and temperature in the northern hemisphere (bottom). The continuous lines represent the results of the model forced by both the variations in CO_2 recorded in the ice cores (Jouzel et al. 1993) and by the variations in the orbital parameters as calculated by Berger (1978). The initial state of the model is an interglacial state, with no ice sheet in the northern hemisphere. Over the last 200,000 years, the model simulates two major glaciations, with a complete freeze-up occurring in steps and a complete deglaciation following the glacial maximum. The coldest temperatures are of course simulated for these glacial maxima. During the last interglacial and the last glaciation (between 126,000 and 80,000 years), despite a high level of recorded variability, temperatures remain sufficiently high so that the frozen-over periods last no longer than 15,000 years.

The red lines correspond to a simulation where the insolation is constant and equal to the current insolation, and where CO_2 levels vary in a similar way to the previous simulation. It can be seen that the volume of ice increases to about 35×10^{15} m^3 and remains at around 30×10^{15} m^3 during the rest of the simulation. Although the average temperature of the northern hemisphere varies in line with the greenhouse gas forcing, it remains too cold to bring about deglaciation. The other curves are the result of simulations where CO_2 remains constant (fixed at 210 ppm for the alternating dash-dot line and at 250 ppm for the dashed curve) and where variations of the orbital parameters are taken into account. This time, alternation between glacial and interglacial periods is obtained, with maximum ice volumes reached for the same periods as in the reference simulation at values inversely related to the level of imposed

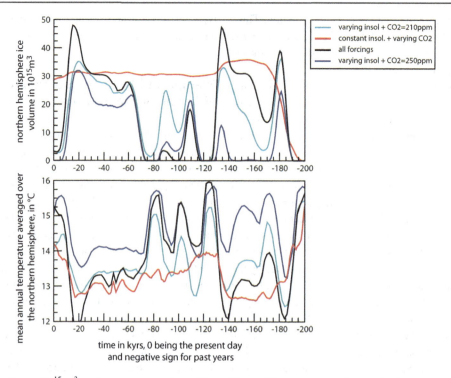

Fig. 25.6 Ice volume (top, in 10^{15} m³) and average temperature of the northern hemisphere (bottom, °C), as a function of time in kyrs, simulated by the Louvain-la-Neuve coupled model ice sheet-northern hemisphere climate (Gallée et al. 1992) over the last 200,000 years. In black, the simulation where CO_2 and insolation vary in line with the Vostok data for CO_2 and according to the forcing by Berger (1978). The red line represents the results for constant insolation, equal to current levels, with variable CO_2. The light and dark blue lines represent the results for variable insolation and constant CO_2, at 210 and 250 ppm respectively. According to Berger et al. (1998), and Loutre and Berger (2000). The authors thank M.-F. Loutre for providing the results of the Louvain-la-Neuve model

CO_2. These sensitivity experiments show that variations in insolation are an essential forcing to explain glacial-interglacial cycles, because forcing the model with CO_2 variations alone does not produce these cycles. On the other hand, experiments with constant levels of CO_2 show that the extent of a glaciation is highly dependent on CO_2.

Examples of other long-range simulations are given in other chapters of this book. In Chap. 29, an example is given of modeling interactions between northern hemisphere ice sheets, the ocean and the atmosphere over a period of 50,000 years (Calov et al. 2002), as well as an example of modeling of ^{18}O variations in the glacial ocean in response to freshwater inputs from the North American ice cap (Roche and Paillard 2005).

Example of the Use of Intermediate Complexity Models to Explore a Multitude of Forcings or Parameters: Exploration of a 'Phase Space'

In the previous section, we showed that, thanks to its efficient use of computing time, a climate model of intermediate complexity allows us to perform long simulations over time scales able to reflect the variations in climate during the Quaternary (glacial-interglacial cycles, abrupt events from the last glacial period). This efficiency also means that it is possible to carry out numerous simulations to explore the sensitivity to the forcings used of a result or to choices made during the construction of the model. An example of this type of use of an intermediate complexity model is given in Chap. 29. The CLIMBER-2 model, incorporating a representation of oxygen isotopes in the ocean, is used to explore the response to freshwater discharges from the North American ice cap in multiple scenarios. The response is then compared to marine records to determine the most probable scenarios in terms of duration and amplitude of the freshwater input.

Another example is given by the work of Schneider von Deimling et al. (2006). This study analyzes ensembles of simulations for the climates of the present day, the future and the Last Glacial Maximum. These ensembles are formed by varying eleven parameters of the model within acceptable ranges, based on our current knowledge of the climate system. The main effect of changes in these parameters is to vary the amplitude of the feedbacks in the climate system and the climate sensitivity of the model (defined as the difference in global temperature due to a doubling of CO_2, see Chap. 31). The current climate, as defined by the observations, makes it possible to make a first choice of parameters so that an acceptable simulated climate can be achieved. The authors show that reconstructions of tropical temperatures during the Last Glacial Maximum help to constrain even further the selection of values of these parameters and so reduce the uncertainties associated with future climate change. This study shows that modeling of past climates followed by comparison with reconstructions can help in the evaluation of climate models used to forecast future climate. This conclusion is also advanced by Hargreaves et al. (2007) using a general circulation model.

Outlook

It might be expected that with the advances in computing, climate models of intermediate complexity would no longer have any reason to exist. In fact, this is not the case because they will always be less time consuming than general circulation models, which incorporate more and more mechanisms and use an increasingly fine resolution. In some ways, general circulation models, such as those used for IPCC simulations, are defined by the capabilities of the most powerful computers. We need to be able to use these models to produce simulations of several hundred years within a reasonable time on available computers. Long simulations, necessary to understand past climate changes reconstructed using multiple indicators and the need to explore different scenarios and model parameters, require faster models. Today's general circulation models will no doubt become the intermediate complexity models of tomorrow, but this concept will continue to exist. In addition, it is important to retain this hierarchy of models because each type of model is established on different assumptions. By comparing the results of different models, it is possible to highlight the relative importance of a particular process which is included in one model but not in the other or which is represented differently in each model.

Conceptual Models

The main objective of the models described above is to try to reproduce the observations we have for the climate system and its variations in the past. As has been highlighted, modeling also aims to improve our understanding of these variations, and it is therefore useful to describe some aspects of the system using extremely simple models, which are intended to illustrate some key processes. These are called conceptual models. There are many varied examples. One example is the Lorenz model (Fig. 25.1d), which often serves as an archetype of the chaotic system. The meteorologist Edward Lorenz proposed a very simple model, based on an idealized thermal convection, which for the first time illustrated that the complexity of the behavior of a dynamic system was absolutely unrelated to the number of degrees of freedom of this system, as many previously imagined. He showed that a very simple system (in this case with only three degrees of freedom) can produce unpredictable behavior, called 'deterministic chaos'. This conceptual model still plays an important educational role, and its mathematical properties are still a subject of active research. Below, some examples directly relevant to the climate system are described in more detail. Other examples, shown in Chap. 28, aim to achieve a better understanding of the glacial-interglacial dynamics (Calder, Imbrie models etc.).

The Budyko/Sellers Model

The Earth's climate is determined above all by its radiative balance. By simulating simplified balances, it is possible to estimate the magnitude of a change in temperature caused by, for example, changes in the incident solar radiation (volcanic dust, changes in the solar constant, nuclear winter, etc.). In 1969, two publications (Budyko 1969, Sellers 1969) came to a somewhat surprising conclusion: if we take account of the feedback between temperature and albedo, a relatively small decrease in the solar constant (−1.5% or −2%) is enough for the Earth to completely freeze over. A similar result also occurs when the greenhouse effect is modified. This indicates that there is a critical threshold towards cooling which causes the climate system to move into a very different state. The meaning of these results has now become even more pertinent with the theory of Snowball Earth (Chap. 26).

In their original versions, Budyko's and Sellers' models are explicitly dependent on latitude and predict a temperature $T(y)$ where y is the latitude. A much simpler version can be formulated to represent the phenomenon of runaway albedo-temperature feedback, with a model with no geographical dimensions. Writing the radiative balance of the Earth as a global average:

$$(1 - \alpha) \times S/4 = (1 - \varepsilon)\sigma T^4 \qquad (25.7)$$

where α is the albedo of the Earth, S is the solar constant; ε is a corrective term to represent the greenhouse effect; σ is the Stefan-Bolztman constant; then the overall global temperature of the planet is easy to calculate.

The problem becomes more interesting with the albedo-temperature feedback. Indeed, if we assume that α is a decreasing function of T, with, for example a constant $\alpha(T)$ (~ 0.3) at high temperatures for a 'blue' planet, a constant $\alpha(T)$ (~ 0.7) at very cold temperatures for a 'white' planet, and a linear α(T) in-between, then we obtain the diagram shown in Fig. 25.7.

The possibility of multiple equilibria leads to the existence of thresholds beyond which the climate system suddenly shifts to a new state of equilibrium. Moreover, this leads to a phenomenon of hysteresis, since it is not possible to easily return to the original state by reversing the disturbance. Thus, to return to the initial state, an inverse perturbation of much greater amplitude is necessary. This is one of the difficulties with the 'snowball' theory: although it is relatively 'easy' for the planet to freeze over completely, as was shown by Budyko and Sellers, it is much more difficult to get out of this cold state.

The Stommel Model (1961)

The existence of multiple equilibria concerns other components of the climate system and an important example in paleoclimatology is the Stommel model (Fig. 25.8).

The model is composed of two well-mixed boxes, of the same volume, representing a mass of cold water with low salinity for high latitudes (with temperature T_1 and salinity S_1) and a mass of warm water with high salinity for low latitudes (with temperature T_2 and of salinity S_2). The difference in density $\Delta\rho = \rho_2 - \rho_1$ between these two boxes is obtained as a function of the positive coefficients of thermal expansion α and saline contraction β assumed to be constant:

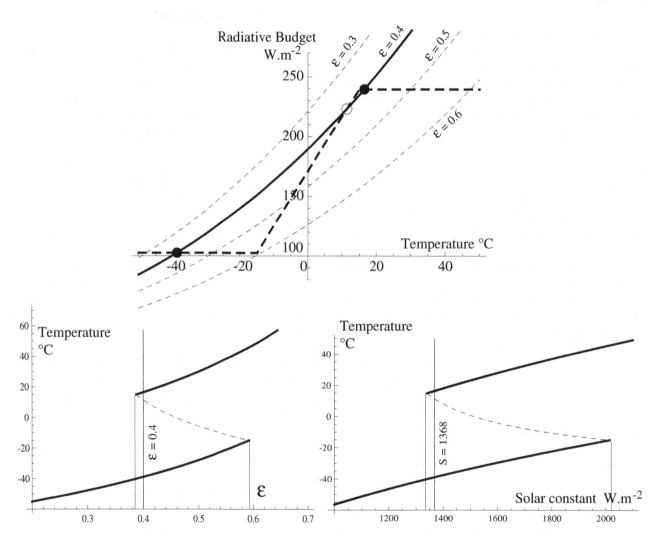

Fig. 25.7 Simplified Budyko and Sellers model. Top: dashed line, the solar term of the radiative balance, i.e. $(1 - \alpha(T)) \times S/4$, with α(T) linear between −15 and +15 °C, and constant beyond this range. Solid line, infrared term, for different values of the greenhouse effect. Balance is achieved when the curves intersect. Note that there are several points of equilibrium, especially for the current situation (ε = 0.4). Bottom: the corresponding points of radiative equilibrium as a function of the greenhouse effect ε (left), or as a function of the solar constant S (right). The equilibrium shown in dotted lines is unstable. For the current parameters, there are therefore two possible stable equilibria, corresponding either to our climate (temperature of around +15 °C) or to a completely frozen planet (temperature of around −40 °C)

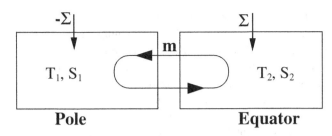

Fig. 25.8 Configuration of the Stommel model

$$\Delta\rho = -\alpha(T_2 - T_1) + \beta(S_2 - S_1) \quad (25.8)$$

A thermohaline circulation m, proportional to the difference in density between the two boxes, mixes the two corresponding bodies of water. This circulation will be positive (Fig. 25.8) when the cold water (box 1) is sufficiently dense to sink below the warm water (box 2), in other words when $\Delta\rho < 0$, and hence:

$$m = -\mu\Delta\rho = \mu[\alpha(T_2 - T_1) + \beta(S_2 - S_1)] \quad (25.9)$$

where μ is an arbitrary constant.

Conversely, if the saline water in box 2 is denser ($\Delta\rho > 0$), the thermohaline circulation m will be negative.

If we impose the temperature gradient $\Delta T = T_2 - T_1$ and force the salinities by applying a constant flow of salt Σ into box 2, and $-\Sigma$ into the box 1, we can infer the evolution in salinity, which are the only variables in the problem. For example, for box 1 (the strict opposite is in box 2):

$$\frac{dS_1}{dt} = -\Sigma + |m|(S_2 - S_1) = -\Sigma + \mu|\alpha\Delta T - \beta\Delta S|\Delta S. \quad (25.10)$$

The equilibrium (or equilibria) of the system is then easily obtained:

$$\Sigma = \mu|\alpha\Delta T - \beta\Delta S|\Delta S, \quad (25.11)$$

which gives a second-degree equation in ΔS, with an absolute value, which can be rewritten as:

$$F = x|1 - x| \text{ by setting: } x = \frac{\beta\Delta S}{\alpha\Delta T}; F = \frac{\Sigma\beta}{\mu(\alpha\Delta T)^2} \quad (25.12)$$

The function $F(x)$ is plotted in Fig. 25.9.

The definition of x shows that x measures the intensity of the salinity gradient relative to the temperature gradient. The sign of m, and consequently, the direction of the thermohaline circulation is positive when x is less than one. The dominant effect is then that of the temperature gradient ΔT. This is a mode of 'thermal' circulation. This is the case for the unique solution x_T when F is negative. Conversely, if the imposed flux of salt Σ is sufficiently strong, and consequently, if F is large enough (greater than 0.25), the solution x_S of the problem is greater than 1, and the thermohaline circulation is reversed because the only equilibrium possible of the system is of the 'saline' type. On the other hand, for an intermediate value of the forcing F, the 'thermal' and 'saline' modes are both possible solutions of the problem.

A third equilibrium point also becomes possible with solution x_I. However, a rapid analysis of the stability of the equilibria obtained shows that this is an unstable equilibrium. In fact:

$$\frac{d\Delta S}{dt} = 2(\Sigma - |m|\Delta S) \quad (25.13)$$

Fig. 25.9 Diagram of stability for the Stommel model (function $F(x)$ defined by Eq. (25.12)). The dotted section corresponds to an unstable equilibrium

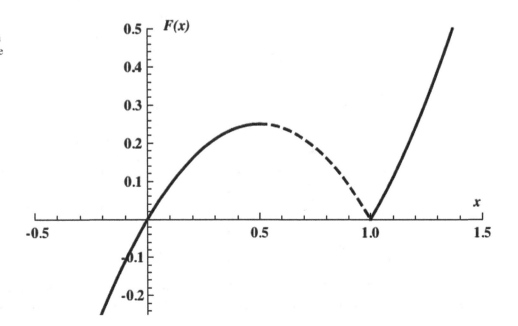

The equilibrium is stable when any infinitesimal increase in the salinity gradient ΔS leads to an increase in the term $|m|\Delta S$, so that the derivative of ΔS relative to time becomes slightly negative. The direction of variation of $|m|\Delta S$ as a function of ΔS is in fact given directly by the figure representing $|1 - x|x$ as a function of x. The descending intermediate branch (dashed line in the figure) indicates a decrease in $|m|\Delta S$ when ΔS increases, leading to an unstable equilibrium. The 'thermal' and 'saline' branches, on the other hand, are perfectly stable. For the same value of the temperatures and the salt flux, there are therefore two stable equilibria possible in this system.

Interestingly, these multiple equilibria are found in much more complex models of the ocean, and even in some ocean-atmosphere coupled models (Rahmstorf 1996; Rahmstorf et al. 2005). It therefore seems that this very simple model captures an important aspect of thermohaline circulation, which explains the existence of sudden variations in the deep ocean circulation. These variations are most likely involved in the sudden climate changes observed during the ice ages (Heinrich events and Dansgaard-Oeschger events, see Chap. 29).

The Welander Model

The multiplicity of equilibria does not explain everything, and of course there are other types of possible behaviors. Another oceanographic example (Fig. 25.10), similar to the Stommel model, concerns convective-advective oscillations (Welander 1982). Although their relevance to climate variations is not established, some have suggested that these oscillations may play a role in the recurrence of Dansgaard-Oeschger events.

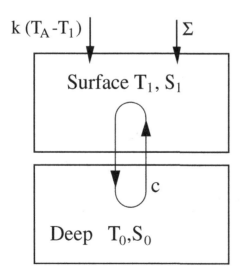

Fig. 25.10 Configuration of the Welander model (1982)

This model is composed of two superimposed boxes, one representing a mass of cold, low-salinity surface waters (temperature T_1 and salinity S_1), and the other a slightly warmer, saltier water mass for the depths (temperature T_0 and salinity S_0). The deep-water box is much larger than the surface water box, and we assume that T_0 and S_0 are constant. The difference in density $\Delta\rho = \rho_1 - \rho_0$ between these two boxes is obtained as a function of the positive coefficients of thermal expansion α and saline contraction β, assumed to be constant:

$$\Delta\rho = -\alpha(T_1 - T_0) + \beta(S_1 - S_0) \quad (25.14)$$

Vertical mixing (convection c) is very small if the column is well stratified (if $\Delta\rho < -\varepsilon < 0$). In the opposite case, it will be large:

$$c = c_0, \quad c_0 \text{ small if } \Delta\rho < -\varepsilon;$$
$$c = c_1, \quad c_1 \text{ large if } \Delta\rho > -\varepsilon.$$

The variables of the problem this time are the temperature and the salinity of the surface water box, and the equations for the corresponding evolution are formulated as follows:

$$\frac{dT_1}{dt} = k(T_A - T_1) + c(T_0 - T_1); \quad \frac{dS_1}{dt} = \Sigma + c(S_0 - S_1).$$
$$(25.15)$$

At equilibrium, the time derivatives in both equations are equal to zero, which yields:

$$T_1^e = \frac{kT_A + c^e T_0}{k + c^e}; \quad S_1^e = S_0 + \frac{\Sigma}{c^e} \quad (25.16)$$

where c^e is the value of the vertical mixing for a temperature T_1^e and a salinity S_1^e, that is, for a density difference of $\Delta\rho^e$ ($c^e = c_0$ if $\Delta\rho^e < -\varepsilon$; $c^e = c_1$ otherwise). This leads to a solution:

$$\Delta\rho^e = -\alpha \frac{k(T_A - T_0)}{k + c^e(\Delta\rho^e)} + \beta \frac{\Sigma}{c^e(\Delta\rho^e)}; \text{ i.e. } \Delta\rho^e = F(\Delta\rho^e)$$
$$(25.17)$$

where F is a constant depending on the sign of $\Delta\rho$ ($F = F_0$ if $\Delta\rho^e < -\varepsilon$; $F = F_1$ otherwise). So, there are zero, one or two solutions, depending on the parameter values, as shown in Fig. 25.11.

In general, if c is a continuous function of $\Delta\rho$ then $F(\Delta\rho)$ will also be a continuous function and there will be an odd number of solutions, alternately stable and unstable, as in the Stommel model. The case of 'zero solution' in Fig. 25.11 (discontinuous case) would in fact correspond to an 'unstable equilibrium' for a continuous model. There is then a boundary cycle (an oscillation): in the absence of intense convection ($c = c_0$), there is a tendency towards an equilibrium in the other domain (with a strong convection,

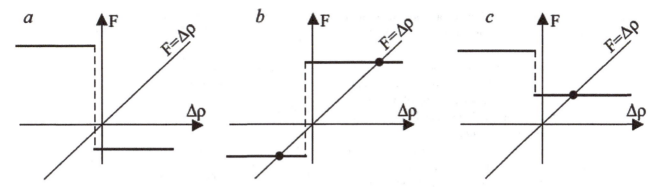

Fig. 25.11 Examples of equilibria given by $F = \Delta\rho$, where F is discontinuous and can only have two constant values. Depending on these values, there are three possible cases: **a** there is no solution (in fact, the system oscillates); **b** there are two equilibria which coexist, one convective and the other diffusive; **c** there is only one equilibrium (convective or diffusive)

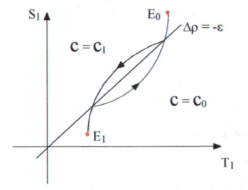

Fig. 25.12 Relaxed oscillation in the Welander model. On a *TS* diagram, a convective zone (when the density ρ_1 is large) and a diffusive zone, separated by an iso-density curve $\Delta\rho = -\varepsilon$, are defined. If the values T, S are located in the convective zone ($c = c_1$), the model tends to bring them back to the point of attraction E_1 (located in the diffusive zone). If they are in the diffusive domain ($c = c_0$), the model tends to bring them back to the point of attraction E_0 (located in the convective zone). The model will thus oscillate from one mode of operation to another, without ever reaching equilibrium

$c = c_1$), as indicated in Fig. 25.12. There is then an oscillation between a 'diffusive' state (c small $= c_0$) and a 'convective' state (c large $= c_1$). These type of oscillations have also been observed, under certain conditions, in three-dimensional ocean models.

Conclusions and Outlook

In this chapter, we explained the basic principles of climate modeling and illustrated their application through examples using models of different levels of complexity to show the strengths of each model type. This hierarchy of climate models is important not only for practical reasons related to the computing time required to study a particular scale of space or time, but above all, because it represents a range of grouped assumptions in each model type, which are necessary to understand the issue at hand. Using models based on different assumptions, it is possible to compare their importance and to better understand the role of a specific process. In principle then, one model is not more reliable than another. A model is always based on a set of assumptions and its relevance must be judged in relation to the question that the modeler has chosen to study. A model generally provides a result, without any uncertainty explicitly associated with this result. The uncertainty is concealed within the assumptions on which the model is based and in the imposed boundary conditions, which, although as realistic as possible, are never as well defined as we would ideally like. It is therefore interesting to compare the results of several models with the same boundary conditions, just as it is interesting to study the sensitivity of the results of a model to certain conditions with poorly constrained boundaries. This may help to determine if a better understanding of these conditions is required or if this is of little importance for the climate being studied.

Here, we mostly described the use of global models. We showed that the resolution of these models, even of the most complex ones, does not always allow comparison with reconstructions which are often representative of regions much smaller than the 'boxes' of a model. We saw that the modeling of paleoclimate indicators such as isotopes permits a more precise comparison and also an analysis of the recorded signal as a function of climate parameters. There are also models with a finer resolution that can be used on a regional scale. The use of this type of model for paleoclimates, still quite limited at the present time, is bound to develop in the future, in parallel with their increasing use for climate forecasts. An example of downscaling methods is given for the Last Glacial Maximum by Jost et al. (2005), which highlights the significant differences between the simulated climates at the European scale using these methods. There is still a lot of work to be done to generate these models. It should be noted that there are also regional models which include a representation of climate indicators such as water isotopes (e.g., REMO-ISO, Sturm et al. 2007).

The evolution of climate models in the near future will involve not only an increasingly fine resolution as computing power increases and an explicit resolution of phenomena previously parametrized or ignored, but also the construction of increasingly comprehensive models of the climate system. Not only will the physical and dynamic aspects of the climate system be addressed, but also the biogeochemical cycles, in particular the carbon cycle. This trend is already emerging and is important not only to understand the evolution of past climates but also to forecast the climates of the future (see Chap. 31).

For Further Information

McGuffie, K., & Henderson-Sellers, A. (2005). *A climate modelling primer* (3rd ed., 280 p). Wiley.

References

Berger, A., (1978). Long-term variations of daily insolation and quaternary climatic changes. *Journal of the Atmospheric Sciences, 35*, 2362–2367.

Berger, A., et al. (1998). Sensitivity of the LLN Climate model to the astronomical and CO_2 forcings over the last 200 Ky. *Climate Dynamics, 14*, 615–629.

Braconnot, P., et al. (2007a). Results of PMIP2 coupled simulations of the Mid-Holocene and Last Glacial Maximum—Part 1: Experiments and large-scale features. *Climate of the Past, 3*, 261–277.

Braconnot, P., et al. (2007b). Results of PMIP2 coupled simulations of the Mid-Holocene and Last Glacial Maximum—Part 2: Feedbacks with emphasis on the location of the ITCZ and mid- and high latitudes heat budget. *Climate of the Past, 3*, 279–296.

Budyko, M. (1969). The effect of solar radiation variations on the climate of the earth. *Tellus, 21*, 611–619.

Calov, R., et al. (2002). Large-scale instabilities of the Laurentide ice sheet simulated in a fully coupled climate-system model. *Geophysical Research Letters, 29*, 2216.

Claussen, M., et al. (2002). Earth system models of Intermediate complexity: Closing the gap in the spectrum of climate system models. *Climate Dynamics, 18*, 579–586.

CLIMAP. (1976). The surface of the Ice-Age earth. *Science, 191*, 1138–1141.

CLIMAP. (1981). *Seasonal reconstructions of the earth's Surface at the Last Glacial Maximum*. Geological Society of America, Map Chart Series MC-36, Boulder, Colorado.

Gates, W. L. (1976). Modeling the Ice-Age climate. *Science, 191*, 1138–1144.

Gallée, H., van Ypersele, J.-P., Fichefet, T., Marsiat, I., Tricot, C., Berger, A. (1992). Simulation of the Last Glacial cycle by a coupled, sectorially averaged climate-ice sheet model. 2. Response to insolation and CO2 variations. *Journal of Geophysical Research, 971*, 15713–15740. https://doi.org/10.1029/92JD01256.

Hargreaves, J. C., et al. (2007). Linking Glacial and future Climates through an ensemble of GCM simulations. *Climate of the Past, 3*, 77–87.

IPCC. (2007). 4e rapport. http://www.ipcc.ch.

Jouzel, J., et al. (1993). Vostok ice cores: Extending the climatic records over the penultimate Glacial Period. *Nature, 364*, 407–412.

Joussaume, S. & Taylor, K. E. (1995). Status of the paleoclimate modeling intercomparison project (PMIP). In *Proceedings of the First International AMIP Scientific Conference*. WCRP Report, (pp. 425–430).

Jost, A., Lunt, D., Kageyama, M., Abe-Ouchi, A., Peyron, O., Valdes, P. J., Ramstein, G. (2005). High resolution simulations of the Last Glacial Maximum climate over Europe: A solution to discrepancies with continental paleoclimatic reconstructions? *Climate Dynamics, 24*, 577–590. https://doi.org/10.1007/s00382-005-0009-4.

Kageyama, M., Braconnot, P., Harrison, S. P., Haywood, A. M., Jungclaus, J. H., Otto-Bliesner, B. L., et al. (2018). The PMIP4 contribution to CMIP6—Part 1: Overview and over-arching analysis plan. *Geoscientific Model Development, 11*, 1033–1057. https://doi.org/10.5194/gmd-11-1033-2018.

Laskar, J., Robutel, P., Joutel, F., Gastineau, M., Correia, A., Levrard, B. A. (2004). A long-term numerical solution for the insolation quantities of the earth. *428*. https://doi.org/10.1051/0004-6361:20041335.

Lorenz, E. (1963). Deterministic nonperiodic flow. *Journal of the Atmospheric Sciences, 20*, 130–141.

Loutre, M.-F., & Berger, A. (2000). No Glacial-Interglacial cycle in the ice volume simulated under a constant astronomical forcing and a variable CO_2. *Geophysical Research Letters, 27*, 783–786.

Manabe, S., & Broccoli, A. J. (1985). A comparison of climate model Sensitivity with data from the Last Glacial Maximum. *Journal of the Atmospheric Sciences, 42*, 2643–2651.

Peyron, O., et al. (1998). Climatic reconstruction in Europe for 18,000 YR B.P. from Pollen Data. *Quaternary Research, 49*, 183–196.

Rahmstorf, S. (1996). On the freshwater forcing and transport of the Atlantic Thermohaline circulation. *Climate Dynamics, 12*, 799–811.

Rahmstorf, S., et al. (2005). Thermohaline circulation hysteresis: A model intercomparison. *Geophysical Research Letters, 32*, L23605.

Ramstein, G., et al. (2007). How cold was Europe at the Last Glacial Maximum? A synthesis of the progress achieved since the first PMIP model-data comparison. *Climate of the Past, 3*, 331–339.

Roche, D., & Paillard, D. (2005). Modelling the oxygen-18 and rapid Glacial climatic events: A data–model comparison. *Comptes Rendus Geoscience, 337*, 928–934.

Schneider von Deimling, T., et al. (2006). Climate sensitivity estimated from ensemble simulations of Glacial Climate. *Climate Dynamics, 27*, 149–163.

Sellers, W. (1969). A global climatic model based on the energy balance of the earth-atmosphere system. *Journal of Applied Meteorology, 8*, 392–400.

Stein, U., & Alpert, P. (1993). Factor separation in numerical simulations. *Journal of the Atmospheric Sciences, 50*, 2107–2115.

Sturm, C. et al. (2007). Simulation of the stable water isotopes in precipitation over South America: Comparing regional to global circulation models. *Journal of Climate, 20*, 3730–3750.

Stommel, H. (1961). Thermohaline convection with two stable regimes of flow. *Tellus, 13*, 224–230.

Welander, P. (1982). A simple heat salt oscillator. *Dynamics of Atmosphere and Oceans, 6*, 233–242.

Wu, H., et al. (2007). Climatic changes in Eurasia and Africa at the Last Glacial Maximum and Mid-Holocene: Reconstruction from Pollen Data using inverse Vegetation modelling. *Climate Dynamics, 29*, 211–229.

The Precambrian Climate

Yves Goddéris, Gilles Ramstein, and Guillaume Le Hir

More than 88% of the history of the Earth occurred in the Precambrian. The Precambrian began with the formation of the Earth 4.6 billion years ago (Ga) and ended 542 million years ago (International Stratigraphic Chart, www.stratigraphy.org). It is subdivided into two large eons: the Archean (between 4 and 2.5 Ga) and the Proterozoic (from 2.5 to 0.542 Ga). The *International Commission on Stratigraphy* is proposing to add an extra eon, the Hadean, covering the first 600 million years of the history of our planet. Notwithstanding, this eon is described as having an *informal* status since no pre-Archean rock has been observed today. In fact, the oldest rocks date back to 4 billion years ago (U/Pb dating on zircon crystals). These are the Acasta gneisses in the Slave Province of Canada. The two formal eons of the Precambrian are subdivided into eras. In particular, the Proterozoic contains three eras: Paleoproterozoic (2.5–1.6 Ga), Mesoproterozoic (1.6–1.0 Ga) and Neoproterozoic (1.0–0.542 Ga).

Today, we find outcrops from the Archean on all the continents. Among the largest, two fragments of continents larger than 0.5×10^6 km^2 were identified: the craters of Kaapvaal (South Africa) and Pilbara (Australia). They are dated at about 3.6–2.9 Ga. Finally, the Precambrian has witnessed several major events in the history of the Earth. These include the onset of plate tectonics, the emergence of the biosphere at least 3.5 billion years ago, the rapid expansion of land surfaces between 3.2 and 2.6 billion years ago and the growth of the partial pressure of oxygen in the atmosphere around 2.3 Ga.

Climate Indicators

Little is known about the evolution of the Earth's climate during the Precambrian. The number of indicators available is very limited. Firstly, sedimentological data are difficult to interpret, given the age of these sediments which have generally been disarranged. Secondly, paleontological data are virtually unusable in terms of climate reconstruction. They are fragmentary and represent only a very simple monocellular biosphere, which is difficult to relate to any environmental evolution. Finally, the isotopic data measured on sediments generally have poorly preserved the original climate signal, having been very often subjected to post-deposit perturbations (diagenesis in particular). From a quantitative perspective, two isotopic signals have been used with varying degrees of success: $\delta^{18}O$ measured on siliceous (cherts) sediments since the 1970s (Knauth and Epstein 1976) and, more recently, the $\delta^{30}Si$ ratio measured on the same cherts (Robert and Chaussidon 2006).

Opal (SiO_2) was precipitated massively during the Archean. The reasons for this level of precipitation are unknown. It may have been caused by biological activity or directly abiotically from the silica-saturated ocean. Finally, this opal may have been produced during the stabilization of clay minerals on the seabed or may be derived from the weathering of volcanic glass. It was subsequently subjected to diagenesis and today appears as siliceous sedimentary rocks called cherts. The isotopic oxygen ($\delta^{18}O$) compositions of these cherts show them to be increasingly depleted in heavy isotopes as we go back in time, reaching a value of 16‰ compared to the international standard SMOW, 3 Ga ago (Fig. 26.1).

Y. Goddéris (✉)
Géoscience Environnement Toulouse, CNRS, Université de Toulouse III, UMR 5563, Toulouse, France
e-mail: yves.godderis@gmail.com

G. Ramstein
Laboratoire des Sciences du Climat et de l'Environnement, LSCE/IPSL, CEA-CNRS-UVSQ, Université Paris-Saclay, 91190 Gif-Sur-Yvette, France

G. Le Hir
Institut de Physique du Globe, CNRS, Université Pierre et Marie Curie, UMR 7154, Paris, France

© Springer Nature Switzerland AG 2021
G. Ramstein et al. (eds.), *Paleoclimatology*, Frontiers in Earth Sciences, https://doi.org/10.1007/978-3-030-24982-3_26

Table 26.1 U-Pb and Re-Os geochronological constraints on Cryogenian glacial onsets and terminations

Paleocontinent	Age (Ma)	Method[a]	Reference
Marinoan deglaciation/cap carbonate: 636.0–634.7 Ma			
Laurentia	>632.3 ± 5.9	Re-Os	(63)
South China	635.2 ± 0.5	U-Pb ID-TIMS	(57)
Southern Australia	636.41 ± 0.45	U-Pb CA-ID-TIMS	(59)
Swakop	635.21 ± 0.59/0.61/0.92	U-Pb CA-ID-TIMS	(56, 83)
Marinoan glacial onset: 649.9–639.0 Ma			
Congo	>639.29 ± 0.26/0.31/0.75	U-Pb CA-ID-TIMS	(83)
Southern Australia	<645.1 ± 4.3	Re-Os	(137)
South China	<654.2 ± 2.7	U-Pb SIMS	(134)
South China	<654.5 ± 3.8	U-Pb SIMS	(58)
Sturtian deglaciation/cap carbonate: 659.3–658.5 Ma			
Southern Australia	>657.2 ± 2.4	Re-Os	(137)
Tuva-Mongol ia	659.0 ± 4.5	Re-Os	(63)
Southern Australia	<659.7 ± 5.3	U-Pb SIMS	(366)
Laurentia	662.4 ± 3.9	Re-Os	(60)
South China	>662.7 ± 6.2	U-Pb SIMS	(65)
Sturtian glacial onset: 717.5–716.3 Ma			
Oman	>713.7 ± 0.5	U-Pb ID-TIMS	(365)
South China	<714.6 ± 5.2	U-Pb SIMS	(64)
South China	<715.9 ± 2.3	U-Pb SIMS	(62)
South China	<716.1 ± 3.4	U-Pb SIMS	(62)
Laurentia	>716.5 ± 0.2	U-Pb CA-ID-TIMS	(32)
Laurentia	<717.4 ± 0.1	U-Pb CA-ID-TIMs	(32)
Laurentia	<719.47 ± 0.29	U-Pb CA-ID-TIMS	(61)

Re-Os isochron ages from sedimentary organic matter. Errors are quoted at the 2σ level of uncertainty. Where multiple uncertainties are given, they represent analytical/analytical + tracer solution/analytical + tracer solution + decay-constant uncertainties

Ca chemical abrasion; *ID6TIMS* isotope-dilution and termal-ionization mass spectrometry; *SIMS* secondary-ion mass spectrometry

Table from Hoffman et al., Sci. Adv. (2017)

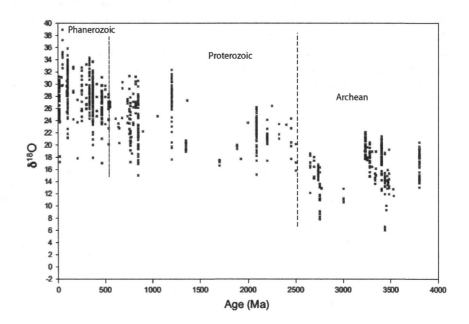

Fig. 26.1 The $\delta^{18}O$ of the Precambrian ocean measured in the cherts

As for carbonates, the $\delta^{18}O$ of silica is a function of the temperature and of the $\delta^{18}O$ of the water in which the silica precipitated:

$$1000 \ln \alpha = \left(3.09 \times \frac{10^6}{T(K)^2}\right) \quad (26.1)$$

where the fractionation factor

$$\alpha = \frac{1000 + \delta^{18}O_{chert}(SMOW)}{1000 + \delta^{18}O_{water}(SMOW)} \quad (26.2)$$

ln denotes its natural logarithm, and $\delta^{18}O_{water}$ is the $\delta^{18}O$ of the water from which the cherts precipitated.

Applied to the cherts, assuming that the isotopic composition of seawater was quite similar to that of today, this paleothermometer predicts the temperatures T of the water from which the cherts precipitated to be close to 85 °C 3 Ga ago, and 50 °C at the Precambrian-Cambrian boundary.

These temperatures are either the sign of very hot oceans or the consequence of an alteration of the cherts after deposition by meteoric or hydrothermal fluids, in which case they provide no information on climate. The question remains open because the $\delta^{18}O$ isotopic signal is particularly sensitive to diagenesis. A third hypothesis has been formulated more recently: it assumes that $\delta^{18}O$ of seawater was much lower than its present value, due to tectonic processes. Under these conditions, the temperature of the sea water could have been similar to the current one. Unfortunately, there is no constraint on the $\delta^{18}O$ of seawater during the Archean. Authors generally assume that this ratio remained constant over time at −1‰ in comparison with the SMOW, calculated from the level of the current ocean which would have received the melt water from all the ice caps on land. This value is the result of the equilibrium that is supposed to exist between the ^{18}O depletion of sea water resulting from the interactions between water and lithosphere at low temperatures and its ^{18}O enrichment during water/lithosphere interactions at high temperature in the hydrothermal systems. Recent work (Kasting et al. 2006) suggests that the $\delta^{18}O$ ratio may have been significantly lower (−9‰ compared to SMOW) during the Archean and Proterozoic periods. This argument is based on the fact that the Archean oceans were probably shallower than the oceans are currently. This results in a shallower water column over the oceanic ridges before 800 Ma, implying a reduction in hydrostatic pressure in the hydrothermal systems. This decrease in pressure limited the penetration of seawater into the ridges in the depths, thus reducing the gain in ^{18}O by sea water through alteration at high-temperature of the oceanic crust ($T > 350$ °C). This reduction in flux at high temperature results in an imbalance in the ^{18}O cycle in the ocean-atmosphere system and its stabilization at lower values than is currently the case. If this scenario proves to be correct, the temperature of the water as inferred from the $\delta^{18}O$ data on cherts, could be significantly lower. Kasting and Howard (2006) argue in favor of 'moderate' climates at the end of the Archean and during the Proterozoic.

An additional element was added to this debate by Robert and Chaussidon (2006), who measured the isotopic composition of silicon ($\delta^{30}Si$) in Precambrian cherts which is distinctly less sensitive to diagenesis than $\delta^{18}O$. These data, when translated into temperatures, (requiring the use of a silicon cycle model and therefore additional assumptions), suggest temperatures of around 70 °C, 3 Ga ago, and 20 °C, 800 million years ago, thus confirming very high temperatures in the distant past.

There remains a potentially major problem: the formation conditions of Precambrian cherts are unknown. Nevertheless, these results showing a gradual cooling of sea water from very high values seem to have been confirmed recently by a totally independent method, based on the resurrection of proteins of unicellular Archean organisms using phylogenetic and statistical methods of analysis (Gaucher et al. 2008).

Finally, irrespective of any debate on the terrestrial temperature during the Archean/Proterozoic, the $\delta^{18}O$ of the cherts show a significant increase at the end of the Archean, between 2.7 and 2.5 Ga (of about 10‰), suggesting a rapid cooling of the oceans of about 20 °C (Fig. 26.2).

The Theory of the Paleothermostat

In 1981, Walker, Hays and Kasting published a ground-breaking article explaining why the climate remained relatively stable (within a temperature range allowing water to remain in the liquid state) for between one million to one billion years (Walker et al. 1981). This study was carried out in order to solve the faint young sun paradox. The models of stellar evolution predict the evolution of the solar constant during the history of the Earth and make it possible to calculate that during the Archean, it would have been 20–30% weaker than today. Under these conditions, the Earth should have totally frozen over although this contradicts the isotopic data which provide an estimate of the temperature of the fluid envelopes of the Earth.

The residence time of the exosphere carbon content (i.e. all the carbon contained in the ocean, the biosphere and the atmosphere) is around 200,000 years which is very short compared to the geological processes of sedimentary carbon burial and continental weathering (François and Goddéris 1998). This measure gives an indication of the average time spent by a carbon atom entering the ocean-atmosphere system via volcanic degassing, for example, before exiting via sedimentary deposits. This response time is very short in the

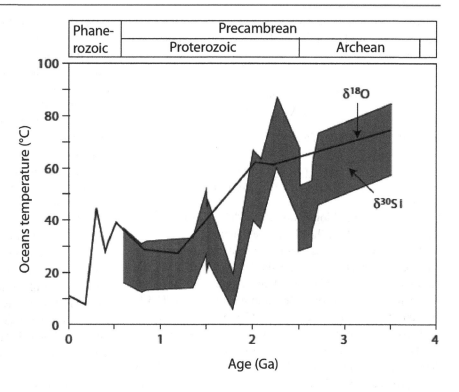

Fig. 26.2 Ocean temperatures reconstructed from δ^{30}Si and from δ^{18}O of Precambrian cherts

context of geological time. If the history of the Earth was condensed into one day, the entire carbon content of the ocean and atmosphere would be recycled about every 3 s. This short response time is obviously linked to the small size of the exosphere carbon reservoir (3×10^{18} mol) and to the fluxes in exchanges of matter (of around 10^{12}–10^{13} mol of carbon per year) between this reservoir and the geological reservoirs. The implications of this short residence time are important for the climate. Any imbalance between the carbon inflows (e.g. solid soil degassing) and outflows (e.g. CO_2 consumption due to silicate weathering) that would have occurred over several million years (a short timescale compared to the history of the Earth) would cause gigantic fluctuations (several orders of magnitude) of the carbon dioxide content in the atmosphere, with dramatic consequences for the climate. Fluctuations of this scale have not occurred in the history of the Earth, apart from some isolated episodes of global glaciations at the end of the Proterozoic. The carbon input and output flows must therefore be close to equilibrium over the geological time scale, from 10^6 to 10^9 years (see François and Goddéris (1998) for a complete mathematical assessment).

The carbon cycle at the million-year scale is described in Fig. 26.3. Only geological flows are taken into account, all rapid recycling (biosphere fluxes and ocean-atmosphere interface) are ignored.

The inflows to the ocean-atmosphere system are soil degassing from volcanoes, F_{vol} and from the oceanic ridges F_{MOR}; the dissolution of the continental carbonates which transfers carbon from the continental crust to the ocean in the dissolved form HCO_3^- F_{cw}; and the oxidation of old sedimentary organic compounds exposed to the atmosphere F_{ow} (black shales, for example, sediments rich in organic matter deposited during episodes of large-scale ocean anoxia and which are exposed to the atmosphere by tectonic activity):

$$CaCO_3 + CO_2 + H_2O \rightarrow Ca^{2+} + 2HCO_3^- \quad (26.3)$$

dissolution of continental carbonates by carbonic acid:

$$CH_2O + O_2 \rightarrow 2HCO_3^- + H^+ \quad (26.4)$$

oxidation of sedimentary organic carbon.

The sinks are represented by the precipitation of carbonate minerals F_{cd} on the ocean floor or on the continental shelves, through the mediation or not of biocalcification, and the burial of organic carbon within sediments F_{od}, both on land and in the ocean.

$$Ca^{2+} + 2HCO_3^- \rightarrow CaCO_3 + CO_2 + H_2O \quad (26.5)$$

precipitation of carbonates

$$CO_2 + H_2O \rightarrow CH_2O + O_2 \quad (26.6)$$

autotrophic productivity and burial of organic carbon.

The long-term equilibrium of the carbon cycle, mathematically required because of its low residence time in the

Fig. 26.3 The exosphere carbon cycle at the scale of a million years

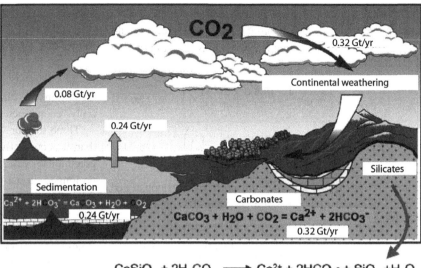

exosphere system, results in the following proximate equation, which needs to be maintained on time scales of a million years to ensure relative stability of the climate:

$$F_{vol} + F_{MOR} + F_{cw} + F_{ow} \approx F_{cd} + F_{od} \quad (26.7)$$

Small deviations from this near equality are likely to explain the climate fluctuations, which may be as significant as the establishment of the Permo-Carboniferous glaciation or the climate optimum in the mid-Cretaceous. In general, under the effect of geological forcings, these imbalances are two orders of magnitude smaller than the fluxes themselves (François and Goddéris 1998). However, this quasi-equilibrium requires a physical basis, which the theory of the paleothermostat described below provides.

The flux of silicate weathering F_{sw} does not appear in the carbon balance because the silicates are not present in the continental crust in significant amounts. Yet, this flow actually consumes carbon from the exosphere:

$$CaSiO_3 + 2CO_2 + 2H_2O \rightarrow Ca^{2+} + 2HCO_3^- + SiO_2 \quad (26.8)$$

This reaction is a generic reaction, the silicate mineral (here the Wollastonite) is scarce on Earth, but it accounts for the alteration budget and therefore illustrates the point. The Ca^{2+} and HCO_3^- ions are carried to the ocean by the rivers. If the ocean is saturated with carbonate minerals, this increased alkalinity will cause the precipitation of calcium carbonates (Eq. 26.5), and thus the storage of exosphere carbon in the sedimentary envelope of the Earth. The contribution of alkalinity to the ocean is thus represented by silicate weathering and carbonates, whereas the loss of alkalinity is related to precipitation of the carbonates. The response time of the alkalinity cycle to any geological perturbation is around 3000 years, and is directly related to the ocean's mixing time which constrains the response time of the alkalinity of the world's oceans to any disturbance (François and Goddéris 1998). So, we again have the quasi-equality:

$$F_{sw} + F_{cw} \approx F_{cd} \quad (26.9)$$

The combination of the Eqs. (26.7) and (26.9) give the following proximate equation:

$$F_{vol} + F_{MOR} + F_{ow} \approx F_{sw} + F_{od} \quad (26.10)$$

If we ignore the existence of imbalances in the organic sub-cycle of carbon ($F_{ow} = F_{od}$, which is a strong assumption), we obtain:

$$F_{vol} + F_{MOR} \approx F_{sw} \quad (26.11)$$

The paleothermostat theory explains the physical reasons for this equality on the basis of two assumptions: firstly, that the weathering of continental silicate rocks depends on the climate and, in particular, on the temperature and the continental runoff. Secondly, excluding water vapor, atmospheric CO_2 is assumed to be the primary greenhouse gas whose variability controls the climate.

Concerning silicate weathering, the intuition of Walker et al. (1981), based on measurements of the dissolution rate of silicate minerals in the laboratory, has been largely verified in the natural environment. The consumption flux of atmospheric CO_2 by silicate weathering was measured on a large number of monolithological silicate drainage basins, both granite (F_{gra}) and basaltic (F_{bas}). It increases with temperature T and with continental runoff R according to the following laws:

$$(F_{gra}) = k_{gra} R \exp\left[-\frac{48200}{R}\left(\frac{1}{T} - \frac{1}{T_0}\right)\right] \quad (26.12)$$

$$(F_{bas}) = k_{bas} R \exp\left[-\frac{42300}{R}\left(\frac{1}{T} - \frac{1}{T_0}\right)\right] \quad (26.13)$$

where the fluxes are expressed in moles of CO_2 consumed per m^2 of land surface area and per year. The approximate Eq. (26.11) can thus be expressed as:

$$F_{vol} + F_{MAR} = A_{gra} k_{gra} R \exp\left[-\frac{48200}{R}\left(\frac{1}{T} - \frac{1}{T_0}\right)\right] \\ + A_{bas} k_{bas} R \exp\left[-\frac{42300}{R}\left(\frac{1}{T} - \frac{1}{T_0}\right)\right] \quad (26.14)$$

where A_{gra} and A_{bas} are the land areas where granite and basalt areas touch. If, for some reason related to the internal geology, soil degassing increases, increases in temperature and runoff are required to maintain the carbon cycle balance. This increase occurs naturally due to the increase in the concentration of CO_2 in the atmosphere, in response to the degassing. Due to its dependence on climate, and because CO_2 is a greenhouse gas, the land silicate weathering on land will therefore track and compensate for the fluctuations in volcanic degassing of CO_2 over time and thus prevent massive fluctuations in the Earth's climate.

The paleothermostat theory is particularly useful to explain the absence of long-term global glaciations during the Archean, when the solar constant was 20–30% lower than it is currently. The colder climate resulting from the lower solar energy consumption led to a slowdown in the consumption of CO_2 by continental silicate weathering and therefore an increase in the CO_2 pressure in the air. The induced warming continued until the silicate weathering balanced out the soil degassing again. However, CO_2 is probably not the only greenhouse gas involved, which complicates the paleothermostat theory.

Methane, for example, is a less abundant but more efficient gas than CO_2 in terms of its greenhouse effect. Methane pressure results from the balance between its production time (methanogenesis) and its destruction time (oxidation by OH radials). In fact, any excess of methane over this balance is not very stable and connects with the CO_2 paleothermostat because a sudden injection of methane oxidizes quite quickly in CO_2.

Major Climate Events in the Precambrian

From 4.5 to 2.4 Ga

In the absence of indicators with the necessary level of resolution, the climate history of the Precambrian is essentially reconstructed through numerical modeling.

Three to four Ga ago, the solar constant was 30% lower than it is currently. In this context, extremely high levels of greenhouse gases would have been required to prevent prolonged global glaciation, which has not been observed in the bedrock from that time or in preserved sediments. NH_3 was suggested initially as a cause but this is not a likely candidate because it is rapidly destroyed by photolysis. This is not true for CO_2. At least 0.3 bar of CO_2 is required in the atmosphere to counteract the low solar constant (Pavlov et al. 2000). The reasons for the presence of very high levels of CO_2 are related to the paleothermostat theory involving the erosion of continental silicates (Walker et al. 1981). The weaker solar constant forces the climate towards colder conditions. This limits the consumption of CO_2 by silicate erosion and allows the CO_2 concentration in the air to grow until the climate becomes sufficiently hot and humid to stabilize the weathering of the silicates so that it reaches the point where it compensates for the soil degassing, leading to a balance in the exosphere cycle of carbon. Moreover, since this degassing is suspected to have been much more intense in this distant past, following the dissipation of the internal heat of the Earth, the CO_2 concentration in the air will be set at a high value, essential to ensure a high level of chemical erosion of the silicates.

However, the absence of siderite ($FeCO_3$) in Archean paleosols suggests an upper ceiling of 0.015 bar of CO_2 in the atmosphere (see references in Catling and Claire 2005). Another greenhouse gas was therefore necessary and methane may have played this role. Methanogenic bacteria are probably among the first organisms to have evolved on the surface of the Earth, and they must have constituted an important part of the primitive biosphere as early as 3.5 Ga. Before 3.5 Ga, probably more methane degassed from the ocean ridges into the prebiotic atmosphere than nowadays, given the lower mantle at this time (Kump et al. 2001). The virtual absence of oxygen in the atmosphere before 2.2 – 2.3 Ga allowed the methane to accumulate in it, probably with mixing ratios of close to 1.6 ppmv, a value one thousand times higher than at present. The residence time of methane in air was also probably a thousand times greater than at present, because, in this weakly oxidized atmosphere, the OH radicals produced by photolysis of water reacted with H_2. As a result, the surface temperature of the Earth could have been as high as 85 °C (Kasting and Howard 2006).

Fig. 26.4 Qualitative curve showing the evolution in the concentration of CO_2 and CH_4 in the atmosphere at the end of the Archean

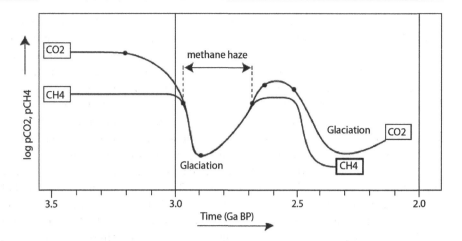

The CH_4/CO_2 ratios probably remained well below 1 up to 3.0 Ga, preventing the formation of a haze of organic compounds in the atmosphere (Lowe and Tice 2004). However, in the 3.0–2.7 Ga range, the level of atmospheric CO_2 could have been considerably reduced, following the growth of the particularly active continental crust at that time. The continental crust had probably reached 60% of its present size, 3–2.9 Ga ago, whereas before 3.2 Ga it was only at 10%. In response to this growth in the surface area of the continents, silicate erosion increased, forcing the level of atmospheric CO_2 to decrease (Goddéris and Veizer 2000; Lowe and Tice 2004). The average global surface temperature dropped by 10 °C (Goddéris and Veizer 2000), at the same time as the CH_4/CO_2 ratio increased, allowing a haze of organic compounds to form around the Earth (Lowe and Tice 2004). This resulted in the cooling being reinforced before the paleothermostat slowly compensated. Traces of glaciations were indeed observed in the supergroups of Pongola and Witwatersrand in South Africa and in the green chists of Berlingue in Zimbabwe, all of these formations dating back to ~2.9 Ga (Fig. 26.4).

The Great Oxidation Event (GOE)

The geochemical and climatological event that marked the beginning of the Proterozoic is the oxygenation of the atmosphere. Biomarkers indicate that the first photosynthetic organisms appeared as early as 2.7 Ga. Atmospheric oxygen probably started to grow around 2.3 Ga. Over 100 or 200 million years, oxygen pressure increased from 10^{-5} bar to 2×10^{-2} bar (Catling and Claire 2005). The $\delta^{13}C$ of sedimentary carbonates (Fig. 26.5) shows a major surge of more than 10‰ at this time (see references in Catling and Claire 2005).

This increase in $\delta^{13}C$ is generally thought to result from the burial of a large amount of organic carbon; this burial caused an imbalance in the carbon cycle: organic matter, low in ^{13}C, was no longer depleted, while photosynthetic organisms continued to pump out carbon dioxide depleted in ^{13}C, causing an increase in the $\delta^{13}C$ of atmospheric CO_2 (reflected in the $\delta^{13}C$ of carbonates) and an increase in the O_2 content of the atmosphere (the oxygen created by photosynthesis not being fully consumed during the decay of

Fig. 26.5 The evolution of $\delta^{13}C$ in Precambrian carbonates

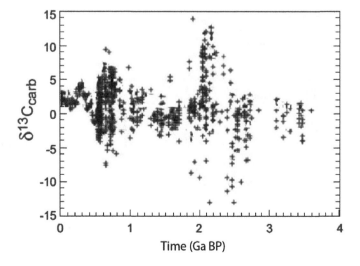

organic matter because of the increased level of conservation). The issue of the oxygen balance during the Paleoproterozoic is still debated to this day, and in particular, the possible role a decrease in the quantity of reduced gas degassed by the mantle is questioned (see references in Catling and Claire 2005). However, there is no doubt about the accumulation of O_2 between 2.3 and 2.1 Ga (Catling and Claire 2005). The GOE is also recorded in the variations in the isotopic composition of iron in sedimentary pyrites between 2.6 and 1.8 Ga, which is suspected to be related to the isotopic composition of seawater (Fig. 26.6). The $\delta^{56}Fe$ of these sediments shows widely dispersed values before 2.3 Ga, but this dispersion decreases considerably after 2.3 Ga. These changes are explained by the existence of an iron rich ocean, fed by hydrothermal springs (containing dissolved iron Fe^{2+}) before 2.3 Ga. This ocean periodically purges itself of a certain amount of its iron during the precipitation of iron oxides during upwelling episodes. Such purges brought water to more or less oxygenated zones. This precipitation extracts the isotope 56 preferentially from the iron; this process of fractional precipitation of iron is thus capable of modifying the $\delta^{56}Fe$ isotopic composition of the deep ocean so long as all the iron carried by the upwellings has not precipitated and a part of it returns to the deep ocean. After 2.3 Ga and the GOE, the Proterozoic ocean became stratified, characterized by a permanently and thoroughly oxygenated surface zone and an anoxic deep ocean. This configuration allows the precipitation of all of the hydrothermal Fe^{2+} brought by the upwellings (see references in Catling and Claire 2005). In this way, the isotopic composition of the deep ocean is no longer affected, since all the iron carried by upwellings precipitated, and stabilized at between 0 and 1‰. Finally, recent data using the independent fractionation of the mass of isotopes of sulfur $\Delta^{33}S$ confirm progressive oxidation of the surface layers from 2.4 to 2.3 Ga onwards (Papineau et al. 2007).

The consequences for the GOE climate were significant. The residence time of methane in the atmosphere decreased strongly and the concentration of CH_4 in the atmosphere probably fell to around 300–100 ppmv, a decrease by a factor of 5–16 compared to Archean values. This caused significant cooling, which could not be immediately compensated for by the low level of CO_2. This triggered the Huronian glaciations during the time it took for the paleothermostat to allow the partial pressure of CO_2 to rise. This major glacial phase may have included at least one episode of total glaciation, but paleomagnetic data on the position of the continental masses remain scanty and difficult to interpret. However, recent paleomagnetic studies show that the Huron glaciation was probably similar to the 'standard' glaciations of the Quaternary. However, the GOE context is completely different in terms of atmospheric composition, paleogeography, solar insolation… and therefore the rhythm of this glaciation is still an open question.

The Proterozoic

Following the Huronian glaciations, the Earth seems to have been subjected to a warm climate persisting over most of the Proterozoic, lasting for about 1 billion years. To date, no trace of glaciation has been found in the timespan 2–0.8 Ga. The measurement of $\delta^{13}C$ from acrytarches (microfossil remains of cysts of photosynthetic eukaryotic organisms) allows a rough estimation of the atmospheric CO_2 level. Indeed, the difference between these ratios and the mean ratio of oceanic carbonates (close to 0‰) defines, with some assumptions, the isotopic fractionation involved in photosynthesis that occurred during the Calvin cycle. This fractionation depends, among other things, on the CO_2 pressure of the water and can therefore be linked to atmospheric CO_2 pressure. At 1.4 Ga, the level of CO_2 was about 10–200

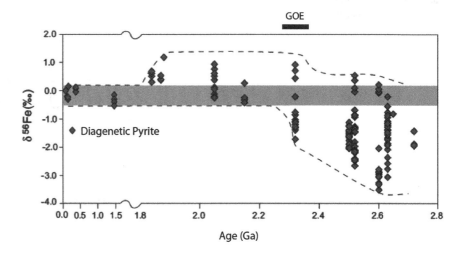

Fig. 26.6 The $\delta^{56}Fe$ of Precambrian sediments

times the pre-industrial level i.e. 2800–56,000 cm^3 of CO_2 per m^3 of air (Kaufman and Xiao 2003).

The level of Proterozoic methane is uncertain. The oxygen level (Fig. 26.7) is estimated to be about 5–18% of its current level, based on the low sulfate concentration of Proterozoic marine carbonates (Pavlov et al. 2003). Since these sulfates are produced by the oxidation of sulphides, their low abundance reflects a low partial pressure of oxygen. Based on these values, Pavlov et al. (2003) calculated a methane pressure in the range of 100–300 ppmv, assuming a CH_4 production in the deep anoxic ocean basins 20-fold higher than current levels. The surface temperature is unknown, the $\delta^{18}O$ of the cherts suggest 50 °C at the end of the Precambrian. As seen above, these high values remain questionable. In order to maintain an average global temperature of 15 °C on the Earth's surface, 300 PAL of CO_2 were needed at 2 Ga, and only 10 PAL at 0.6 Ga (1 PAL = present atmospheric level). Log_{10}.

The End of the Proterozoic: Global Glaciations

After more than a billion years of no glaciations, the end of the Proterozoic (the Neoproterozoic, from 900 to 543 Ma) was marked by the strongest glaciations in the history of the Earth. These are suspected to have been global, hence the name 'snowball' glaciations. Two events are acknowledged: the first between 723 and 667 Ma (the Sturtian glaciation) and the second between 667 and 634 million years (the Marinoan glaciation). They were followed by a glacial episode of lower intensity around 583 Ma, comparable to the glaciations of the Phanerozoic (the Gaskiers glaciation).

The Sturtian and Marinoan glaciations have a particular set of characteristics. (1) The paleolatitude of the glacial deposits was located mostly in the intertropical zone, which implies a major glaciation, since the ice reached the equator (Evans 2000). The histogram showing the presence of glacial deposits as a function of paleolatitude, is totally atypical for snowball glaciations. Although this histogram shows a peak at high latitudes for all of the Phanerozoic glaciations, this peak is displaced to the lower latitudes for snowball glaciations (Evans 2000). (2) Glacial deposits are directly overlaid with atypical carbonate deposits (cap carbonates) with no break, suggesting a transition of extraordinary rapidity on a geological scale from a very cold climate to a very hot climate (Hoffman et al. 1998). If the Earth was covered with ice, the hydrological cycle would have almost completely shut down, thereby allowing CO_2 to accumulate in the atmosphere as a result of volcanic degassing. When more than 0.29 bar of CO_2 has accumulated in the atmosphere (Pierrehumbert 2004), the greenhouse effect intensifies and deglaciation is suddenly initiated. A climate characterized by a strong greenhouse effect followed the very cold climate. Continental weathering recommenced and quickly reached a high level. The massive surge of alkalinity in the ocean became predominant over carbonates, which explains the presence of carbonate deposits directly on top of glacial deposits. (3) Banded iron formations (BIF) reappeared during and just after glaciation, although these had disappeared during the Proterozoic, around 1.8 Ga. The return of the BIFs is qualitatively compatible with the installation of sea ice over the whole of the oceans, greatly reducing the vertical mixing of the ocean and favoring the development of anoxic conditions in the deep ocean. The Fe^{2+} emitted at the ridges can therefore be transported by upwellings to the surface waters, where it precipitates as BIF in contact with oxygen. (4) The presence of an iridium peak in the basal cap carbonates suggests an accumulation of

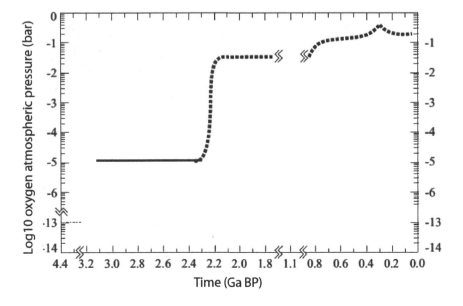

Fig. 26.7 History of the level of oxygen in the atmosphere

iridium (brought by cosmic materials falling to the ground) on pack ice for 3–12 million years and its inclusion in sediments at the time of an ice flood corresponding to the end of Marinoan glaciation (Bodiselitsch et al. 2005). These last decades the chronology of Neoproterozoic glaciations has been considerably improved. The chronology is now well established with two dissymmetric glaciations: a long lasting Sturtian episode for more that 50 Ma and a shorter Marinoan episode lasting around 15 Ma separated by a short interglacial period only lasting 10 Ma (Hoffman et al. 2017). (5) The $\delta^{13}C$ isotopic ratio of the cap carbonates is well documented, especially for the most recent snowball glaciation (Fig. 26.8). It is particularly low at around −3‰ towards the end of the glaciation, reaching very low values (−5‰) at the top of the cap carbonates (Halverson et al. 2005).

As early as 1998, Hoffman et al. interpreted this low value as a sign of shallow burial of organic carbon, in turn, an indicator of significantly slower biological productivity in the oceans. By taking the isotopic fractionation of carbon during photosynthesis at −20‰, the proportion of organic carbon buried relative to total carbon sedimentation increases from 10% at the base of the cap carbonates to 0% at the top (compared to 25% today). These low values suggest an oceanic biosphere very affected by the glacial episode, which is indicative of its magnitude. Nevertheless, there are still many uncertainties concerning the interpretation of the cap carbonates. They are largely related to the fact that cap carbonates are in fact dolomites and that the kinetics of their precipitation is still not fully understood although it appears to be associated with general conditions of anoxia and probably with a bacterial activity reducing the amount of sulphates at the water-sediment interface.

Glaciation Onset

Using models coupling climate and the carbon cycle, Donnadieu et al. (2004) showed that glacial triggering was closely linked to the configuration of the continents during the late Proterozoic. From 800 Ma onwards, the supercontinent Rodinia (Fig. 26.9), located at the equator, began to drift.

Numerous basaltic effusions mark the beginning of rifting. The formation of these highly weathered basaltic surfaces led to an increase in the consumption of atmospheric CO_2 (Goddéris et al. 2003). Moreover, the continental blocks began to drift, but remained between the latitudes 60° S and 60° N. The dislocation of Rodinia increased the supply of moisture to the continents, which favored continental weathering. This weathering occurred even more rapidly on large continental areas located in the hot and humid intertropical convergence zone. CO_2 consumption increased and atmospheric pressure of CO_2 plunged. The dislocation of Rodinia alone (Fig. 26.10) and the consequent increase in weathering (Fig. 26.11) explain a decrease in radiative forcing of 6.85 W/m^2 and an overall cooling of 8 °C. To this effect is added the intensified weathering of the fresh basaltic surfaces brought into the hot and humid climatic zones, conditions favoring their weathering (Goddéris et al. 2003).

This is particularly true of the magmatic Laurentian province (Donnadieu et al. 2004). The combined effects of lowering continental masses and weathered basaltic surfaces caused the system to tip into global glaciation, bringing sea ice as far south as 30°, at which point the positive feedback between albedo, ice cover and cooling accelerated. Atmospheric CO_2 increased from over 1800 ppmv, 800 Ma ago at the time of Rodinia, to a level below the threshold for global glaciation of 250 ppmv.

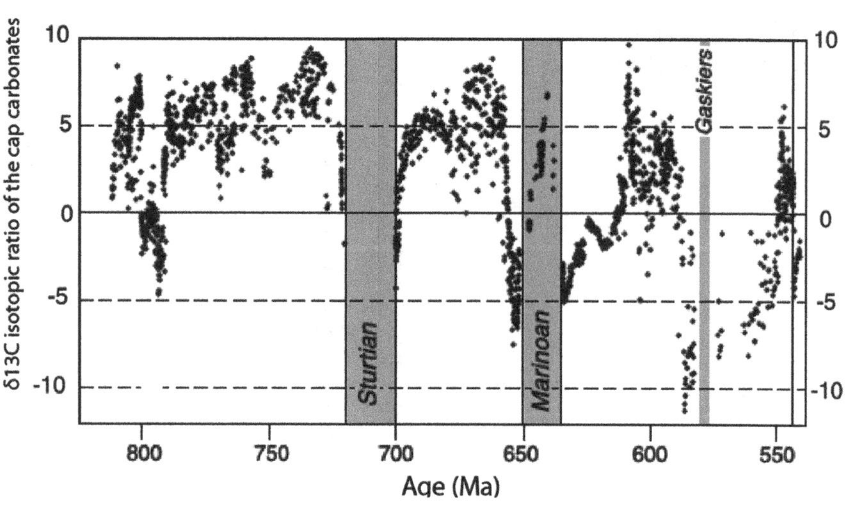

Fig. 26.8 Evolution of the $\delta^{13}C$ of sedimentary carbonates at the end of the Proterozoic. The two gray bands indicate the glaciations thought to be global

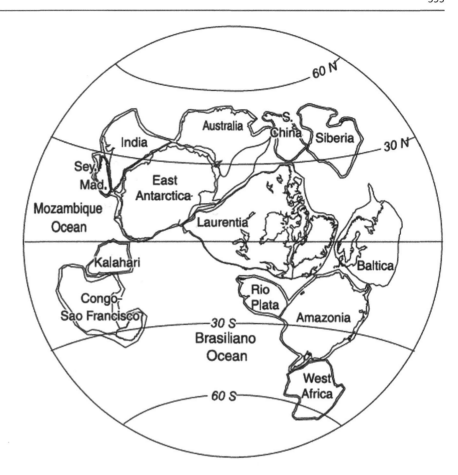

Fig. 26.9 The supercontinent Rodinia, about 800 million years ago

The tectonic theory for glacial inception has two advantages:

1. It explains the persistence of cold climates, even outside the episodes of 'snowball' glaciation for the duration of the Neoproterozoic, as long as the continents are dispersed along the equator. This configuration is observed in a window from 750 to 600 Ma (Torsvik et al. 2001) and thus includes the two extreme episodes.
2. It also explains why such global glaciations did not occur in more recent times. The equatorial configuration of the continents, the driver of global glaciation, never occurred during the Phanerozoic.

The possibility that clathrates (complexes formed from several molecules of methane and present today in certain marine sediments) played a destabilizing role has been proposed to explain the triggering of the Marinoan glaciation and the decrease of the $\delta^{13}C$ of the carbonate sediments just before glaciation (Fig. 26.8). This hypothesis suggests an ad hoc increase in CH_4 degassing from sediments just before glaciation. The result is a decrease of $\delta^{13}C$ in the sea water and an increase in the level of atmospheric CH_4. The resulting global warming disturbs the paleothermostat and CO_2 is consumed faster by the weathering of the silicates.

The methane valve then closes for reasons that have still to be discovered and the remaining CO_2 level is no longer sufficient to prevent tipping into global glaciation. Finally, Pavlov et al. (2003) suggest that atmospheric methane levels remained high throughout the Proterozoic (100–300 ppmv) and that an additional episode of oxygenation of the surface layers created the conditions favorable to glaciation in the Neoproterozoic. This scenario is similar to the one elaborated to explain the Huronian glaciations. Rapid oxidation of methane to CO_2 considerably reduces the greenhouse effect in a time window too short for the paleothermostat to rebalance the temperature of the Earth's surface.

During Glaciation

During the main glacial period, the near absence of sedimentation does not allow data-based reconstruction. As a result, all theories are based on climate modeling. The average annual temperature over land during a complete glaciation is about −25 °C in the equatorial zone, with the coldest point being reached on the West African Craton with −110 °C (Donnadieu et al. 2003). The oceans become covered over with ice very quickly. The thickness of the sea ice is subject of debate. Model-based simulations suggest

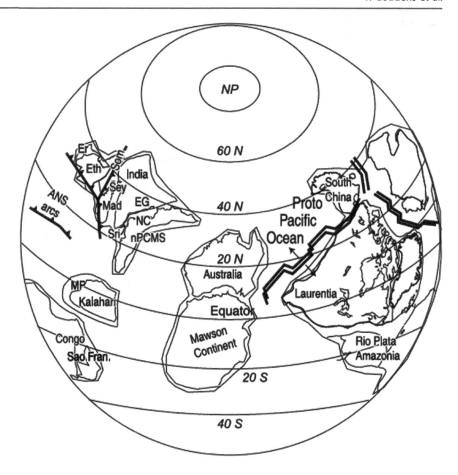

Fig. 26.10 The Rodinia dislocated about 750 million years ago

that this thickness must have rapidly exceeded a kilometer (Goodman 2006). However, for photosynthetic activity to persist during glaciation, the existence of thinner ice allowing light to filter through, at least in some areas, would be required. In contrast to the simulation suggesting thick ice everywhere, Hyde et al. (2000) presented a possibility where open water areas persisted in the tropical zone, while the rest of the ocean was completely covered with ice. This simulation was performed by coupling an energy balance model with an ice model. Nevertheless, simulations carried out with climate models taking into account the general circulation of the atmosphere and the ocean show that this solution is not stable. The ocean was probably entirely covered with ice with no real "oasis zones" where life could exist. By contrast, the presence of thin ice (of around 10 m thick) in the equatorial zone may have been possible. This was suggested by McKay (2000) and then modeled by Pollard and Kasting (2005). As for land masses, ice sheets became quickly established. Donnadieu et al. (2003) showed that by setting the CO_2 pressure at the pre-industrial value of 280 ppmv and by reducing the solar constant by 6%, the continental ice sheet would have taken 400,000 years to reach its equilibrium size of 190 million km^3 of ice covering 90% of continental surfaces (Fig. 26.12). Its maximum thickness would have been 5000 m. Even when the ice cover became total, light snowfall would have continued on the continents due to the sublimation of ice from the sea ice. In addition, the ice sheet would have remained dynamic as its base was wet. Ice flow rates of 5–10 m/year have been calculated for the most continental parts of the ice sheet and a rate of 50 m/year for coastal areas (Donnadieu et al. 2003). An active water cycle is therefore maintained during total glaciation, even if it is greatly reduced.

Exiting from Glaciation

In order to exit from glaciation, very high levels of CO_2 are required. Caldeira and Kasting (1992) calculated that a fully engulfed Earth would require 0.12 bar of CO_2 to initiate deglaciation. According to these authors, the greenhouse effect then becomes sufficiently powerful to counteract the high albedo of a totally frozen Earth. Unfortunately, the energy balance model used to produce this estimate was probably too simple for this particular environment but, even more importantly, the solar constant was fixed at its present value. Taking a solar constant of 6%, the deglaciation threshold becomes 0.29 bar. Pierrehumbert (2004) has also shown that the use of a GCM further pushes back this threshold for deglaciation and that 0.29 bar must be

Fig. 26.11 Weathering rate of continental silicate rock (10^4 molCO$_2$/km^2/year) during the Neoproterozoic for the Rodinia configuration and for the dislocated configuration at 1800 ppmv of CO_2

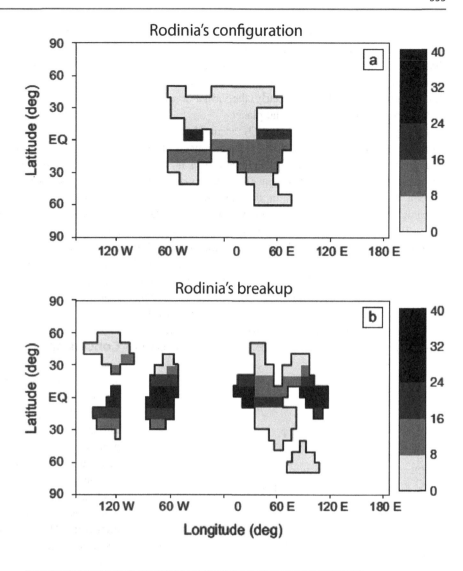

Fig. 26.12 Numerical simulation of the thickness of continental ice under Neoproterozoic conditions, at 280 ppmv of CO_2

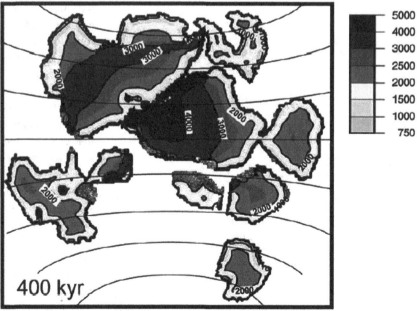

considered to be a low limit. The reasons it is difficult to melt the 'snowball' Earth are multiple. One of these is that the dryness of the atmosphere resulting from the very low temperatures limits the formation of high-level clouds and thus the greenhouse effect. Similarly, the virtual absence of water vapor limits warming of the planet.

During the glaciation period, CO_2 accumulates in the atmosphere. Indeed, the virtual disappearance of the hydrological cycle means that atmospheric CO_2 is no longer absorbed by continental weathering (Hoffman et al. 1998). Carbon therefore accumulates in the atmosphere through volcanic degassing. By taking the current degassing rate of 6.8×10^{12} mol of CO_2 per year, it takes 8 million years for 0.29 bar of CO_2 to accumulate in the atmosphere. This timeframe is compatible with the estimates of the duration of 'Snowball' Earth (Bodiselitsch et al. 2005). However, this only applies if all of the carbon sinks are stopped during glaciation and this is not the case, since the existence of thin ice allows a fracture at the ocean-atmosphere interface. A surface area of 3000 km^2 of open water is sufficient to ensure a massive diffusion of CO_2 into the ocean, ensuring the balance between the ocean and the atmosphere (Le Hir et al. 2007). Under these conditions, the ocean undergoes major acidification (the pH drops to 6) and the weathering of the oceanic crust becomes an efficient carbon sink, prolonging glaciation by counteracting the growth of atmospheric CO_2. If this process is included, it appears that even over 30 million years, atmospheric CO_2 does not reach the threshold for deglaciation (Fig. 26.13)

The most recent studies show that an important problem with the hypothesis of 'Snowball' Earth is understanding the conditions under which it melts. Nevertheless, processes able to reduce the duration of the glaciation have never been tested. For example, prolonged volcanic activity during glaciation probably led to the accumulation of ash on the ice, decreasing its albedo, and therefore, the CO_2 threshold required to initiate deglaciation.

Conclusion

Twenty years ago, the Precambrian was a large unknown gap in our understanding of the geochemical and climate history of our planet. This period, lasting approximately 4 billion years, saw the emergence of the major regulatory processes in our environment (the emergence of the continents, the establishment of plate tectonics, the appearance of life and of the modern atmosphere). Our understanding gradually unfolded as the experimental techniques (in particular isotopic techniques) advanced and as modeling improved. This has been a major leap forward that has occurred in the past ten years. Very significant progress has been made in describing the evolution of the environment at the Archean-Proterozoic boundary. Today, thanks to the ever more refined study of the isotopic ratios of sedimentary rocks of this period ($\delta^{13}C$, $\delta^{56}Fe$, $\Delta^{33}S$), the Great Oxygenation Event is one of the best documented events of the Precambrian. Similarly, the description of the major, possibly global, glaciations that marked the end of this very long period has improved dramatically in recent years. We now know the number, age and probable extent of the glaciations that preceded the Cambrian explosion of life, and we have an increasingly coherent image of their modalities.

However, four billion years is a long time and many areas of uncertainty still remain. We have progressed to the point where certain key periods in the Precambrian have been clearly illuminated. For instance, the description of Neoproterozoic glaciation has improved on many issues. The most striking one of the new Cryogenian chronology is the grossly unequal duration of the cryochrons (Fig. 5.14a). The Sturtian lasted four time longer than the Marinoan. Another surprising caracteristic of the new chronology is the brevity of the nonglacial interlude between the cryochrons. When

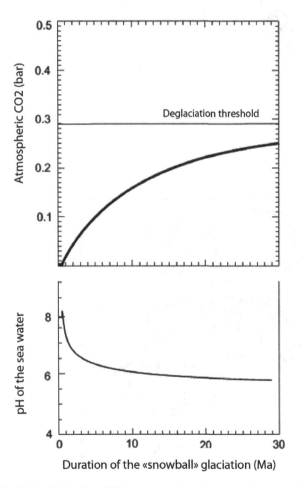

Fig. 26.13 Evolution of CO_2 concentration in the atmosphere and pH of the ocean during a global glaciation, assuming contact between the ocean and the atmosphere via fractures in sea ice

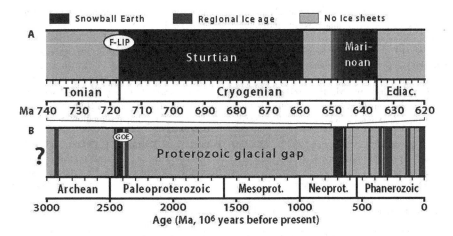

Fig. 26.14 Glacial epochs on Earth since 3.0 Ga. **a** Black bands indicate duration of the Sturtian and Marinoan cyrochrons (Table 26.1). The graded start to the Marinoan cryochron denotes chronometric uncertainty, not gradual onset. **B** Snowball Earth chrons (black), regional-scale ice ages (medium gray), and nonglacial intervals (light gray) since 3.0 Ga. Ellipse GOE is centered on the Great Oxidation Event, as recorded by the disappearance of mass-independent S isotope fractionations ≥ 0.3 per mil (‰) in sedimentary sulfide and sulfate mineral. The dashed gray line indicates questionable glaciation. Figure from Hoffman et al., Sci. Adv. 2017

the Marinoan began, a Snowball Earth had terminated less than 20 Ma earlier. When the Sturtian began, no low-latitude glaciation had occurred for 1.7 Ga (Fig. 5.14b). Nevertheless, more obscure periods, marked by slower, more progressive changes and yet which cover very long portions of the Precambrian, have still to be elucidated. In particular, the reasons why most of the Proterozoic did not experience any glaciation need to be explored. Is this truly the case or is there a bias in our observation of Precambrian formations? If this is true, then this would be the longest persistent hot period (1.2 billion years!) in the history of our planet. Why are the great quasi-periodic glacial advances observed since the emergence of multicellular life not active in the Paleo and Meso-Proterozoic? The climate history of the Precambrian that we have so far reconstructed concerns essentially that of its glacial crises (the GOE, the Neoproterozoic), but these represent only a small fraction of the immensely long Precambrian. Reconstructing a less disjointed image of this period is a major challenge for the coming decades.

References

Bodiselitsch, B., Koeberl, C., Master, S., & Reimold, W. U. (2005). Estimating duration and intensity of Neoproterozoic snowball Glaciations from Ir Anomalies. *Science, 308*(5719), 239–242.

Caldeira, K., & Kasting, J. F. (1992). Susceptibility of the early earth to irreversible Glaciation caused by carbon dioxide clouds. *Nature, 359,* 226–228.

Catling, D. C., & Claire, M. W. (2005). How Earth's atmosphere evolved to an oxic state: A status report. *Earth and Planetary Science Letters, 237,* 1–20.

Donnadieu, Y., Fluteau, F., Ramstein, G., Ritz, C., & Besse, J. (2003). Is there a conflict between the Neoproterozoic Glacial deposits and the Snowball Earth interpretation: An improved understanding with numerical modeling. *Earth and Planetary Science Letters, 08,* 101–112.

Donnadieu, Y., Goddéris, Y., Ramstein, G., Nédelec, A., & Meert, J. G. (2004). Snowball EARTH triggered by continental break-up through changes in runoff. *Nature, 428,* 303–306.

Evans, D. A. D. (2000). Stratigraphic, Geochronological, and paleomagnetic constraints upon the Neoproterozoic climatic paradox. *American Journal of Science, 300,* 347–433.

François, L. M., & Goddéris, Y. (1998). Isotopic constraints on the cenozoic evolution of the carbon cycle. *Chemical Geology, 145,* 177–212.

Gaucher, E. A., Govindarajan, S., & Ganesh, O. K. (2008). Paleotemperature trend for precambrian life inferred from resurrected proteins. *Nature, 451,* 704–708.

Goddéris, Y., et al. (2003). The Sturtian Glaciation: Fire and ice. *Earth and Planetary Science Letters, 211,* 1–12.

Goddéris, Y., & Veizer, J. (2000). Tectonic control of chemical and isotopic composition of ancient oceans: The impact of continental growth. *American Journal of Science, 300,* 434–461.

Goodman, J. C. (2006). Through thick and thin: Marine and meteoric ice in a 'Snowball Earth' climate. *Geophysical Research Letters, 33.* https://doi.org/10.1029/2006gl026840.

Halverson, G. P., Hoffman, P. F., Schrag, D. P., Maloof, A. C., & Rice, A. H. N. (2005). Towards a Neoproterozoic composite carbon isotope record. *Geological Society of America Bulletin, 117,* 1181–1207.

Hoffman, P. F., Abbot D. S., Ashkenazy, Y., Benn, D. I., Brocks, J. J., Cohen, P. A., et al. (2017). Snowball Earth climate dynamics and Cryogenian geology geobiology. *Science Advances.*

Hoffman, P. F., Kaufman, A. J., Halverson, G. P., & Schrag, D. P. (1998). A Neoproterozoic Snowball Earth. *Science, 281,* 1342–1346.

Hyde, W. T., Crowley, T. J., Baum, S. K., & Peltier, W. R. (2000). Neoproterozoic 'Snowball Earth' simulations with a coupled climate/ice sheet model. *Nature, 405,* 425–429.

Kasting, J. F., & Howard, M. T. (2006). Atmospheric composition and climate on the early earth. *Philosophical Transaction Royal Society London B, 361,* 1733–1742.

Kasting, J. F., et al. (2006). Paleoclimates, ocean depth, and the oxygen isotopic composition of seawater. *Earth and Planetary Science Letters, 252,* 82–93.

Kaufman, A. J., & Xiao, S. H. (2003). High CO_2 levels in the proterozoic atmosphere estimated from analyses of individual microfossils. *Nature, 425*, 279–282.

Knauth, L. P., & Epstein, S. (1976). Hydrogen and oxygen isotope ratios in nodular and bedded cherts. *Geochimica Cosmochimica Acta, 40*, 1095–1108.

Kump, L. R., Kasting, J. F., & Barley, M. E. (2001). Rise of atmospheric oxygen and the 'upside-down' Archean Mantle. *Geochemistry Geophysics Geosystems, 2*(1). https://doi.org/10.1029/2000gc000114.

Le Hir, G., Goddéris, Y., Ramstein, G., & Donnadieu, Y. (2007). A scenario for the evolution of the atmospheric $p$$CO_2$ during a Snowball Earth. *Geology* (in press).

Lowe, D. R., & Tice, M. M. (2004). Geologic evidence for Archean atmospheric and climatic evolution: Fluctuating levels of CO_2, CH_4, and O_2 with an overriding tectonic control. *Geology, 32*(6), 493–496.

McKay, C. P. (2000). Thickness of tropical ice and photosynthesis on a Snowball Earth. *Geophysical Research Letters, 27*, 2153–2156.

Papineau, D., Mojzsis, S. J., & Schmitt, A. K. (2007). Multiple sulfur isotopes from paleoproterozoic Huronian interglacial sediments and the rise of atmospheric oxygen. *Earth and Planetary Science Letters, 255*, 188–212.

Pavlov, A. A., Hurtgen, M. T., Kasting, J. F., & Arthur, M. A. (2003). Methane-rich proterozoic atmosphere. *Geology, 31*, 87–90.

Pavlov, A. A., Kasting, J. F., Brown, L. L., Rages, K. A., & Freedman, R. (2000). Greenhouse warming by CH_4 in the atmosphere of Early Earth. *Journal of Geophysical Research, 105*(E5), 11981–11990.

Pierrehumbert, R. T. (2004). High levels of atmospheric carbon dioxide necessary for the termination of global Glaciation. *Nature, 429*, 646–649.

Pollard, D., & Kasting, J. F. (2005). Snowball Earth: A thin-ice solution with flowing sea glaciers. *Journal of Geophysical Research, 111*.

Robert, F., & Chaussidon, M. (2006). A palaeotemperature curve for the precambrian oceans based on silicon isotopes in cherts. *Nature, 443*, 969–972.

Torsvik, T. H., et al. (2001). Rodinia refined or obscured: Paleomagnetism of the Malani igneous suite (NW India). *Precambrian Research, 108*, 319–333.

Walker, J. C. G., Hays, P. B., & Kasting, J. F. (1981). A negative feedback mechanism for the long-term stabilization of earth's surface temperature. *Journal of Geophysical Research, 86*, 9776–9782.

The Phanerozoic Climate

Yves Goddéris, Yannick Donnadieu, and Alexandre Pohl

The Phanerozoic period covers the last 542 million years of Earth's history, about 12% of the history of our planet. With regard to the evolution of life, the Phanerozoic experienced major events such as the rapid diversification of multicellular organisms which first appeared in the Cambrian (541–485 Ma), the colonization of continental surfaces by living organisms during the Ordovician (485–444 Ma) and the appearance of the first hominids about 8 million years ago. Over this same period, the appearance of the Earth's surface changed considerably: the continental drift during the Phanerozoic moved all of the continental masses situated in the southern hemisphere and along the equator since the Cambrian to join together around 280 million years ago to form a supercontinent: the Pangea. This would then disintegrate during the Jurassic around 180 million years ago with the emergence of the Atlantic Ocean.

Our understanding of the Phanerozoic climate, even if it is still incomplete, has been rapidly evolving in recent years. Data accumulated over the last few decades has improved consistently in quality and the numerical models used to reconstruct past climates have evolved considerably: they have a better spatial resolution and include an increasing number of processes. In general, it is true to say our conception of how and why the Earth's climate evolved during this key period in the history of life on Earth is now undergoing a true revolution.

The Proxies for the Phanerozoic Climate

There are no direct indicators of climate conditions in the geological past. However, qualitative reconstructions can be produced from sedimentological and paleontological data, and, in general, are more reliable than they are for the Precambrian. Geochemical data, in particular the isotopic indicators measured in marine sediments, provide some quantification, but interpretation of them is rarely straightforward. Below we provide a non-exhaustive list of examples of indicators. Finally, we will highlight the indicators that allow the reconstruction of CO_2 concentration in the atmosphere over geological time.

Sedimentological Indicators

A compilation of sedimentological data indicative of glacial climate was carried out in 1992 by Frakes et al. (1992). It consists of an inventory of tillite-type glacial deposits (clays formed from erosion products resulting from the friction of glaciers on their bedrock) and, on the other hand, a reconstruction of the minimum paleolatitude attained by rock debris carried by sea ice. The result shows a fluctuation of hot and cold modes over a period of approximately 135 Ma. The coldest climate mode was identified during the Permo-Carboniferous glaciation.

A more recent study provides results in agreement with those of Frakes et al. (1992). Boucot et al. (2004) compiled data on continental coal deposits, indicators of an arid climate, as a function of paleogeography and time. They constructed a qualitative curve of variations on the equator-pole climate gradient for the entire Phanerozoic. They interpret the existence of weak gradients as the sign of a warm climate.

Y. Goddéris (✉)
Géosciences Environnement Toulouse, CNRS-Université de Toulouse III, UMR 5563, Toulouse, France
e-mail: yves.godderis@gmail.com

Y. Donnadieu · A. Pohl
Aix-Marseille Université, CNRS, IRD, Coll France, CEREGE, Aix-en-Provence, France

Isotopic Indicators

The $\delta^{18}O$ of Carbonates

A detailed study of the isotopic composition of oxygen ($\delta^{18}O$) in carbonate sediments was carried out mainly on fossilized brachiopod shells (Veizer et al. 1999). It shows two trends (Fig. 27.1). The first is a slow, almost linear, increase of $\delta^{18}O$ from the Cambrian, from values of around −10‰ (relative to the standard Pee Dee Belemnite) up to current values close to 0‰. This increase is still difficult to interpret. If the $\delta^{18}O$ of the ocean has remained close to its present value and if the evolution of the $\delta^{18}O$ of the brachiopods is interpreted in terms of temperature over the last 540 million years, then the temperature of seawater must have reached 70 °C in the Cambrian, a level which is lethal to most marine organisms and therefore difficult to reconcile with the very large phase of diversification of marine organisms documented at this time (Zhuravlev and Riding 2001). Two possibilities have been proposed to solve this paradox: either the decrease of $\delta^{18}O$ in the past reflects a diagenetic alteration of the brachiopod shells, in which case the signal is irrelevant, or the $\delta^{18}O$ of seawater was lower in the past. Seawater is, in fact, influenced by the tectonic processes: as silicate rocks are transformed into $\delta^{18}O$-depleted clay sediments, continental and hydrothermal alteration at low temperatures tend to increase the $\delta^{18}O$ of the water in contact with the minerals. A fractionation of 20% is observed for the low-temperature alteration at the ridges, 12.5% for the continental alteration, while the high-temperature hydrothermal alteration decreases the $\delta^{18}O$ of seawater by enriching the alteration production with a fractionation of −18‰. The role of these geological processes on the $\delta^{18}O$ of seawater is not yet clearly understood. There is as yet no consensus on this issue especially since recent studies suggest that the value of the $\delta^{18}O$ of the ocean has remained constant since 760 Ma (Bergmann et al. 2018; Hodel et al. 2018).

The second trend highlighted in the long-term recordings of the $\delta^{18}O$ of the brachiopods is the periodic oscillations superimposed on the long-term linear trend described previously. If this is subtracted, the oscillations have an amplitude of 3–5‰ (Veizer et al. 2000). The most surprising aspect is that the period of these oscillations is in agreement with the periodicity of the hot and cold modes determined by Frakes et al. (1992), suggesting the presence of a true climate signal. The use of a paleothermometer, linking the isotopic fractionation between calcite and seawater to the precipitation temperature of the carbonate, makes it possible to reconstruct the temperature variations of the water in which the brachiopods lived, provided that the $\delta^{18}O$ of the seawater is known, a fact dependent on the volume of continental ice. It should be noted, however, that examples of diagenetic alteration have been identified in which isotopic exchange with runoff leads to values for $\delta^{18}O$ very different from the original values, but in which seasonal oscillations seem to be preserved. This is merely an artifact. Finding a pseudo-climatic periodicity in a diagenesis signal is not impossible and does not constitute proper evidence of the preservation and the consistency of the isotopic signal.

The fractionation α between calcite and water is expressed by the relationship:

$$T(K) = \frac{18.03 \times 10^3}{1000 \ln \alpha + 32.42} \quad (1)$$

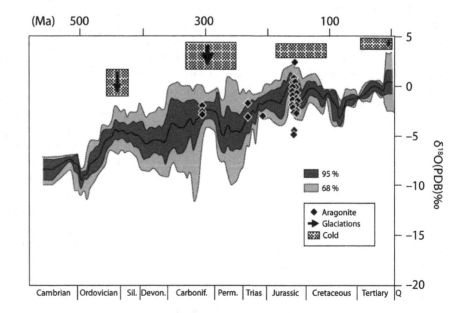

Fig. 27.1 $\delta^{18}O$ of carbonate sediments (calcite) measured over the whole Phanerozoic. The two light and dark shaded ranges contain 68% and 95% respectively of the data points. The lozenges represent measurements made on aragonitic fossils

where

$$\alpha = \frac{1000 + \sigma^{18}O^{PDB}_{sample}}{1000 + \sigma^{18}O^{PDB}_{solution}} \quad (2)$$

is the $\delta^{18}O$ (PDB) of the carbonate sample, and $\delta^{18}O$ is the $\delta^{18}O$ of the solution from which the carbonate precipitated. A $\delta^{18}O$ of seawater of 0‰ compared to SMOW is equivalent to −30‰ compared to PDB. A variation of −1‰ in the $\delta^{18}O_{sample}$ corresponds to an increase of +4 °C in the temperature at which precipitation occurred.

The volume of continental ice and the $\delta^{18}O$ of seawater are far from being well known through the geological past of the Earth. To simplify, Veizer et al. (2000) assumed the volume of continental ice to be twice the current volume at the glacial period peaks and to be zero during a warming period. An oscillation in the ice volume of this scale could account for a 2‰ change in the $\delta^{18}O$ of seawater, which presumes that the remaining 1–3‰ are attributable to temperature changes. Making these assumptions, Veizer et al. (2000) propose that seawater in the equatorial zone (where all fossil brachiopods have been found) was up to 3.5 °C colder during the Ordovician glacial maximum than now, 3 °C colder during the Permo-Carboniferous and 2 °C colder during the Jurassic, without taking into account rapid shifts which could reach amplitudes of 9 °C (Fig. 27.2).

It should be noted that the majority of the brachiopod fossils used are from the Paleozoic (Veizer et al. 1999) and that the resolution for the Mesozoic and Cenozoic ages is weak within this Phanerozoic database. The Mesozoic was covered more precisely by measurements of $\delta^{18}O$ from benthic and planktonic foraminiferal shells (e.g., Bice et Norris 2002) and on belemnite rostra (Dera et al. 2011). One of the most remarkable results is the estimation of the deep-water temperature during the Cretaceous. This temperature was approximately 10 °C at the end of the Cretaceous and could have reached 15 °C around 100 Ma (Friedrich et al. 2012). Measurements for the Mesozoic are also covered by $\delta^{18}O$ on phosphates (see next section).

Finally, the Cenozoic is covered by a high-resolution database (Zachos et al. 2008), which is the most significant step forward in terms of climate reconstruction (Fig. 27.3). The $\delta^{18}O$ measurements carried out in forty ODP and DSDP drill sites, were conducted on benthic foraminifera that once lived in the deep ocean. They are generally considered to be indicators of changes in surface water temperature at high latitudes (where dense surface waters sink to produce the deep waters of the global ocean) and of changes in the isotopic composition of the ocean on average, which are a function of the ice cap volume. They are particularly good at recording the phases of rapid growth of the Antarctic ice sheet.

The $\delta^{18}O$ of Phosphates

Another particularly promising approach is based on the study of $\delta^{18}O$ measured in phosphates, particularly in fish teeth or conodonts, small tooth-shaped structures of 0.25–2 mm, consisting of apatite and having belonged to vermiform animals that disappeared at the end of the Triassic. The paleothermometer is expressed as follows:

$$T(°C) = 112.2 - 4.2 \left(\sigma^{18}O^{SMOW}_{sample} - \sigma^{18}O^{SMOW}_{water} \right) \quad (3)$$

$\sigma^{18}O^{SMOW}_{sample}$ is the $\delta^{18}O$ (SMOW) of the phosphate sample, and $\sigma^{18}O^{SMOW}_{water}$ is the $\delta^{18}O$ of the solution from which the phosphate precipitated.

The advantage of phosphates, especially the enamel of fossil teeth and conodonts, is their greater resistance to diagenetic alteration than carbonates. In general, the $\delta^{18}O$ measured on phosphate does not appear to show significant

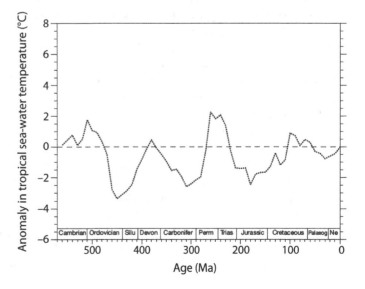

Fig. 27.2 Temperature anomalies of tropical waters reconstructed from $\delta^{18}O$ carbonate data

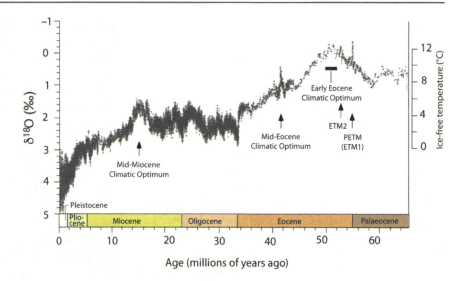

Fig. 27.3 $\delta^{18}O$ of benthic foraminifera during the Cenozoic

decay as one goes back in time, suggesting a lower sensitivity to diagenesis and therefore better reliability. This has been clearly demonstrated for the Devonian. The $\delta^{18}O$ measured on apatite from conodonts indicates a seawater temperature of about 25 °C for the end of Givetien and Frasnian (391–374 Ma), taking the $\delta^{18}O$ of the seawater to be −1‰ due to the probable absence of ice caps, as these only developed during the Famennian (Caputo et al. 2008). The $\delta^{18}O$ measured on the calcite of the brachiopod shells from the same period indicates significantly higher temperatures of between 30 and 40 °C (Veizer et al. 1999). Similar differences are observed in the amplitudes of the temperature changes between the two methods. For example, measurements of $\delta^{18}O$ on phosphates suggest a drop in tropical water temperature from 40 to 32 °C between 490 and 465 million years which seems to be correlated with a major acceleration in the expansion of biodiversity (Trotter et al. 2008). The $\delta^{18}O$ data from brachiopods suggest a temperature drop of only 4 °C over the same period. It appears that the isotopic composition of brachiopod shells depends largely on kinetic fractionation processes (typical of diagenesis) and, to a lesser extent, on the metabolism of these animals. Consequently, the $\delta^{18}O$ measured on the brachiopods could be a weak reflection of the environmental conditions that prevailed at the time of the formation of the shell. Nevertheless, the debate on the validity of the brachiopod data is still ongoing, especially since the recent publication of a new paleothermometer which revises upwards the temperatures reconstructed from phosphates (Pucéat et al. 2010).

The use of fish teeth from various parts of the world supports the reconstruction of latitudinal gradients of water temperatures, which provides essential clues to climates in the distant past. Finally, data from the teeth of fossil vertebrates offer immense opportunities in terms of the measurement of temperatures and their latitudinal gradients in continental environments.

The 'Clumped' Carbonate Isotope Method or the Δ_{47} Method

The major problem with using oxygen isotopes for the reconstruction of seawater temperatures in the past is the lack of knowledge of the $\delta^{18}O$ ratio of the seawater in which the carbonates and phosphates formed. A new technique has recently been proposed, which makes it possible to overcome this limitation. This involves essentially counting the number of bonds between rare isotopes in the $CaCO_3$ molecules, in particular, the ^{13}C–^{18}O bonds. The difference between the actual number of rare bonds and the number of bonds there would be if the bonds were stochastically distributed depends entirely on the temperature and not at all on the isotopic composition of the water in which the carbonate was formed. It is measured with the assistance of Δ_{47}:

$$\Delta_{47} = \left(\frac{R^{47}_{measured}}{R^{47}_{stochastic}} - 1\right) \times 1000 \qquad (4)$$

where $R^{47}_{measured}$ is the ratio of the mass of $^{18}O^{13}C^{16}O$ molecules to the mass of light $^{16}O^{12}C^{16}O$ molecules measured in the CO_2 emitted from the attack on carbonate by phosphoric acid. $R^{47}_{stochasitc}$ is the same as the ratio for a stochastic distribution of the molecules. The Δ_{47} depends on the temperature of the medium in which the carbonate formed (Ghosh et al. 2006) according to the formula:

$$\Delta_{47} = 0.0592(10^6 \times T^{-2}) - 0.02 \qquad (5)$$

This technique also has the advantage of being impervious to diagenesis within a temperature range of 0–200 °C. The first use of this technique was devoted to the study of samples from the Lower Silurian (around 435 million years) and from the Middle Pennsylvanian (Carboniferous, around 310 million years) (Came et al. 2007). It produces contradictory results: for the Carboniferous samples, they are, for example, in agreement with the $\delta^{18}O$ measurements on

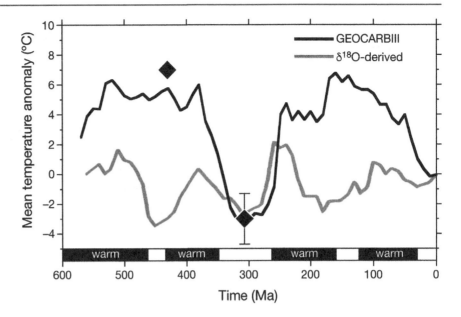

Fig. 27.4 Phanerozoic temperature anomalies. The black line tracks the output of a numerical model and represents the mean global temperature anomalies (Berner 1994). The gray line represents the $\delta^{18}O$ data on calcite (Veizer et al. 2000), and the two diamonds show the Δ_{47} data (Came et al. 2007)

brachiopods, but for the Silurien samples, they are completely opposite in that they indicate temperatures around 8 °C higher than the current ones, whereas the $\delta^{18}O$ data indicates temperatures around 2.5 °C lower than currently (Fig. 27.4).

By combining the $\delta^{18}O$ and Δ_{47} data from Ordovician sedimentary calcites, a recent study showed for the first time that it was possible to calculate the $\delta^{18}O$ ratio of seawater over this period of time and thus to trace back the volume of ice present on the continents during the glacial peak at the end of the Ordovician (Finnegan et al. 2011) the estimate of which is in agreement with the reconstruction of sea level variations for that time (Loi et al. 2010).

Indirect Isotopic Indicators

The $\delta^{13}C$ of Carbonate Sediments

The $^{13}C/^{12}C$ ratio ($\delta^{13}C$ expressed with respect to the PDB standard) of the carbonate sediments recorded the isotopic composition of the total carbon dissolved in seawater (in the form of dissolved CO_2, bicarbonate and carbonate ions, denoted by $\sum CO_2$ or by the acronym DIC—Dissolved Inorganic Carbon) at the time of deposition. The main trend of this isotopic indicator is a general increase during the Paleozoic, from $-2‰$ during the Cambrian to $+4‰$ at the end of the Carboniferous. Veizer et al. (2000) (Fig. 27.5).

This geological stage presents the highest value for this signal for the entire Phanerozoic. After a rapid decrease during the Permian, the $\delta^{13}C$ of carbonates registered minimal fluctuations around the present value of $+1.5‰$. This $\delta^{13}C$ is an indicator of the behavior of the carbon cycle, but unfortunately it is not very clear how exactly to interpret it.

The simplified budget of the $\delta^{13}C$ of the oceanic DIC δ_{oc} is written as:

$$C_{oc}\frac{d\delta_{oc}}{dt} = F_{cw}(\delta_{cw} - \delta_{oc}) + F_{ow}(\delta_{ow} - \delta_{oc}) + F_{cw}(\delta_{cw} - \delta_{oc})$$
$$+ F_{MOR}(\delta_{MOR} - \delta_{oc})$$
$$- F_{cd}(\delta_{oc} - \varepsilon_{carb} - \delta_{oc}) - F_{od}(\delta_{oc} - \varepsilon_{MO} - \delta_{oc})$$
(6)

where C_{oc} is the DIC content of the ocean and F_{cw}, F_{ow}, F_{vol} and F_{MOR} are the carbon fluxes transferred from the lithosphere to the ocean by dissolving continental carbonates, by the oxidation of sedimentary organic carbon, by degassing linked to the volcanic activity and by the oceanic ridges, respectively. F_{cd} and F_{od} are the fluxes of carbonate deposits from all environments, and the burial of organic carbon respectively. The δ are the $\delta^{13}C$ corresponding to each of these fluxes: δ_{cw} is close to 0‰, δ_{ow} to $-25‰$; δ_{MOR} is estimated at -5 or even $-6‰$. δ_{vol} is less well known, but its value is certainly located between the mantle value and that of the carbonates deposited on the abyssal sea floor, that is ±0‰ on average over a long-time scale. ε_{carb} is the isotopic fractionation between the DIC of seawater and the carbonate minerals. This fractionation is low (around 1.2‰) (Hayes et al. 1999), indicating that carbonate deposits cannot be responsible for the temporal evolution of δ_{oc}. However, the fractionation between the buried organic matter and the oceanic DIC ε_{MO} is very high (±20‰ today). The organic flows F_{ow} and F_{od} dominate the budget because this fractionation means that their combined flow is multiplied by $\delta^{13}C$ and so is an order of magnitude greater than the other terms. The flux variations most influencing the temporal evolution of the $\delta^{13}C$ of the ocean are therefore those that

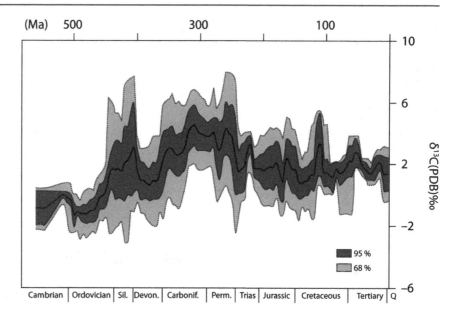

Fig. 27.5 $\delta^{13}C$ of calcite sediment from the Phanerozoic. % of data points

affect the processes with a $\delta^{13}C$ signature furthest away from the oceanic value, in other words, the oxidation of the organic sedimentary carbon exposed on land or the burial of organic carbon in sediment. However, it is impossible to discriminate between them using the $\delta^{13}C$ signal alone.

However, geochemical interpretation is possible for some major excursions. In particular, the positive and very large Carboniferous excursion (a positive excursion of 4‰ throughout the Carboniferous (Veizer et al. 1999) is interpreted as the recording of the burial of a very large amount of organic carbon (this was the time large coal beds were being deposited in Europe, Russia and North America), which extracted preferentially the ^{12}C from the ocean-atmosphere system. This resulted in a higher consumption of atmospheric CO_2 and a cooling of the climate which coincides reasonably well with the Permo-Carboniferous ice age. This observation encouraged the association of any positive excursion of $\delta^{13}C$ with a cooling of the climate, especially since many positive excursions coincide with episodes of extensive burial of organic matter (episodes of global anoxia and black shale formation), such as the end of the Devonian (Frasnian-Famennian episode, circa 375 Ma) or the global anoxic events of the Cretaceous (Aptian circa 120 Ma and Turonian circa 90 Ma). This interpretation, however, is not conclusive. To illustrate the difficulty of interpreting the $\delta^{13}C$ signal in geochemical and climate terms, it should be noted that certain anoxic episodes, during the formation of gray or black shales, are accompanied by negative excursions of the oceanic $\delta^{13}C$, such as the Toarcian anoxic episode. These events probably indicate favorable conditions (anoxic environment) for the preservation of organic matter, but the total amounts of organic matter ultimately buried are low. The Toarcian negative excursion can be interpreted as the result of a large CO_2 outgassing from the mantle when the Karoo-Ferrar traps were established in South Africa and could therefore be linked to a warming of the climate by the greenhouse effect. It should also be pointed out that, at shorter intervals during glacial-interglacial Quaternary oscillations, the $\delta^{13}C$ of the ocean decreased during glacial periods due to a reduction in the biosphere, a consequence of aridity on the continents (Part III, Chaps. 1 and 2).

Finally, some very rapid and pronounced negative excursions could be attributed to the destabilization of methane hydrates contained in marine sediments. The methane released by the sediments is characterized by a $\delta^{13}C$ of −60‰ and these negative values allow very pronounced excursions of $\delta^{13}C$. In general, it is assumed that the CH_4 contribution to the exosphere is short (10^4 to 10^5 years) and intense, resulting in rapid negative excursions (e.g. the thermal maximum of the Palaeocene-Eocene transition (McInerney and Wing 2011).

It is even possible to imagine a combination of effects: the Karoo traps were established in the coal-rich sediments of Gondwana 183 million years ago. As a result, a massive degassing of reduced carbon, low in ^{13}C, towards the atmosphere, causing the negative excursion in $\delta^{13}C$ observed during the Toarcian (McElwain et al. 2005).

Finally, a database of all the $\delta^{13}C$ values measured on benthic foraminifera from 40 ODP and DSDP drillings provides high-resolution coverage of the entire Cenozoic (Zachos et al. 2008). The dominant signal from this curve is the decrease in $\delta^{13}C$ of about 2% from mid-Miocene (15 Ma) onwards. The reasons for this reduction remain obscure. It could be a sign of a global decrease in the burial of organic carbon over the last 15 million years. However, over the same period, isotopic fractionation ε_{MO} decreased by approximately 8‰, whereas it remained relatively constant over the rest of the Phanerozoic. This fractionation was reconstructed over the whole Phanerozoic by measuring the $\delta^{13}C$ of organic carbon in marine sediments over many different periods and comparing it to the $\delta^{13}C$ of sedimentary carbonates (Hayes et al. 1999). Since this fractionation is dependent on the concentration of H_2CO_3 in the waters, it can be inferred that its decline is linked to a drop in atmospheric CO_2 since the Miocene. Yet this simple interpretation is challenged by independent estimates of the CO_2 level suggesting pressures below 300 ppmv during the Miocene (see section on atmospheric CO_2). Nevertheless, the combination of the drop in ε_{MO} fractionation and the drop in oceanic $\delta^{13}C$ suggests an increase in CO_2 being buried in the form of organic carbon in sediments since 15 Ma, probably linked to the establishment of the Himalayan orogeny.

The $^{87}Sr/^{86}Sr$ Isotope Ratio of Carbonate Sediments

Marine carbonates record the strontium isotopic ratio, $^{87}Sr/^{86}Sr$, of seawater without fractionation. The residence time of Sr in seawater (2–5 million years) ensures an even value throughout the ocean and therefore low dispersion of the data. A high-resolution curve (1 million years) has been published by Veizer et al. (1999). This signal has been widely used to constrain the extent to which CO_2 is consumed by alteration of continental silicates during the Cenozoic, particularly in response to the Himalayan orogeny (Raymo 1991). Two main types of Sr intake to the ocean are identified: an exchange flux at the level of the ocean ridges, which doesn't affect the Sr concentration of the water but modifies its isotopic ratio. Today, the water enters the ocean ridges with a $^{87}Sr/^{86}Sr$ ratio of 0.709 and exits after contact with mantellic rocks with a typical value of 0.703. This process therefore tends to reduce the $^{87}Sr/^{86}Sr$ ratio of seawater and bring it closer to the mantle value. Conversely, the isotopic ratio of the rivers, inherited from the weathering of continental rocks is now equal to 0.712. There is some correlation between periods with high $^{87}Sr/^{86}Sr$ seawater ratios and glaciation episodes. This has been interpreted as a sign of greater weathering during cold climate periods, in response to intensified physical erosion, which in turn promotes chemical weathering. The signal is particularly clear for the last 40 million years, and the rapid increase in the $^{87}Sr/^{86}Sr$ ratio of seawater has been interpreted as the signature of increased continental weathering during the uplift of the Himalayas. It has been suggested that there is a correlation between this increase in the $^{87}Sr/^{86}Sr$ ratio and a decrease in the CO_2 concentration in the atmosphere. In this model, the Himalayas are considered to have triggered the cooling in the Cenozoic.

This hypothesis has been extensively developed. It is found again in the more recent literature, linking orogeny with global cooling of the climate in response to intensified weathering. Nevertheless, this hypothesis is in contradiction with the paleothermostat theory. In fact, weathering becomes a function of climate in a positive feedback loop. The colder it gets, the more erosion increases, forcing an uptake in CO_2 consumption by weathering, in turn forcing increased cooling. The silicate weathering and volcanic degassing are uncoupled, and the carbon content of the ocean and the atmosphere is consumed in less than a few million years (Goddéris and François 1996), which would lead to unregistered climate disasters during the Tertiary. It should be underlined, however, that it is possible that orogens pump CO_2 while respecting the paleothermostat theory. The consequences of the uplift of a mountain range are much more complex than a simple uplift causing increased weathering. We shall see later that they lead to a number of geological phenomena which ultimately link orogenesis to cooling.

Moreover, the evolution of the $^{87}Sr/^{86}Sr$ isotopic ratio of seawater cannot be interpreted solely in terms of changes in the relative importance of mantellic and continental flows. The isotopic ratio of the source rocks of the weathering may also have changed over time, and, particularly in the orogenic zones, which considerably complicates the interpretation of the isotopic sign of strontium. Nevertheless, the quality of the Phanerozoic signal should motivate further analysis of this indicator in the future.

Along with the $^{87}Sr/^{86}Sr$ isotopic signal, the $^{187}Os/^{188}Os$ osmium isotopic ratio of seawater measured from sediments taken during ocean drilling programs is also used to constrain the evolution of continental and hydrothermal weathering fluxes. The methodology is very similar to that of Sr, but the major advantage of osmium is its short residence time in the ocean, around 10–30 kyr, although it is sufficiently longer than the mixing time of the water masses to ensure values that are representative of the global ocean. River contributions are the dominant factor, with a $^{187}Os/^{188}Os$ ratio of 1.3 and a flux of 1800 mol per year^{-1}. Hydrothermal inputs have an isotopic signature of 0.13 and a flux of around 100 mol per year^{-1}. A flow linked to cosmic dust, with a ratio of 0.13 and a flux of 80 mol per year^{-1}, must also be added. The main sinks are ocean sediment deposits. The ratio can therefore be indirectly linked to the evolution of the Earth's climate and of greenhouse gases, through the characterization of the geological flows of the carbon cycle. Nevertheless, osmium is a very scarce element, which makes it difficult to measure. The mean concentration of Os is 10 fg/g (fentogram gram^{-1}, 1 fg = 10^{-15} g) in seawater and 9 fg g^{-1} in rivers.

The Level of Atmospheric CO_2

There is clearly no direct measurement of this level for the distant past beyond 900,000 years, the period covered by ice cores drilled in the Antarctic ice. All methods are therefore indirect and depend on a number of assumptions. This inevitably produces significant uncertainties. In addition, the low residence time of carbon in the ocean-atmosphere system (200,000 years) is an inherent cause of scattering of data points. Temporal resolution for the geological past very rarely reaches this level of precision. It follows that two points, attributed to the same geological moment, may have very different values.

An excellent overview was carried out by Royer et al. (2001) using several reconstruction methods. For the very distant past, the counting of the stomata of the fossil leaves is commonly used. For modern species, there is a positive correlation between the number of stomata and the ambient CO_2 level. These correlations are applied to old geological samples. This is probably the least precise method, but it has the advantage of being able to go back very far in the past (as far back as the Devonian) and does not encounter the problems experienced by isotopic systems, such as diagenesis.

A second method allows measurements to be traced back to the Paleozoic. It involves measuring the $\delta^{13}C$ of pedogenic carbonates in paleosols and is based on the fact that the level of CO_2 in modern soils results from a mixing of the atmospheric CO_2 in the atmosphere and the CO_2 in the soil through respiration. The $\delta^{13}C$ of this C_2^{atm} mixture is registered in the pedogenic carbonates:

$$CO_2^{atm} = S(z) \frac{\delta_{sample} - 1.0044\delta_{resp} - 4.4}{\delta_{atm} - \delta_{sample}} \quad (7)$$

where δ_{sample} is the $\delta^{13}C$ of the pedogenic carbonate. The method requires setting the $\delta^{13}C$ of the breathed CO_2, δ_{resp}, at typical values. It therefore depends on the proportion of C_4 plants to C_3 plants, for which the isotopic fractionations are very different. In practice, it is not applicable after the emergence of C_4 plants, 15 Ma ago. The method also requires knowledge of the $\delta^{13}C$ of the atmosphere δ_{atm} and $S(z)$, the amount of breathed CO_2 at the estimated depth z of the pedogenic carbonate in the paleosol. The $\delta^{13}C$ of the atmosphere is estimated by measuring the $\delta^{13}C$ of marine carbonates of the same age and by imposing the fractionation value between the carbonates and the atmospheric CO_2. As for the fraction of breathed CO_2, it is calculated by making major assumptions about soil temperature, porosity and biological productivity. In fact, if CO_2 production in soils is a function of productivity and temperature (which partially controls the degradation of organic matter by bacteria), its diffusion to the atmosphere depends largely on the physical structure of soils. The level of CO_2 at a given depth is therefore dependent on the relative importance of the production and loss by diffusion. This is by far the most uncertain method. A recent recalibration of the method has led to a considerable reduction in past reconstructed CO_2 pressures (Breecker et al. 2010; Foster et al. 2017). A similar method consists of studying the trace $\delta^{13}C$ of pedogenic carbonates contained in goethite, a mineral formed during soil alteration reactions (Royer et al. 2001).

A third approach uses the measurement of isotopic fractionation of carbon by phytoplankton and its relationship with the dissolved CO_2 content in seawater. Initially, the difference between the $\delta^{13}C$ of carbonates and that of total organic carbon was used. It was subsequently found to be error-prone, in particular due to the presence of organic matter of various origins in both continental and marine sediments. This fractionation is now measured by directly using biomarkers in the organic matter, such as alkenones (Pagani et al. 2005). The link between isotopic fractionation and the level of CO_2 dissolved in water is based on correlations established for the present. For example, this one is based on a compilation of GEOSECS campaign data:

$$\varepsilon_P = 12.03[CO_{2aq}] - 3.56$$
$$10 \leq [CO_{2aq}] \leq 90 \, \mu M \quad (8)$$

where ε_p is the photosynthetic fractionation of phytoplankton, and $[CO_{2aq}]$ is the concentration of gaseous CO_2 dissolved in water.

The use of correlations established for current conditions is the main weakness of this method, since they are extrapolated to CO_2 ranges that are significantly higher than the current level, using compounds made by organisms with an unknown metabolism. In addition, isotopic fractionation is also a function of the growth rate of these organisms, which complicates reconstruction.

A final method is based on the measurement of the ratio of boron isotopes $^{11}B/^{10}B$ ($\delta^{11}B$) in carbonate sediments (Royer et al. 2001). The relative abundance of the two dissolved borate species (H_4BO_4 and $H_3BO_4^-$) depends on the pH of the sea water. There is an isotopic fractionation of about 19‰ between the two species. The carbonates are mainly made up of the H_4BO_4 species and the isotopic $\delta^{11}B$ composition of the carbonates will therefore depend on the pH. Nevertheless, the link with atmospheric CO_2 is not clear. First, this requires assumptions that $\delta^{11}B$ of seawater remains constant over time, that the isotopes are shared between the two species and that the relative abundances of ^{11}B and ^{10}B remain the same. It was shown that this was probably not the case and that $\delta^{11}B$ of total borate probably changed in the past. Finally, we must make strong assumptions about the alkalinity of seawater, to bring the pH up to the pressure of atmospheric CO_2.

The compilation of all these reconstructions inevitably shows a large dispersion of points (Royer 2006; Foster et al. 2017) (Fig. 27.6). Nevertheless, some trends may emerge. The level of atmospheric CO_2 seems to have been high before the Devonian (with values generally in excess of 2000 ppmv). This period is followed by a time interval covering the end of the Carboniferous and the beginning of the Permian, during which time atmospheric CO_2 remained between 200 and 500 ppmv. The data for the Mesozoic are more confusing and show no clear trend. For the same time period, it is common to have atmospheric CO_2 estimates varying by a factor of 5–10. Finally, the Cenozoic appears to be marked by a general decrease in atmospheric CO_2 pressure, reflected above all in a general reduction in the maximum reconstructed values. The most precise record (i.e., with the best temporal resolution and lowest dispersion) is the one obtained by reconstructing the isotopic fractionation in carbon of the oceanic biosphere based on the measurement of the $\delta^{13}C$ of the alkenones (Pagani et al. 2005). It shows a rapid decrease in atmospheric CO_2 from the beginning of the Eocene until the end of the Oligocene: around 50 million years ago, the CO_2 content is estimated to have been 1500 ppmv and fell to between 200 and 300 ppmv 23 million years ago. CO_2 levels then remained constant throughout the Miocene at values slightly below 250 ppmv. Finally, CO_2 levels during the Pliocene were explored using two methods: through isotopic fractionation in carbon and by the counting the stomata of fossil leaves. Both methods suggest that CO_2 levels have risen: between 350 and 450 ppmv from 2.9 to 3.3 million years for the first method, and between 370 and 250 ppmv from 5.3 to 2.6 million years for the second.

The Great Climate Modes of the Phanerozoic and Their Possible Causes

The climate reconstructions of the Phanerozoic show a succession of modes warmer than currently and of cold modes similar to currently, with the emergence of ice caps. This succession is observed in sedimentological records of glacial sedimentary deposits, including tillites, and in ice rafted debris (IRD), debris carried by sea ice (Frakes et al.

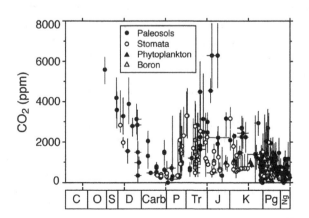

Fig. 27.6 Levels of phanerozoic CO_2 reconstructed by various methods based on proxies

1992). However, these climate oscillations are also observed in isotopic data, such as in the $\delta^{18}O$ of carbonates deposited on the seabed, which reflect, at least partially, the $\delta^{18}O$ of seawater at the time of deposition (Veizer et al. 2000). These isotopic data remain difficult to interpret because they combine not only climate indicators (seawater temperature, continental ice volume), but also geochemical data such as salinity of the seawater, the speciation of carbonates, the $\delta^{18}O$ of seawater which is itself influenced by continental and hydrothermal alteration flows. Nevertheless, the $\delta^{18}O$ during the Phanerozoic shows oscillations with a periodicity of 135 million years, in line with sedimentological reconstructions, thereby reinforcing its validity as a good climate indicator. To date, this periodicity of 135 million years remains largely unexplained, but its length indicates that it might have to do with the tectonic processes that shaped the Earth's surface or with astronomical movements (Shaviv and Veizer 2003). The accumulation of recently obtained isotopic data on phosphates (including fossil fish teeth and conodonts) has greatly improved the resolution of the alternation of hot and cold modes, especially during the Mesozoic and Devonian periods (Dromart et al. 2003; Pucéat et al. 2003; Joachimski et al. 2004).

The cold climate modes of the Phanerozoic occur during the Ordovician (from about 470–440 Ma), the Permo-Carboniferous (from about 330–270 Ma), the Jurassic and the Cretaceous. This period is marked by a succession of short cold events: at the end of the Toarcian around 176 Ma, at the Callovian-Oxfordian boundary around 161 Ma, at the transition from the Lower Valanginian to the Middle Valanginian towards 140 Ma, at the beginning of the Aptian around 125 Ma and at the Cenomanian-Turonian boundary around 94 Ma. Finally, the end of the Cenozoic, when Antarctica first started to freeze over 34 million years ago up to the current period, was in cold mode with ice first appearing in the southern polar regions and later in the northern hemisphere.

The Causes of Cold Climate Modes

These are generally subjected to more study than the causes of warm modes. The following processes have been suggested:

1. Orogenesis
   The establishment of mountain chains causes an increase in physical weathering following the establishment of glaciers, steep slopes and alternating freeze-thaw regime (Raymo 1991). This results in a greater fracturation of the rock and thus greater sensitivity to chemical weathering which consumes CO_2. Therefore, locally, this process increases the vulnerability of the continental surfaces to

weathering. The result would be a drop in atmospheric CO_2 and an overall cooling. This theory is based on measurements of current weathering rates in mountain ranges, on the strong correlation between physical erosion and chemical weathering in modern watersheds, and on the $^{87}Sr/^{86}Sr$ isotopic ratio of the water run-off from the Himalayas. A second effect should also be taken into account: the very high sedimentation rates in the seas at the foot of the orogens lead to very efficient burial of organic carbon and thus to increased consumption of atmospheric CO_2 thanks to the action of the biological pump. This process is responsible for trapping two to three times more CO_2 than the silicate weathering in the Himalayan orogeny (Galy et al. 2007).

2. The development of vascular vegetation on land. Tall vegetation with a developed root system acts at three levels on weathering rates. The roots stabilize the soils and thus increase the contact time between inland waters and the silicate minerals. In addition, root and microbial respiration in soils increases the partial pressure of CO_2, and thus acidifies the water which percolates towards the bedrock. Finally, the plants secrete organic acids which also contribute to the acidification of the waters. As a result of these three effects, there is an increase in the consumption of atmospheric CO_2 through dissolution of the continental silicates. This hypothesis is based on studies carried out in particular in Iceland on lava flows on slopes covered and uncovered with stemmed vegetation. It appears that weathering rates are eight to ten times larger under dense vegetation cover (Berner 2004). A small-scale laboratory study suggests that non-vascular plants (lichens and mosses) could have a similarly accelerate chemical weathering of continental surfaces (Lenton et al. 2012).

3. Increased burial of organic carbon during the anoxic phase of the ocean. This hypothesis is often proposed to explain positive excursions in the $\delta^{13}C$ ratio of oceanic carbonates correlated with climate cooling. It requires particular environmental conditions: either conditions favorable to maintaining water stratification in large ocean basins and preventing the ventilation of the deep waters in these basins, or conditions of oceanic hyperproductivity leading to the absorption of oxygen in the deep waters through the recycling of organic matter produced in the euphotic zone. This burial may also occur on land-based environments, as has happened during the Carboniferous period.

4. The movement of the solar system into a galactic arm. This recent hypothesis attempts to explain the periodicity of 135 million years in cold modes. The galactic arm is an area of formation of intense stars and of emission of galactic cosmic rays. Reaching the atmosphere, these are thought to participate in the nucleation of low-level clouds, increasing the albedo of the atmosphere and cooling the climate (Shaviv and Veizer 2003). To date, there is no experimental evidence of the validity of this mechanism, which remains purely speculative.

5. The fragmentation of a supercontinent. The resulting increase in rainfall activates the consumption of CO_2 by silicate weathering and thus cools the climate. This effect is particularly important if the supercontinent breaks up along the equator, the site of intense rainfall (Goddéris et al. 2014).

6. The migration of continents towards the low latitudes, characterized by climatic conditions favoring the weathering of continental silicates and thus an increased consumption of atmospheric CO_2 (Nardin et al. 2011).

7. Any reduction in degassing of greenhouse gases from the mantle or sediments towards the atmosphere.

8. The establishment and subsequent weathering of basaltic provinces on the continents. Basalts weather much more efficiently than the average continental crust on which they spread (in equivalent conditions, basalt weathers eight times faster than granite). This finding was established from a study of weathering in basaltic watersheds (Dessert et al. 2001). The weathering of new basalt thus produces a long-term decrease in the partial pressure of atmospheric CO_2. The question remains as to the weathering of submarine basaltic plateaus. Do they contribute to the cooling of the climate system or not? The pH buffer imposed by carbonate speciation in seawater nevertheless suggests that weathering of oceanic basalts is a minor phenomenon, with basalt dissolution being minimal at around pH 8, a value for seawater which probably didn't change much over the course of the Phanerozoic.

The Causes of Warm Climate Modes

Curiously, cold modes have always been considered to be accidents in a prolonged warm state. This is probably the reason why the suggested causes of warm modes are fewer and less discussed in the literature, with the exception of the thermal event of the Palaeocene-Eocene transition.

The following mechanisms have been proposed:

1. Any increase in degassing of greenhouse gases from the mantle or sediments to the atmosphere. This could be due to increased volcanic activity releasing massive amounts of CO_2, basaltic effusion events over land (Dessert et al. 2001) or methane degassing from gas hydrates accumulated in sediments (McInerney and Wing 2011).

2. The creation of a supercontinent, reducing rainfall and thus weathering of the continental silicates, allowing an

increase in pressure of CO_2 in the air and a warming of the climate (Goddéris et al. 2014).

The Phanerozoic Terrestrial Paleothermostat

The short residence time (200,000 years) of carbon in the exosphere, as well as the reaction time of ocean alkalinity (3000 years) impose the following near-equal relationship between the fluxes of inorganic carbon, ignoring any possible imbalances in the organic carbon cycle (see Chap. 5):

$$F_{vol} + F_{MOR} \approx F_{sw} \quad (9)$$

where F_{vol} is volcanic degassing, F_{MOR} ocean ridge degassing and F_{sw} half the CO_2 consumption by silicate weathering. This factor of a half stems from the fact that two moles of atmospheric carbon are consumed for two equivalents of alkalinity produced by the dissolution reaction of the continental silicates. Only one of these two moles will finally be buried in the form of ocean carbonate. The other mole of carbon remains in the ocean-atmosphere system (see Chap. 5). The flux of CO_2 consumption by silicate weathering is a function of temperature and of continental runoff. Generally, it increases as the CO_2 content increases. But its response to an increase in CO_2 is also a function of the continental plant cover, the presence of orogens and of intense physical weathering, the configuration of the continents, the modification of the superficial lithology, following, for example, the establishment of basaltic surfaces on land during major magma events. F_{sw} can therefore be expressed in the following way:

$$F_{sw} \alpha f_1(T) \times f_2(R) \times f_3(\text{erosion}) \\ \times f_4(\text{vegetation}) \times f_5(\text{litho}) \quad (10)$$

where T is the continental temperature, and R is the runoff. The functions f_1 and f_2 are known: the first is an exponential function of the temperature, the second a linear function of the runoff. f_1 and f_2 have been determined for granites and basalts. The function f_3 is unknown. The only indicator available is that there is a very strong positive correlation between physical erosion fluxes and chemical weathering fluxes for both large and small watersheds. We can deduce from this that f_3 is an increasing function of the rate of erosion, but its precise mathematical expression had yet to be defined. As for f_4, studies of lava flows in Iceland suggest that the rate of weathering increases by a factor of 8 when vascular vegetation develops (Berner 2004). f_4 increases with vegetation cover but also when mosses and lichens cede to vascular plants with a well-developed root system (see discussion on the Devonian in the following section).

Finally, f_5 expresses the level of dependence on the lithological type. It can be expressed as a constant factor equal to 8 or 10 for new (rapidly deteriorating) basaltic surfaces and equal to 1 for granite surfaces (Dessert et al. 2001).

For example, this simple formalism shows that the establishment of an orogen leads to an increase in the consumption of atmospheric CO_2 through silicate weathering (f_3 increases), but that the conditions of the paleothermostat are always verified: the climate cools globally, and the decrease in the f_1 and f_2 factors compensates for the increase in f_3. It can then be said that the vulnerability of continental surfaces to weathering has changed. Indeed, if the degassing of the solid Earth does not change, Eq. (9) dictates that CO_2 consumption by silicate weathering remains virtually constant on the scale of several million years. However, Eq. (10) dictates a decrease in f_1 and f_2 to compensate for the increase in f_3. We can say that the weathering of continental surfaces has increased, whereas the total silicate weathering flux has remained unchanged. However, to allow f_1 and f_2 to adapt to the new conditions, the equilibrium level of CO_2 is lower, and the climate is colder and dryer. Similarly, the establishment of a basaltic province increases the factor f_5 and the climate will cool in compensation. The same applies to the colonization of the continental surfaces which are described below.

It should nevertheless be noted that the relation 10 is a simplification. The relationship between CO_2, temperature, and continental runoff is complex and is largely dependent on the paleogeographic configuration, which complicates the problem considerably.

Finally, if the possibility of imbalance in the organic carbon cycle is taken into account, the thermostat equation is written as:

$$F_{vol} + F_{MOR} + F_{ow} \approx F_{sw} + F_{od} \quad (12)$$

where F_{ow} is the oxidation of sedimentary organic carbon and F_{od} is the overall burial of organic carbon. If deposits were to increase, for example, due to the development of exceptional conditions for the preservation of organic matter (such as the appearance of large-scale anoxia), while the oxidation of exposed sedimentary organic carbon on land remained constant, this would give the following inequality:

$$(F_{vol} + F_{MOR}) - F_{sw} \approx F_{od} - F_{ow} \geq 0. \quad (13)$$

In this case, silicate weathering must be less than the total degassing of the solid Earth to maintain the paleothermostat balance. This condition will be verified because the increase in buried organic carbon reduces the CO_2 pressure in the air which, in the first order, causes a decrease in the f_1 and f_2 factors.

Finally, it should be noted that the paleothermostat constitutes a very powerful stabilizing force of the Earth's climate. If causes external to the CO_2 cycle disturb the climate (cosmic rays and cloud nucleation, the passage of the Earth into a cloud of galactic dust, methane degassing), the resulting change in f_1 and f_2 factors would cause an imbalance in the carbon cycle, which would find a new balance over a few million years, by adjusting the pressure of atmospheric CO_2 to re-establish climate conditions verifying the paleothermostat.

The Paleozoic Climate: The Chronology of Major Trends and Their Causes

In general, the Paleozoic climate is described as warmer than the current one, except for two glacial events with very different characteristics.

The reasons for this warm climate state are not clearly understood, but several hypotheses have been put forward. On the one hand, the degassing of the solid Earth seems to have been generally greater by about 60% than it is at present. This assertion is based on the fact that the sea level was generally higher in the geological past than it is today, except for the Permo-Carboniferous transition. This high sea level can be explained firstly by the larger volume occupied by the ocean ridges and therefore a supposedly greater degassing of the solid Earth. This result has never been confirmed by other methods apart from sea level and remains questionable. On the other hand, the absence of abundant vascular vegetation until the end of the Devonian prevented the development of modern soils on land surfaces. This absence of soil reduced the contact time between the inland water and minerals, thus limiting their weathering. Similarly, the absence of a root system reduced the acidity of soil solutions and thus the consumption of atmospheric CO_2 by continental silicate weathering, which in turn promotes high CO_2 levels. Berner (2004) estimates that the average global temperature was ±6 °C higher than it is currently, based on a numerical modeling study. These very high values are confirmed by measurements using the paleothermometer made up of the number of rare molecules $Ca^{18}O^{13}C^{16}O_2$ in the carbonates of the Lower Silurian (Came et al. 2007).

Whatever the causes, this warm climate state was interrupted by several glacial events of very different durations and amplitudes. The Atlas Fig. chapter 3.9 shows the paleogeographic maps of the Earth during the main periods of the Phanerozoic.

The Ordovician Glaciation

The $\delta^{18}O$ data on apatite show a long-term cooling trend during the Lower and Middle Ordovician and a sudden drop in temperatures during the Hirnantian in the Late Ordovician (Trotter et al. 2008), which is independently confirmed by the $\Delta^{47}CO_2$ analysis (Finnegan et al. 2011, Fig. 27.7). Glacial sediments, which are the only direct evidence of Ordovician glaciation, are only documented during this very short cool interval, which has long suggested that glaciation is a short-term cold accident punctuating an otherwise very hot period of geological time. Geochemical studies reconstructing the composition of oceanic $\delta^{18}O$ (Finnegan et al. 2011) and glacio-eustatic variations (Loi et al. 2011) suggest that the ice cap at the South Pole would have reached a volume almost twice as large as during the Last Glacial Maximum. Indirect indices such as variations in sea level (Dabard et al. 2015) or $\delta^{18}O$ excursions (Rasmussen et al. 2016) today suggest that the first ice caps could have been in place since the Middle Darriwilien (about 470 Ma) in the Ordovician. In addition, it appears that glacial events also punctuated the Lower Silurian. The Ordovician glaciation is

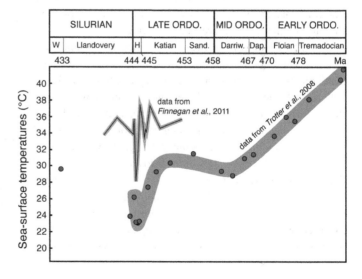

Fig. 27.7 Tropical sea-surface temperatures reconstructed based on $\delta^{18}O$ data measured on apatite (Trotter et al. 2008) and based on $\Delta^{47}CO_2$ proxy data (Finnegan et al. 2011)

therefore increasingly considered to have been a long cold period (about 470–425 Ma), sometimes referred to as 'Early Paleozoic Ice Age' (Page et al. 2007), and within which the Hirnantien only represents a glacial maximum. This vision is supported by the most recent climate models (Pohl et al. 2016).

The causes of this glaciation are still poorly understood. Nardin et al. (2011) showed that the long-term cooling of the climate can be explained by the paleogeographic evolution occurring throughout the Ordovician, and in particular the migration of continents in the intertropical zone conducive to weathering, which brings about a fall in the atmospheric concentration of CO_2. Regarding the Hirnantien glacial peak, the best explanations also suggest a fall in atmospheric CO_2, but the mechanisms to achieve this are subject to debate. Kump et al. (1999) proposed an interesting hypothesis: the fall in CO_2 level could have been a consequence of the establishment of New Caledonian and Appalachian orogens during the Middle and Upper Ordovician, which would have increased the vulnerability of the continental surfaces to weathering. Other mechanisms have also been proposed, including the establishment of the first plants on land (Lenton et al. 2012). The difficulty in explaining this event lies in the magnitude of the cooling, which is around −7 °C at tropical latitudes, whereas sea surface temperatures appear to have varied by only 1–2 °C at the same latitudes during the last glacial-interglacial cycle (CLIMAP Project 1981). However, numerical modeling of the ocean-atmosphere coupled system during the Ordovician revealed climatic instability associated with the sudden development of sea ice, which explains a sharp fall in temperatures in response to a moderate decrease in atmospheric CO_2 concentration, thus loosening the constraints that would be placed on CO_2 sinks to explain the geochemical data (Pohl et al. 2016). To explain the emergence from the Hirnantian glacial maximum, Kump et al. (1999) proposed the following mechanism: as the ice cover on the supercontinent Gondwana increased, the available surface of continental silicates exposed to weathering falls, thus causing an accumulation of CO_2 in the atmosphere. This persuasive scenario was tested with a simple climate model.

The Devonian Climate

The Devonian (419–359 Ma) is marked by numerous biological disturbances. Although the first traces of vegetation appeared during the Middle Ordovician (Rubinstein et al. 2010) and the existence of vegetation fires during the Silurian are suggested based on the presence of charcoal in the sedimentary record, the development of a long-stemmed biosphere begins in the Lower Devonian. Plants reaching heights of up to 2 to 3 meters were identified during the Eifelian (390 Ma; Stein et al. 2007). Trees, 8–10 m high, began to colonize the land towards the end of the Givetien (385 Ma), with, among others, giant ferns such as *Archeopteris* and *Cladoxyopsides* (Anderson et al. 1995). True large forests are likely to have become established towards the end of the Devonian (Frasnian; 380 Ma, Scott and Glaspool, 2006). As trees appeared and flourished, weathering of the continental silicates rapidly accelerated, while degassing from the solid Earth remained almost constant (Berner 2004). Soils developed along with root systems, increasing the acidification of the water in contact with the minerals as well as increasing the contact time between inland water and silicate rocks. This resulted in a rapid decrease in the partial pressure of atmospheric CO_2, from 2000 ppmv in the early Devonian to 1000 ppmv at the end (Foster et al. 2017). The extent to which the climate cooled as a result of this colonization is uncertain however because the change of the albedo of continental surfaces, following the replacement of bare soils by forests, compensates at least partially for the fall in atmospheric CO_2 (Le Hir et al. 2011). It is also interesting to note that the overall cooling of the climate may have been beneficial to the development of modern leaves, which are large in size and have many stomata, encouraging primary production on land to the detriment of more primitive plants. We can thus infer the establishment of a positive feedback between cooling and the colonization of land surfaces by ever more efficient plants.

Another mechanism that could explain the drop in CO_2 during the Devonian is an increase in the amount of CO_2 trapped in sediment, also as a result of colonization of land by continental plants. Indeed, the appearance of lignin in plant tissues from 410 Ma onwards increased the amount of organic carbon preserved in continental environments and on the margins. Lignin is indeed much more resistant to mineralization than marine organic matter. It was first thought that lignin appeared before the development of organisms capable of decomposing it, causing an increase in the burial of carbon and the reduction of CO_2, which could have contributed to the establishment of the glaciation of the Late Paleozoic (Nelsen et al. 2016). Nevertheless, recent studies have shown that decomposers evolved in parallel to lignin, which calls into question an 'organic' trigger for the Permo-Carboniferous glaciation (Nelsen et al. 2016).

Finally, the end of the Devonian is characterized by a mass extinction event, affecting tropical marine environments in particular. This event lasted approximately 1–3 million years, culminating at the Frasnian-Famennian boundary. It is accompanied by the deposition of anoxic sediments (black shales), accompanied by two positive excursions of the $\delta^{13}C$ of carbonate sediments. It has been proposed that these events are the consequence of the emergence of pulses as vascular plants colonized the land

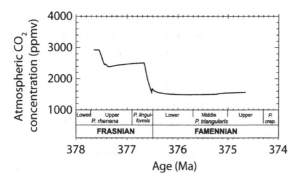

Fig. 27.8 Calculated levels of CO_2 in the atmosphere at the end of the Devonian

and of increased weathering bringing additional nutrients to the ocean and causing intense blooms in the oceanic biosphere. These would have led to increased oxygen consumption in the surface waters and the establishment of large-scale anoxia. Moreover, these anoxic events are correlated with second-order fluctuations of the sea level (a few tens of meters), thereby disturbing the weathering of carbonate platforms which alternate between being emerged and flooded. A numerical study, coupling a 1D energy balance climate model and a geochemical model of ocean and continental surfaces predicts a fall of 1500 ppmv across the Frasnian-Famennian boundary, over an interval of only 3 million years, as well as a cooling reaching more than 2 °C in the equatorial zones (Fig. 27.8; Goddéris et Joachimski 2004). These results are in agreement with the development of the first glaciers at the end of the Devonian.

The Permo-Carboniferous Glaciation

The cooling trend which started in the Devonian continued and reached its peak during the Permo-Carboniferous glaciation (330–270 Ma) also known as the '*Late Paleozoic Ice Age*' (Montañez and Poulsen 2013). It is the most important glacial event of the Phanerozoic. Debris carried by drifting icebergs reached the paleolatitude of 30° (Frakes et al. 1992). The temperature of the tropical waters was probably 2 °C below present values (Veizer et al. 2000). At the same time, the continents were covered with dense forests in low latitude areas (in the latitude band 15°N–15°S).

This major and prolonged cooling is traditionally considered to have been the final consequence of the colonization of continental surfaces by vascular plants, favoring continental weathering. A second major effect of this accelerated growth also occurred: the burial of organic carbon in continental environments reached a level during the Carboniferous never before attained during the whole Phanerozoic. Approximately 31×10^{15} mol Ma^{-1} of carbon were buried on the continents during the Carboniferous, while in the Devonian, this carbon sink was thirty times lower (Berner 2004). This is nearly double the estimated burial rate for the Cenozoic. The reason for the efficiency of this burial during the Carboniferous remains partly obscure, but it is probably linked to two interconnected factors: (i) the abundance of vegetation in certain zones, (ii) a low sea level during the Carboniferous (although it was high during the Devonian), creating low-lying coastal lands and extensive swamps. This efficient burial on the continents of reduced carbon depleted in ^{13}C, instigated the longest and most extensive positive excursion of the $\delta^{13}C$ in carbonate sediments (5‰, spread out over almost all of the Carboniferous and Permian).

Although this model of the forcing of the Carboniferous-Permian glaciation by biotic factors is commonly accepted, several elements today are causing us to question it. First, the colonization of the continents by vascular plants of large growth form was completed at the end of the Devonian, several tens of millions of years before the establishment of the Carboniferous-Permian glaciation (Davies and Gibling 2013). In addition, the increase in continental organic carbon burial seems relatively uncorrelated both with the $\delta^{13}C$ signal, which shows uniformly high values between 360 and 260 Ma (Fig. 27.5), and with the chronology of the colonization of continental surfaces by plants. A recent study coupling a climate model with a long-term carbon cycle model, highlighted the key role of tectonic changes in the entry into as well as the exit from the Carboniferous-Permian glaciation (Goddéris et al. 2017). The Hercynian orogen, by creating steep topographic slopes, would have allowed physical erosion to increase causing the destruction of the superficial soil formations, which had previously protected the bedrock from chemical weathering. The weathering of the newly exposed continental rocks would then have induced a drop in atmospheric CO_2 to levels allowing entry into glaciation (Fig. 27.9 carb). This scenario of tectonic forcing leading to glaciation is not in opposition to the biotic hypothesis, since the increased weathering of the continents would increase the nutrient flows to the ocean and thus necessarily favor an increase in primary productivity in the ocean.

The sequestration of organic carbon in continental sediments may have forced a decrease in CO_2 but it also led to a considerable increase in O_2 pressure in the atmosphere, from 15% at the end of the Devonian to 32% of the air content around 290 Ma (Berner 2004). This increase is also recorded in the $\delta^{13}C$ of fossil remains of continental vegetation. The fractionation of carbon isotopes reached 23‰ 290 Ma ago and was therefore about 5‰ higher than the average of the estimated values for the rest of the Paleozoic and Mesozoic. In modern plants, this fractionation is a function of the O_2/CO_2 ratio in the atmosphere. A value of 23‰ indicates a ratio of 1000. While the CO_2 pressure was around 300 ppmv

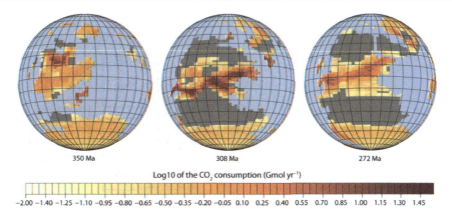

Fig. 27.9 carb—Paleogeographic pattern of CO_2 consumption by silicate weathering immediately before, during and after the Carboniferous-Permian glaciation. Weathering is inhibited by the development of a thick saprolith along the equator, except when steep slopes and high runoff maintain a high flow of erosion (308 Ma). The gray continental surfaces represent arid areas with no runoff and no weathering

(close to the current value), O_2 reached a pressure close to 30% of the total pressure of the atmosphere at ground level.

The End of the Paleozoic

The end of the Paleozoic is marked by the exit from the Permo-Carboniferous glaciation. The climate of the end of Permian is a much drier and warmer climate.

The transition from a cold mode to a warm mode is triggered during the final aggregation of the Pangea in the middle of the Permian. The formation of a supercontinent reduced the amount of precipitation on the continents, which partially inhibited the consumption of CO_2 by silicate weathering. In response to this imbalance between the volcanic source of CO_2 and the sinks through weathering and deposition of carbonates, atmospheric CO_2 accumulated, heating the system until the increased weathering due to increasing temperature compensated for the lack of precipitation, restoring the balance between sources and sinks of CO_2. This era is explored by much more powerful numerical models than the simple models used for the Paleozoic. The use of atmospheric general circulation models coupled with biogeochemical cycle models allows a more detailed investigation of the role of paleogeography.

Simulations carried out by atmospheric general circulation models coupled with biogeochemical cycle models calculate that 250 Ma ago, CO_2 pressure was only 2400 ppmv in response to the coming together of Pangea (Donnadieu et al. 2006) and that continental temperatures reached 19 °C on average. By comparison, today, a similar average, strongly influenced by the very cold Antarctic continent, would be only 6.7 °C. The end of the Permian is marked by a major negative excursion of the $\delta^{13}C$ of marine carbonates (−3‰ over about ten million years. See Fig. 27.5). This is interpreted as indicating a drastic reduction in the sequestration of continental organic carbon in response to the decline in productivity by the biosphere, due to a major reduction in precipitation (Berner 2004). Moreover, the end of the Permian is marked by intense volcanic activity, with the establishment of large fissural eruptions (traps) in Siberia. These three factors reinforced the global warming trend from the end of Permian onwards.

The Mesozoic

The three geological stages of the Mesozoic have long been considered to be typical examples of hot climates, especially the Cretaceous. Long-term carbon cycle models such as GEOCARB and all subsequent generations estimate very high values for atmospheric CO_2 pressure over the entire Mesoozoic, between 4 and 10 times the current value, supported by strong degassing of the solid Earth (Berner 2004). The only significant event was the appearance of flowering plants (angiosperms) in the Cretaceous, thought to further increase the efficiency of the consumption of atmospheric CO_2 through silicate weathering. This brought about a significant drop in CO_2 after 130 million years ago.

New perspectives on the climate trends of the Mesozoic have emerged recently and suggest other important long-term changes. The first notable feature (Fig. 27.10) is the steady increase in the $\delta^{13}C$ of carbonate sediments by about 1‰ from its lowest point in the Jurassic to the middle Miocene in the Cenozoic (Katz et al. 2005). This increase is typically explained by an increase in the ratio of organic carbon to total carbon buried in marine sediments, reaching ±20%, as the dislocation of Pangea increased the surface area available for the accumulation of organic carbon. Many sedimentological studies show a significant increase in

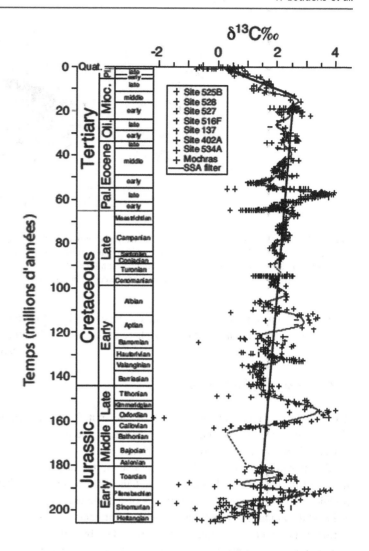

Fig. 27.10 $\delta^{13}C$ of carbonate sediments over the last 200 million years. (according to Katz et al. 2005)

the amount of organic carbon preserved on the shores of the Atlantic Ocean as it opened up during the Jurassic and Cretaceous periods.

This slow increase in the amount of buried organic carbon is also responsible for a net supply of oxygen to the ocean-atmosphere system (3×10^6 Gt O_2). This process should also consume atmospheric CO_2 by storing an increasing share of photosynthesized carbon in sediments. The paleothermostat (see Chap. 5, Volume 2) also allows the carbon cycle to be maintained close to equilibrium. Indeed, the cooling initiated by the increase in sequestration of organic carbon will be compensated for by a reduction in the consumption of CO_2 by the weathering of continental silicates. The response of CO_2 pressure in the atmosphere to this long-term evolution still needs to be quantified.

A second notable feature of the Mesozoic is the fragmentation of Pangea, which began as early as 250 Ma. An event of this magnitude has a major impact on the Earth's climate and the carbon cycle. Using a numerical model coupling an atmospheric general circulation model and a model of global biogeochemical cycles, Donnadieu et al. (2006) explored the climate and biogeochemical consequences of this dislocation. A configuration such as that of Pangea implies a weak continental runoff, due to its large continental nature. This causes a partial inhibition of continental silicate weathering. According to the theory of the paleothermostat, atmospheric CO_2 will increase, forcing the temperature to rise until the consumption of CO_2 by silicate weathering compensates once more for the degassing of the solid Earth. On the other hand, a configuration where the continents are dispersed leads to an increase in runoff and thus greater efficiency of silicate weathering. This results in increased CO_2 consumption due to weathering (Fig. 27.11), and the climate cools down until the paleothermostat is again balanced. At a constant rate of degassing of the Earth, the average annual temperature of the continents would have decreased from 19 °C at the beginning of the Triassic to 10 °C at the end of the Cretaceous (Donnadieu et al. 2006).

An interesting finding is that this simulated global cooling is not linear over the whole Mesozoic. The main episode of

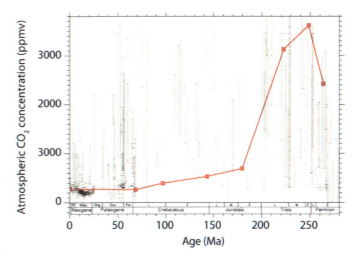

Fig. 27.11 Levels of atmospheric CO_2, calculated for the last 250 million years (red line), in response to continental drift. The vertical lines represent the available data

cooling and humidification of continental surfaces occurred during the Upper Triassic, between the Carnian and the Rhetian (237–201 Ma). The Lower and Middle Triassic are characterized by very high continental temperatures and extensive aridity (average annual continental runoff of only 23.5 cm/year), consistent with sedimentological data from the establishment of redbeds and massive deposits of evaporites. The corresponding atmospheric CO_2 pressures are close to 3000 ppmv, in agreement with the current reconstructions (Royer 2006), suggesting values between 2000 and 4000 ppmv for the same period.

An abrupt change occurred in the last stage of the Triassic (Rhetian 209–201 Ma), during which time 50% of the Mesozoic cooling occurred in the model. Goddéris et al. (2008) calculated CO_2 pressures of around 900 ppmv and global average temperatures lower by 4.6 °C. The CO_2 levels estimated based on the count of stomata on fossil leaves confirm these low levels of CO_2 between 500 and 1000 ppmv (Royer et al. 2004). The $\delta^{18}O$ measured on brachiopod shells show an increase of 0.8‰ between the Carnian and the Rhaetian, i.e. an overall cooling of more than 3 °C, (Korte et al. 2005) in line with modeling results. Similarly, sedimentological data clearly show an increase in moisture and a decrease in temperature during the Rhaetian (Fig. 27.12). Changes in clay mineralogy and in the conditions for pedogenesis are signs of the installation of cooler and wetter climate regimes around 209 Ma ago, during the Norian-Rhaetian transition (Ahlberg et al. 2002).

The causes of this rapid cooling may be found in the general drift of the Pangea towards the north. During the Middle Triassic, large continental areas were located in the southern zone of the inter-tropical divergence, a very arid area and therefore not conducive to weathering. The shift of the Pangea to the north brought these large areas into the humid equatorial zone, allowing increased atmospheric CO_2 consumption through increased runoff. Thus, the world became colder, but more humid, allowing the paleothermostat equilibrium to be maintained (equilibrium degassing of the solid Earth—silicate weathering), but at a lower level of CO_2 than in the middle of the Triassic. It is remarkable that these cooler conditions (but nevertheless up to 4 °C warmer than is currently the case on the continents) persisted after the Triassic, driven by the break-up of the Pangea rather than by its general latitudinal movement.

The Cenozoic

The overall climate evolution of the Cenozoic is better understood than that of the preceding epochs. Nevertheless, the causes of this evolution are still widely disputed. The climate history of the last 65 million years is that of a transition from the warmer Cretaceous climate, characterized by little or no polar ice caps, towards the current glacial climate.

The oldest stage of the Cenozoic, the Paleocene, is characterized by a climate similar to that of the late Cretaceous. The first break with the Mesozoic is at the Paleocene-Eocene transition (56 Ma, Fig. 27.3). This transition is marked by an extremely intense global warming. The deep waters of the ocean warmed up to about 5–7 °C in response to global warming and to a reorganization of ocean circulation. Similarly, the surface waters heated up by 8 °C (Thomas et al. 1999; Zachos et al. 2003; Sluijs et al. 2006). This warming, probably reinforced by the destabilization of methane hydrates in the sediments (McInerney and Wing 2011) was of short duration, spanning just 200,000 years.

This brief episode was followed by the Eocene climate, which lasted about 5 million years. (Fig. 27.3).

From the time of entry into the Middle Eocene, the climate began to cool down globally, leading to the appearance of small temporary ice sheets that developed on the Antarctic

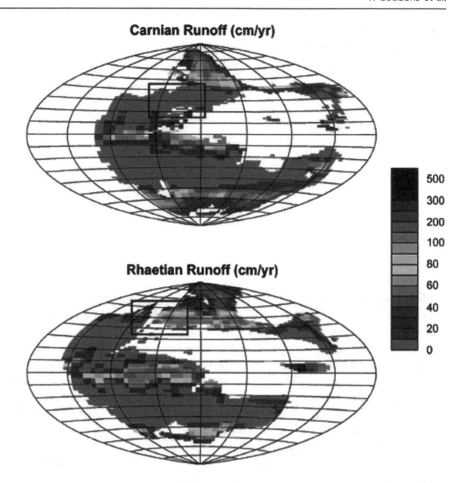

Fig. 27.12 Continental runoff calculated for the Carnian and Rhaetian, two stages of the Triassic, at 3122 ppmv of CO_2. The red rectangle shows the position of Europe and indicates increased humidification of the area between the Carnian and the Rhaetian

in the last phase of the Eocene (Zachos et al. 2008). At the Eocene-Oligocene transition (34 Ma) a distinct, cold climate pulse was experienced. The $\delta^{18}O$ data obtained from benthic foraminifera show a deep-water cooling of about 4 °C (Fig. 27.3). A permanent Antarctic cap began to emerge (Zachos et al. 2008).

The second step towards global cooling occurred at the Oligocene-Miocene boundary (23 Ma). But this cold episode is followed by the Miocene climate optimum between 23 and 15 Ma. The latter is particularly problematic because, according to the $\delta^{11}B$ and $\delta^{13}C$ data from alkenones, atmospheric CO_2 pressure should have been low, at around the present value, or even lower, between 200 and 300 ppmv (Pagani et al. 1999). The reasons for this climate optimum are still unknown.

From 15 Ma onwards, the climate cooled rapidly and the East Antarctic sheet developed. The last stage of cooling, 5 Ma ago, was marked by the establishment of the West Antarctic ice sheet (Fig. 27.3).

The reasons for this cooling are still in dispute. There are two opposing theories. According to one, the climate evolution of the Cenozoic was largely driven by the opening of key ocean passages and the closure of others. According to the other, it was the drop in CO_2 level that was responsible for this cooling. From 50 Ma onwards, ocean basins began to become established in the Drake Passage area, allowing shallow water exchanges between the Atlantic and Pacific Oceans. Isotopic analyses of neodymium in deep sediments indicate that around 41 Ma, the flow of exchanges between ocean basins intensified around Antarctica (Scher and Martin 2006). The intensification of the rate of expansion of the seabed in the Drake and the Tasmanian Passages 34 Ma ago, allowed the Antarctic circumpolar current to become established, definitively isolating the South Pole continent. This date coincides with a major development phase of the Antarctic ice cap.

The role of ocean passages in global climate change has been called into question by modeling studies, which tend to show that glaciation and the appearance of an ice cap over Antarctica are mainly associated with a decrease in the level of CO_2 Lefebvre et al. (2012), see also Chap. 3. It should be noted, however, that this premise is very poorly documented. Indeed, this is a period for which $\delta^{11}B$ data are non-existent. At best, we know that CO_2 levels were around 1000 ppmv 40 Ma ago, and about 300 ppmv 24 Ma ago. These data suggest decreasing levels of CO_2 although it is not possible to document the evolution precisely. Nevertheless, DeConto and Pollard (2003) have shown in a simulation that even where the Drake Passage is kept artificially closed, the decrease in atmospheric CO_2 concentration causes glaciation to occur over the Antarctic. Glaciation is simply delayed by about 2 to 3 million years.

Changes in the climate caused by the opening or non-opening of the Drake Passage will also affect the global carbon cycle, potentially increasing the climate response of the system. One possible consequence of the opening of the Drake Passage is an upheaval of the thermohaline circulation facilitating the formation of deep waters in the Antarctic and triggering the plunging of waters in the North Atlantic (Huber and Nof 2006). The result is a warming of the northern hemisphere of approximately 3 °C and a severe cooling of the Antarctic. As most of the continental area is located in the northern hemisphere, an increase in global consumption of CO_2 by continental silicate weathering is to be expected and thus a reduction in the amount of CO_2 in the atmosphere, reinforcing the cold climate mode being established (Elsworth et al. 2017). In conclusion, although the overall climate effect of the opening of the Drake Passage remains weak, there may have been positive feedbacks in the carbon cycle which substantially amplified the response. These have yet to be documented with precision.

A second driver of the evolution of climate also took place during the Cenozoic: the Himalayan orogen.

How orogeny affects the carbon cycle is complex. We have identified two effects: one is the chemical weathering of exposed silicates in the mountain chains, the other is the sequestration of organic carbon at the foot of the mountains. Take first the increase in weathering of continental surfaces through increased erosion. The development of glaciers, the alternating freezing and thawing patterns at high altitudes and steep slopes all favor the break-up of rocks and increase the area of contact with solutions. This results in increased weathering and increased consumption of CO_2. This increased weathering is seen in an increase in the erosion factor f_3 in Eq. (10), and the level of CO_2 is lowered until the weathering of the silicates again compensates for the degassing of the solid Earth (Goddéris and François 1996). Currently, 4×10^{12} kg yr^{-1} of suspended solids are transported to the ocean from the Himalayan zone, representing 17% of the world's erosion flow, whereas the ratio of the Himalayan surface to the total continental area is only 4%. It is therefore to be expected that a major orogen would considerably increase the consumption of atmospheric CO_2 through chemical weathering of the exposed rocks (increase in factor f_3). However, this result is not confirmed by current data of fluxes of dissolved elements in the rivers from the Himalayas. They suggest a modest consumption of 0.7×10^{12} mol yr^{-1} of CO_2 by weathering of Himalayan silicates, only 6% of the world total of 11.7×10^{12} mol yr^{-1}. One of the reasons for this low rate of chemical weathering may be that the erosion motor is too efficient in the Himalayas and that the discharge of the debris produced by mechanical erosion is too fast to allow the progress of effective chemical weathering. This would produce a weathering system which would be very limited by the very slow kinetics of the dissolution of minerals. In addition, the lithology is such that calcium silicates are scarce in the Himalayas and the weathering fluxes are mostly of sodium and potassium silicates. Since these chemical reactions do not lead to precipitation of carbonates, their effect on the carbon cycle is minimal over the long term (France-Lanord and Derry 1997).

However, the rate of sedimentation, which is extremely high in the Bay of Bengal, is responsible for the preservation of very large quantities of organic matter, of both continental and marine origin. It is estimated that the sequestration of carbon at the foot of the orogen is two to three times higher than the consumption of CO_2 by weathering of the Himalayan silicates. A recent study shows that 100% of the organic carbon of continental origin transported by the Himalayan rivers is preserved in the sediments of the Gulf of Bengal (Galy et al. 2007). France-Lanord and Derry (1997) estimated that the sedimentary organic carbon reservoir grew to 0.6×10^{12} mol yr^{-1}. This value is of a similar order of magnitude to estimates from numerical simulations, carried out using a carbon cycle model reversing the records of $\delta^{13}C$ in carbonates during the Cenozoic period (Goddéris and François 1996). The Himalayas consume carbon (Fig. 27.13), but in organic form, and therefore, they are, at least partially, responsible for the cooling of the climate during the Cenozoic. The quantification of the impact of this mechanism on atmospheric CO_2 has yet to be completed.

Abrupt Climate Events During the Phanerozoic

As well as defining the major climate modes of the Phanerozoic, recent efforts have defined episodes of rapid climate change that have punctuated the history of the Earth at a $100,000^{-year}$ scale. We outline below three of these events, discussing their causes.

The Callovian-Oxfordian Transition (Middle Jurassic-Upper Jurassic)

This was a brief cooling episode during the Jurassic. Such events occurred several times during the Jurassic and Cretaceous periods.

The $\delta^{18}O$ values measured in fish teeth and belemnites suggest an abrupt drop in temperature of 8 °C starting in the Upper Callovian and remaining until the middle Oxfordian. At the same time, boreal ammonite fauna invaded the

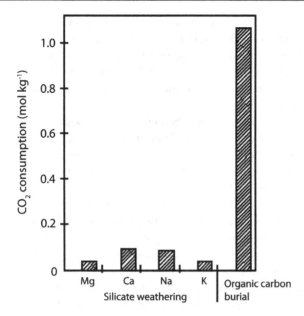

Fig. 27.13 Atmospheric CO_2 consumption by the Himalayas, per kg of sediment exported. On the left, the contribution of silicate weathering, reconstructed from the balance of each cation exported by the mountain range. On the right, the contribution associated with the sequestration of organic carbon in the Bay of Bengal (France-Lanord and Derry 1997)

Tethyan domain, suggesting widespread cooling at all latitudes (Dromart et al. 2003). The total duration of the event was about 3 million years. This episode is also marked by the reduction, by a factor of about 10, of carbonate deposits. The $\delta^{13}C$ values also suggest a positive excursion of about 0.5‰ over the same period. Finally, a drop in the sea level of several tens of meters is supported by evidence (Dromart et al. 2003), suggesting the establishment of temporary ice caps (Fig. 27.14).

The reasons for this cooling are not completely understood, nor, indeed are those of all the abrupt cooling episodes of the Jurassic and the Cretaceous. However, several avenues of enquiry have been opened. In particular, for this specific event, the amount of organic matter buried in marine sediments increased greatly during the middle Callovian just before the cooling. The percentage of organic matter increased from less than 1% during the early Callovian to 5%, and even 10% in the middle Callovian. This increased rate of burial of organic matter may have led to an increase in consumption of atmospheric CO_2 on a temporal scale sufficiently short, relative to the response time of the terrestrial paleothermostat so that it was unable to intervene as a stabilizer. This would have resulted in a reduction in CO_2 pressure which could have initiated the subsequent cooling. This cooling, accompanied by a drop in sea level caused by the development of glaciers on the coldest continents, was responsible for the near halting of carbonate sedimentation on the continental shelves.

Another hypothesis advanced recently connects the massive reduction in carbonate deposits on the continental shelves to the cooling episode. To date, it was assumed that the arrival of colder climate conditions resulted in a reduction in bioconstruction activity in the reef areas. Conversely, Donnadieu et al. (2011) suggest that the strong decrease in carbonate reef activity due to external causes (tectonic reasons for example) caused an accumulation of alkalinity in the oceans. In fact, this alkalinity continued to be supplied by rivers (continental weathering), whereas the alkalinity sink by deposition of carbonates was greatly reduced. The result was a massive dissolution of atmospheric CO_2 in the oceans. This caused CO_2 pressure to drop from 800 to 200 ppmv during crises in the carbonate production (lasting a few hundred thousand years), resulting in a global average cooling of 9 °C.

The Cretaceous-Tertiary Boundary, Meteorite and the Deccan Traps

The Cretaceous-Tertiary boundary (K–T), dating back to 66 Ma, has been studied in detail because it corresponds to a mass extinction event, which eliminated, among other species, the dinosaurs. Two major events occurred at the K–T boundary: the collision of a meteorite with the Earth and the establishment of the Deccan traps. This latter is a major magmatic event which may have had a major impact on the biosphere, but certainly had on the Earth's climate from 10^5 to 10^6 years.

Dessert et al. (2001) have simulated the impact of the Deccan Traps on the geochemistry and climate of the Earth. The total volume of lava put in place is 3×10^6 km^3, corresponding to the emission of 1.6×10^{18} mol of CO_2, or half the current carbon content of the exosphere. This emission could have occurred within a timeframe of about 10^5 years. This time scale is shorter than the response time of the geological carbon cycle. This is thus far beyond the capacity of the Earth's paleothermostat to respond. This produced a very rapid increase in the partial pressure of CO_2 which was increased to more than 3.5 times its initial level in 100 000 years (i.e. 1000 ppmv, assuming that the pre-disturbance CO_2 levels were at the pre-industrial level of 280 ppmv: Dessert et al. 2001) The global average temperature was thus increased by 4 °C a hundred thousand years after the establishment of the Traps.

Once the eruption ended and time passed, the Earth's paleothermostat could then take on its stabilizing role. The surplus CO_2 was slowly consumed by silicate weathering, which was itself accelerated by the increased greenhouse effect. Over 2 million years, the level of CO_2 returned to a stable level, one that was lower than the pre-disturbance level by 60 ppmv, corresponding to an overall cooling

Fig. 27.14 Summary of events at the Middle Jurassic-Upper Jurassic transition. On the left, the $\delta^{13}C$ of the carbonates, measured at different locations. On the right, layers rich in organic carbon, and the accumulation rate of carbonates

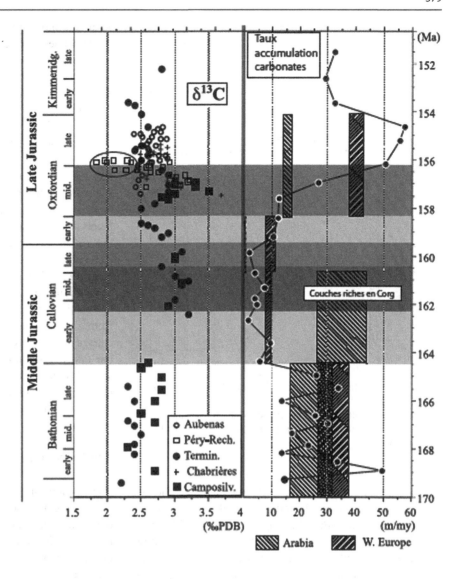

of ±0.5 °C compared to the temperature prior to the establishment of the Deccan Traps (Fig. 27.15). This cooling was the result of continental surfaces (continental shelf) that were very resistant to weathering, being replaced by 500,000 km² of fresh basaltic surfaces, eight to ten times more prone to weathering. Mathematically, it is the factor f_5 (Eq. 10) of the paleothermostat which has increased globally (the whole of the continental surfaces being slightly more vulnerable to weathering) and, for a degassing which returned to its pre-disturbance level, the CO_2 must stabilize at a lower level in order to correct the imbalance due to continental weathering.

The result of a magma episode, such as the establishment of continental traps, is initially a short-lived warming episode (10^5 years), followed by a global cooling that persists for several million years, as long as the basaltic surfaces exposed to the atmosphere are not entirely destroyed by weathering. A similar study carried out on Siberian traps (Permo-Triassic boundary) shows that the atmospheric CO_2 level stabilized a few million years after the end of the event at 750 ppmv below its pre-disturbance level of 4500 ppmv, which caused a global cooling of more than 1 °C.

At the time of the K–T limit, another major phenomenon occurred: a large meteorite with a diameter estimated at about ten kilometers collided with the Earth and fell into the Yucatan Peninsula. The impact created a large crater, identified by geophysics, which is currently buried under a thousand meters of sediment (the Chicxulub crater). This event was catastrophic, much shorter than the great fissure eruptions of the Deccan, which date from the same period, but which have had a prolonged impact for several hundreds of thousands of years.

The impact of the meteorite is easily identified because sediments from the K–T boundary are composed of a thin layer, rich in iridium, a very rare metal on Earth, a sign of contribution of cosmic origin. This layer also contains minerals (spinels) whose chemical composition indicates that they could not have been formed on Earth. Analysis of

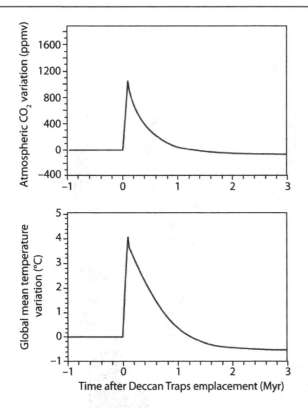

Fig. 27.15 Quantification of the impact of the establishment of the Deccan traps on the global average temperature and the level of atmospheric CO_2

the spinels showed that the meteorite had the chemical composition of carbon chondrites, with the particularity of being very rich in sulfur. The impact of such a collision was considerable. This impact, out of all proportion with current observations, is difficult to simulate because of the extent of the disturbance to the chemistry of the atmosphere. Thus, the simulations are based on assumptions made in the context of studies on a 'nuclear winter', where climatologists have calculated the impact on the global climate of a large-scale nuclear conflict. This can only be considered to be a very simplified approach.

The collision of a meteorite with the Earth has many consequences, although it is difficult to quantify them precisely:

- it releases an enormous quantity of aerosols (sulfates, nitrates) which reach the upper atmosphere where they can remain for several years;
- the aerosols cause an attenuation of about 50% of the solar radiation, resulting in a cooling of about ten degrees on the ground for a decade. Agronomists estimate that half the vegetation of the northern hemisphere could have been killed in the first years;
- the disruption is greatest if the impact occurs in the spring when the vegetation most needs solar radiation;
- an enormous quantity of water vapor is emitted into the atmosphere, which becomes charged with nitrates and sulfates, and falls back in the form of highly acid rain, toxic to plants;
- finally, chondrites contain many toxic heavy metals, particularly nickel, which inhibits chlorophyll activity.

We are obviously far from a full comprehension of all the events that marked the end of the Cretaceous, with the disappearance of many animal and vegetal species. Continental sediments testify to the appearance of widespread fires and to the pioneering return of the ferns, the most resistant of plants and the first to colonize the areas devastated by fire. In the ocean, sedimentological, biological and geochemical data show a considerable decrease in primary production by algae; this only returned to its former level after about five or six million years.

It is likely that the collision with the meteorite and fumes from the Deccan fissures both contributed to the major changes in the environment that marked the KT boundary, the first event through considerable sudden effects lasting several years and the second through geochemical effects that persisted over long periods relative to the time constants of the biosphere.

The Paleocene–Eocene Thermal Maximum (PETM)

The destabilization of methane hydrates (very active greenhouse gases) in sediments can cause climate fluctuations over short timescales (10^5 years). Several events of this type have been identified during the Phanerozoic, but the best documented is located at the Paleocene-Eocene transition. In the space of 20 000 years (Fig. 27.16), the $\delta^{13}C$ of the ocean decreased by 3‰, before returning to its initial value 240 000 years later (McInerney and Wing 2011). Over the same time, the temperature of the deep ocean waters increased from 5 to 7 °C. Similarly, a warming of 8 °C of surface waters was recorded. This warming is attributed to a sudden destabilization of methane hydrates in ocean sediments (Dickens 2003). The methane released by sediments is characterized by a $\delta^{13}C$ of −60‰. As a result, a flow of 2500 Gt of carbon spread over 20,000 years is sufficient to explain the observed isotopic excursion. This event caused a significant but temporary warming of the atmosphere. The $\delta^{13}C$ excursion is then reabsorbed over 200,000 years by the 'conventional' processes of the carbon cycle: continental weathering and sequestration in sediments.

The reason for the destabilization of the gas hydrates has yet to be explained. These can be released into the ocean and atmosphere if the water temperature rises or the pressure decreases. For example, regional eruptions occurred in the North Atlantic 55 Ma ago, shortly before the PETM

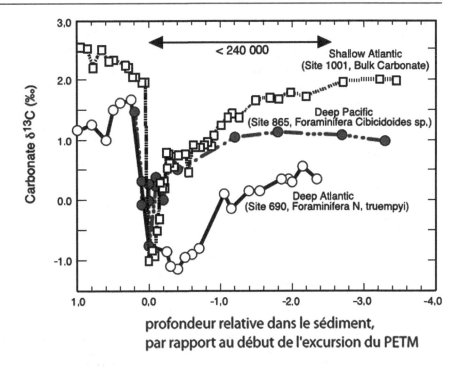

Fig. 27.16 Temporal evolution of $\delta^{13}C$ of benthic foraminifera at the Paleocene-Eocene transition at three distant oceanic sites. The arrow represents a period of 240,000 years

(McLennan and Jones 2006). The reduction in the height of the water column could have led to a drop in pressure and a destabilization of the methane hydrates in sediments. There may be a link between these regional eruptions and the PETM, although this assumption needs to be tested by more precise dating. Another hypothesis is that a significant change in ocean circulation could have significantly warmed the deep waters (by 4–5 °C), causing the destabilization of gas hydrates stored in sediments and initiating the PETM.

Conclusions

The reconstruction of climates on the scale of geological time is still open to discussion. The main difficulty lies in the fact that climate and biogeochemical cycles cannot be dissociated. There are numerous indicators of geochemical and climatic changes, but they are always difficult to interpret because of their indirect nature. The numerical models used are often very simple, taking a global average approach in most studies, which does not explicitly take into consideration the many parameters of the climate system. Yet ancient climates also represent an amazing testing-ground where new techniques can be developed and innovative ideas can be explored. In this field of study, climate models, initially developed to understand the evolution of current climate, are applied to extreme conditions, and the extent to which they are suitable is questionable. One limitation is the simplistic way the changing geographical configuration of the past is taken into account due to the lack of precise information. This is a climate factor of the highest importance and we are

not sure that complex models, such as atmospheric general circulation models coupled with ocean-atmospheric models, react correctly when boundary conditions are changed in such a drastic manner.

Nevertheless, the refining of analytical techniques and models, the process of trial and error, the successes and the failures allow us to discover a general history of the Earth's climate at the same time as multicellular organisms evolved. One of the major debates of recent years has been around the link between the level of atmospheric CO_2 and the evolution of climate during the Phanerozoic, which appear to have been decoupled during certain major events. Given the enormous uncertainties that exist in the reconstruction of CO_2 levels using isotopic methods or based on paleontological data (Royer 2006) and taking into account the uncertainties around isotope-based climate reconstructions, it is not possible to claim the existence or non-existence of a decorrelation between CO_2 level on one side and climate on the other. However, the use of a new generation of models that closely couples the carbon cycle with climate by taking into account the spatial variability of the processes suggests a coherence in the joint history of CO_2 and climate, in line with the major climate trends of the Mesozoic (Donnadieu et al. 2006). Similarly, the emergence of isotopic techniques allowing the reconstruction of the climate with increasingly fine latitudinal resolution makes it possible to reconsider the commonly studied events in the climate history of our planet. New consolidated images appear, in which atmospheric CO_2 is a key driver of climate change but is modulated by first order factors, such as the paleogeographic configuration, largely ignored for a long time, or the more or less periodic variations in the Earth's orbit, the role of which is explained in Chap. 7.

One of the major challenges for the near future is understanding the relationship between these climate and geochemical evolutions on the one hand and biological evolution on the other.

References

Ahlberg, A., Arndorff, L., & Guy-Ohlson, D. (2002). Onshore climate change during the Late Triassic marine inundation of the Central European Basin. *Terra Nova, 14*(4), 241–248.

Anderson, H. M., Hiller, N., & Gess, R. W. (1995). Archeopteris (progymnospermopsida) from the Devonian of Southern Africa. *Journal of the Linnean Society, 117*(4), 305–320.

Bergmann, K. D. et al. (2018). A 600-million-year carbonate clumped-isotope record from the Sultanate of Oman. *Journal of Sedimentary Research, 88,* 960–979

Berner, R. A. (2004). *The phanerozoic carbon cycle* (p. 150p). New York: Oxford University Press.

Bice, K. L., & Norris, R. D. (2002). Possible atmospheric CO_2 extremes of the Middle Cretaceous (late Albian-Turonian). *Paleoceanography, 17,* 1070.

Boucot, A. J., Xu, C., & Scotese, C. R. (2004). Phanerozoic climatic zones and paleogeography with a consideration of atmospheric CO_2 level. *Paleontological Journal, 38,* 115–122.

Breecker, D. O., Sharp, Z. D., & McFadden, L. D. (2010). Atmospheric CO_2 concentrations during ancient greenhouse climates were similar to those predicted for AD 2100. *Proceedings of the National Academy of Science, 107,* 576–580.

Came, R. E., et al. (2007). Coupling of surface temperatures and atmospheric CO_2 concentrations during the paleozoic era. *Nature, 449,* 198–201.

Caputo M. V. et al. (2008). Late devonian and early carboniferous glacial records of South America. In *Resolving the late Paleozoic ice age in time and space* (Vol. 441). Geological Society of America Special Papers, (pp. 161–173).

Dabard M. P. et al. (2015). Sea-level curve for the Middle to early Late Ordovician in the Armorican Massif (western France): icehouse third-order glacio-eustatic cycles.*Palaeogeography Palaeoclimatology Palaeoecology, 436,* 96–111.

Davies N. S., & Gibling, M. R. (2013). The sedimentary record of Carboniferous rivers: continuing influence of land plant evolution on alluvial processes and Palaeozoic ecosystems. *Earth-Science Reviews, 120,* 40–79.

DeConto, R. M., & Pollard, D. (2003). Rapid cenozoic glaciation of antarctica induced by declining atmospheric CO_2. *Nature, 412,* 245–248.

Dera, G. et al. (2011). Climatic ups and downs in a disturbed Jurassic world. *Geology, 39*(3), 215–218.

Dessert, C., et al. (2001). Erosion of deccan traps determined by river geochemistry: Impact on the global climate and the $^{87}Sr/^{86}Sr$ ratio of seawater. *Earth and Planetary Science Letters, 188*(3/4), 459–474.

Dickens, G. R. (2003). Rethinking the global carbon cycle with a large, dynamic and microbially mediated gas hydrate capacitor. *Earth and Planetary Science Letters, 213*(3–4), 169–183.

Donnadieu, Y., et al. (2011). A mechanism for brief glacial episodes in the mesozoic greenhouse. *Paleoceanography, 26.* https://doi.org/10.1029/2010pa002100.

Donnadieu, Y., et al. (2006). A GEOCLIM simulation of climatic and biogeochemical consequences of Pangea breakup. *Geochemistry Geophysics Geosystems, 7*(11). https://doi.org/10.1029/2006gc001278.

Dromart, G., et al. (2003). Ice age at the middle-late jurassic transition? *Earth and Planetary Science Letters, 213,* 205–220.

Elsworth, G. et al. (2017). Enhanced weathering and CO_2 drawdown caused by the latest Eocene strengthening of the Atlantic meridional overturning circulation. *Nature Geoscience, 10*(3), 213–216.

Finnegan, S., et al. (2011). The magnitude and duration of late ordovician-early silurian glaciation. *Science, 331,* 903–906.

Foster, G. L., Royer, D. L., Lunt, D. J. (2017) Future climate forcing potentially without precedent in the last 420 million years. *Nature Communications.* https://doi.org/10.1038/ncomms14845.

Frakes, L. A., Francis, J. E., & Syktus, J. I. (1992). *Climate modes of the phanerozoic.* Cambridge: Cambridge University Press.

France-Lanord, C., & Derry, L. A. (1997). Organic carbon burial forcing of the carbon cycle from Himalaya erosion. *Nature, 390,* 65–67.

Friedrich, O. et al. (2012). Evolution of middle to Late Cretaceous oceans - a 55 m.y. record of Earth's temperature and carbon cycle. *Geology, 40*(2), 107–110.

Galy, V., et al. (2007). Efficient organic carbon burial in the Bengal fan sustained by the Himalayan erosional system. *Nature, 450,* 407–410.

Ghosh, P., Garzione, C. N., & Eiler, J. M. (2006). Rapid uplift of the Altiplano revealed through C-13-O-18 bonds in paleosol carbonates. *Science, 311*(5760), 511–515.

Goddéris, Y., et al. (2008). Causal of casual link between the rise of nannoplankton calcification and a tectonically-driven massive decrease in the Late Triassic Atmospheric CO_2? *Earth and Planetary Science Letters, 267,* 247–255.

Goddéris, Y., & François, L. M. (1996). Balancing the cenozoic Carbon and Alkalinity Cycles: Constraints from isotopic records. *Geophysical research letters, 23*(25), 3743–3746.

Goddéris, Y. et al. (2014). The role of palaeogeography in the Phanerozoic history of atmospheric CO_2 and climate. *Earth-Science Reviews, 128,* 122–138.

Goddéris, Y. et al. (2017). Onset and ending of the late Palaeozoic ice age triggered by tectonically paced rock weathering. *Nature Geoscience, 10*(5), 382–385.

Hayes, J. M., Strauss, H., & Kaufman, A. J. (1999). The abundance of ^{13}C in marine organic matter and isotopic fractionation in the global biogeochemical cycle of carbon during the past 800 Ma. *Chemical Geology, 161,* 103–125.

Hodel, F. et al. (2018). Fossil black smoker yields oxygen isotopic composition of Neoproterozoic seawater. *Nature Communications, 9,* Article Number: 1453.

Huber, M., & Nof, D. (2006). The ocean circulation in the Southern Hemisphere and its climatic impacts in the eocene. *Palaeogeography, Palaeoclimatology, Palaeoecology, 231,* 9–28.

Joachimski, M. M. et al. (2004). Oxygen isotope evolution of biogenic calcite and apatite during the Middle and Late Devonian. *International Journal of Earth Sciences,93*(4): 542–553.

Katz, M. E. et al. (2005). Biological overprint of the geological carbon cycle. *Marine Geology, 217,* 323–338.

Kump, L. R., et al. (1999). A weathering hypothesis for glaciation at high atmospheric pCO_2 during the late ordovician. *Palaeogeography, Palaeoclimatology, Palaeoecology, 152,* 173–187.

Le Hir, et al. (2011). The climate change caused by the land plant invasion in the Devonian. *Earth and Planetary Science Letters, 310* (3–4), 203–212.

Lefebvre, V. et al. (2013). Was the Antarctic glaciation delayed by a high degassing rate during the Early Cenozoic? *Earth and Planetary Science Letters, 371–372,* 203–211.

Lenton, T. M. et al. (2012). First plants cooled the Ordovician. *Nature Geoscience, 5,* 86–89.

Loi, A. et al. (2010). The Late Ordovician glacio-eustatic record from a high-latitude storm-dominated shelf succession: The Bou Ingarf section (Anti-Atlas, Southern Morocco). *Palaeogeography Palaeoclimatology Palaeoecology, 296*(3–4), 332–358.

McElwain, J. C., Wade-Murphy, J., Hesselbo, S. P. (2005). Changes in carbon dioxide during an oceanic anoxic event linked to intrusion into Gondwana coals. *Nature, 435*(7041), 479–482.

McInerney, F. A., & Wing, S. L. (2011). The paleocene-eocene thermal maximum: A perturbation of carbon cycle, climate, and biosphere with implications for the future. *Annual Review of Earth and Planetary Sciences, 39*, 489–516.

Mountanez, I. P., & Poulsen, C. J. (2013). The late paleozoic ice age: An evolving paradigm. *Annual Review of Earth and Planetary Sciences, 41*, 629–656.

Nardin, E., et al. (2011). Modeling the early paleozoic long-term climatic trend. *Geological Society of America Bulletin, 123*, 1181–1192.

Nelsen, M. P. et al. (2016). Delayed fungal evolution did not cause the Paleozoic peak in coal production. *Proceedings of the National Academy of Science, 113*(9), 2442–2447.

Pagani, M., Zachos, J. C., Freeman, K. H., Tipple, B., & Bohaty, S. (2005). Marked decline in atmospheric carbon dioxide concentrations during the Paleogene. *Science, 309*, 600–603.

Page, A. A. et al. (2007). Were transgressive black shales a negative feedback modulating glacioeustasy in the Early Palaeozoic icehouse? In *Deep-time perspectives on climate change: marrying the signal from computer models and biological proxies*. pp. 123–156.

Pohl, A. et al. (2016). Glacial onset predated Late Ordovician climate cooling. *Paleoceanography, 31*(6), 800–821

Pucéat, E., et al. (2010). Revised phosphate-water fractionation equation reassessing paleotemperatures derived from biogenic apatite. *Earth and Planetary Science Letters, 298*, 135–142.

Pucéat, E. et al. (2003). Thermal evolution of Cretaceous Tethyan marine waters inferred from oxygen isotope composition of fish tooth enamels. *Paleoceanography, 18*,(2), Article Number: 1029.

Rasmussen, C. M. O. et al. (2016). Onset of main Phanerozoic marine radiation sparked by emerging Mid Ordovician icehouse. *Scientific Reports, 6*, Article Number: 18884.

Raymo, M. E. (1991). Geochemical evidence supporting T.C. chamberlin's theory of glaciation. *Geology, 19*, 344–347.

Royer, D. L. (2006). CO_2-forced climate thresholds during the phanerozoic. *Geochimica and Cosmochimica Acta, 70*, 5665–5675.

Royer, D. L., Berner, R. A., & Beerling, D. J. (2001). Phanerozoic atmospheric CO_2 change: Evaluating geochemical and paleobiological approaches. *Earth-Science Reviews, 54*, 349–392.

Rubinstein, C. V. et al. (2010). Early middle ordovician evidence for land plants in Argentina (eastern Gondwana). *New Phytologist, 188* (2), 365–369.

Scher, H. D., & Martin, E. E. (2006). Timing and climatic consequences of the opening of drake passage. *Science, 312*, 428–431.

Scott, A. C., & Glasspool, J. (2006). The diversification of Paleozoic fire systems and fluctuations in atmospheric oxygen concentration. *Proceedings of the National Academy of Science, 103*(29), 10861–10865.

Shaviv, N. J., & Veizer, J. (2003). Celestial driver of phanerozoic climate? *GSA Today, 13*(7), 4–10.

Sluijs, H. et al. (2006). Subtropical arctic ocean temperatures during the Palaeocene/Eocene thermal maximum. *Nature, 441*(7093), 610–613.

Stein, W. E. et al. (2007). Giant cladoxylopsid trees resolve the enigma of the Earth's earliest forest stumps at Gilboa. *Nature, 446*(7138), 904–907.

Thomas, D. J. et al. (1999). New evidence for subtropical warming during the late Paleocene thermal maximum: stable isotopes from the Deep Sea Drilling Project Site 527, Walvis Ridge. *Paleoceanography, 14*, 561–570.

Trotter, J. A. et al. (2008). Did cooling oceans trigger Ordovician biodiversification? Evidence from conodont thermometry. *Science, 321*, 550–554

Veizer, J., et al. (1999). $^{87}Sr/^{86}Sr$, $\delta^{13}C$ and $\delta^{18}O$ evolution of phanerozoic seawater. *Chemical Geology, 161*, 59–88.

Veizer, J., Goddéris, Y., & François, L. M. (2000). Evidence for decoupling of atmospheric CO_2 and global climate during the phanerozoic eon. *Nature, 408*, 698–701.

Zachos, J. C. et al. (2003). A transient rise in tropical sea surface temperature during the Paleocene-Eocene Thermal Maximum. *Science, 302*(5650), 1551–1554.

Zachos, J. C., Dickens, G. R., & et Zeebe, R. E. (2008). An early Cenozoic perspective on greenhouse warming and carbon-cycle dynamics. *Nature, 451*(17). https://doi.org/10.1038/nature.

Zhuravlev, A. Y., & Riding, R., (2001). The ecology of the Cambrian radiation - Introduction. Perspectives in paleobiology and Earth history series, pp. 1–7.

Climate and Astronomical Cycles

Didier Paillard

A Little History

From the Discovery of Ice Ages to the First Climate Theories

The scientific notion that climate could evolve over time first appeared along with the birth of paleoclimatology in the nineteenth century when the existence of ice ages was discovered. Although the scientists of the time were aware that the Earth had undergone many upheavals, notably through the successions of various animal and vegetable fossil species, it was only when evidence of previous glacial periods emerged that the idea of a changing climate really took shape. The traces left behind by the movement of glaciers during the last glacial period: moraines, erratic boulders, glacial striations and features, were clearly identified in many regions of Europe and North America and promoted the idea that the climate, at least in the northern hemisphere, may have been considerably colder in the past. The nineteenth-century scientists' point of view on glaciations was quite different from ours, deeply rooted as it was in a 'catastrophic' perspective on the evolution of the Earth, which was at that time imagined to have been punctuated by 'deluges' and other cataclysms, as highlighted in the excerpt below.

> The appearance of these large sheets of ice must have led to the annihilation of all organic life on the surface of the earth. The soil of Europe, formerly ornamented with tropical vegetation and inhabited by troops of great elephants, enormous hippopotamuses and gigantic predators, was suddenly buried under a vast mantle of ice covering the plains, lakes, seas and plateaus. The movement of a powerful creation was succeeded by the silence of death. The springs dried up, the rivers ceased to flow, and the rays of the sun, rising on this frozen beach (that's if they reached it), were only greeted by the whistles of the northern winds and the thunder of the crevasses that opened on the surface of this vast ocean of ice.
> (L. Agassiz, *Studies on glaciers*, 1840)

For geologists of that era, climate changes were primarily related to changes in topography. Although the continental drift was not yet well established, the scientists tried to explain their field observations by means of vertical movements of the continents: uplift of the mountains, erosion, change in sea level etc. Under the influence of Charles Lyell, these changes came to be seen as slow and progressive. The assumption of catastrophism faded away to be replaced with gradualism which presupposes that the modifications of the Earth's surface obey the physical laws that now apply but over immensely long durations. This new point of view was opposed to any external (especially heavenly) influence, which no doubt explains the reluctance of geologists when faced with the first astronomical theories of glaciations:

> But though I am inclined to profit by Croll's maximum eccentricity for the glacial period, I consider it quite subordinate to geographical causes or the relative position of land and sea and abnormal excess of land in polar regions.
> (C. Lyell to C. Darwin, 1866)

The idea that the climate stems above all from geography and topography has indeed merit. One need only look at the etymology of the word climate (κλιμα = inclination, i.e. the height of the Sun above the horizon, in other words, the latitude of the location) or look for 'climatology' in an academic organization chart to be convinced that this idea is still valid. The physical principles underlying the functioning of the climate system were also updated in the nineteenth century, notably by Joseph Fourier, who established the laws of heat diffusion, explaining how heat is redistributed by the surface fluids of the atmosphere and the ocean, and also discussed the essential role of the greenhouse effect:

D. Paillard (✉)
Laboratoire des Sciences du Climat et de l'Environnement, LSCE/IPSL, CEA-CNRS-UVSQ, Université Paris-Saclay, 91190 Gif-sur-Yvette, France
e-mail: Didier.Paillard@lsce.ipsl.fr

© Springer Nature Switzerland AG 2021
G. Ramstein et al. (eds.), *Paleoclimatology*, Frontiers in Earth Sciences,
https://doi.org/10.1007/978-3-030-24982-3_28

> The temperature can be increased by the interposition of the atmosphere, because the heat finds less obstacle to penetrate the air, being in the state of light, than it finds it to pass through the air when converted into dark heat.
>
> (J. Fourier, 1824)

It is in this context that the two main physical theories are presented, which are still relevant today, and which make it possible to explain the existence of glacial periods: the astronomical theory and the variations in the atmospheric concentration of CO_2.

From Adhémar to Milankovitch: The Role of Insolation

Although the notion that the climate is influenced by the stars has undoubtedly been around for a very long time, the first astronomical scientific theory of the ice ages was formulated by Joseph Adhemar in 1842. It was based simply on common sense: since ancient times, astronomers have highlighted the 'three movements of the Earth': the diurnal cycle, the annual cycle, and the precession of the equinoxes (Hipparchus, about 130 before J.-C.). While it is clear that the annual and diurnal cycles generate temperature variations, the same must be true of the third movement of the Earth. As will be explained a little later, the precession of equinoxes has the consequence of modifying the position of the perihelion (point of the Earth's orbit closest to the Sun) in relation to the seasons: today, the Earth is closest to the Sun around January 4 but this date changes slowly to cover the whole of the year over about 21,000 years. In contrast to today, 10,500 years ago, the Earth was far from the Sun in January and close to it in July. Adhemar suggested that this mechanism could modify the climate. More specifically, the winters of the northern hemisphere now occur when the Earth is close to the Sun and, conversely, those of the southern hemisphere when the Earth is far from the Sun. Adhemar proposed that this explains the absence of a large ice cap in the north, due to milder and shorter winters, and conversely, the presence of a large Antarctic cap. The situation would have been exactly the opposite 10,500 years ago, which allowed him to explain the periods of great glacial expansion that had just been revealed by geologists.

Adhemar's theory was criticized for many reasons, some largely unfounded, but it was on the very foundations of his theory that his detractors, Charles Lyell and Alexander von Humbolt, would find compelling arguments. In fact, the mechanism of precession works in an anti-symmetrical way between the poles, but also between the seasons. It can easily be shown that although, for example, less energy is received in winter, this is compensated for by an equivalent excess of energy received in summer. If the seasonal contrast varies with the precession, the full complement of energy received does not change. How then could this have any effect on climate? According to Adhemar, although the astronomical forcing is effectively anti-symmetric with respect to the seasons and zero for the annual average, the climate processes are probably not.

In 1864, James Croll clarified this concept. According to him, the accumulation of snow occurs chiefly in winter, with melting occurring in summer. Croll emphasized the role of the winter accumulation which essentially supports Adhemar's argument: longer or colder winters favor a greater accumulation of ice allows the initiation of a glaciation. In addition, aware of the progress that had been made in celestial mechanics notably by Pierre Simon de Laplace and Urbain le Verrier, Croll went on to introduce the effect of variations in the eccentricity of the Earth's orbit. These variations modulate the intensity of seasonal contrasts. Indeed, in the case of a circular orbit, the effect of the precession on climate is zero, since there would be neither a perihelion (nearest point) nor an aphelion (farthest point). The greater the eccentricity, the greater the climate effects of the precession. Croll therefore linked the great glaciations with eccentricity maxima. Hence, he pushed back the estimate of the last glaciation to 80,000 years ago, and suggested an even more intense glaciation 240,000 years ago. Although Croll proposed a much more solid and evolved astronomical theory, he did not succeed in convincing the scientific community of his time. However, the interglacial-glacial alternations discovered in some sediments argued in favor of a more or less periodic mechanism. Yet the first dating of elements available from that time, extrapolating the rates of erosion or counting lake varves, indicated a much more recent glaciation. Croll went on to introduce a third important astronomical parameter for the calculation of the variations of the solar energy received in a given place: the obliquity of the terrestrial axis, that is to say its inclination with respect to the plane of the Earth's orbit. However, obliquity has little effect on winter insolation, and therefore was of little importance in Croll's theory.

Milankovitch (1941) formulated the astronomical theory which still applies today. The main criticism that can be levelled at Croll's theory is that it considered winter to be the most pertinent season. This objection was already expressed by Joseph Murphy during Croll's time, as observations of the eternal snow and mountain glaciers showed that summer melting had much more impact on the ice mass balance than snow accumulation. However, the precession was often advanced to explain the current asymmetry of temperatures between the northern hemisphere and the southern hemisphere, and the presence of Antarctica, since today, the austral winter is longer and further from the Sun than the northern winter. Croll's theory was therefore probably based on this nineteenth century misconception of the current climate.

Milankovitch resolved this problem by deducing that the current north-south asymmetry was linked to geography and not to astronomical forcing. The critical season for the evolution of ice caps is therefore summer, which completely reverses the reasoning of Adhemar and Croll. The obliquity of the Earth's axis, that is to say its inclination with respect to the plane of Earth's orbit, becomes the most important parameter for glacial-interglacial evolution. The foundations of modern astronomical theory were thus laid down.

From Tyndall to Arrhénius: The Role of Carbon Dioxide

The importance of the role of the greenhouse effect was well understood by the nineteenth century, notably through the work of Joseph Fourier. As early as 1845, Jacques Joseph Ebelmen, a French chemist, first suggested that changes in the atmospheric concentration of CO_2 might have consequences for the climate (Bard 2004). Indeed, by focusing on the chemistry of minerals, he established the bases of carbon geochemistry: a source primarily of volcanic origin and sinks related to the erosion of silicates and the burial of the organic material. Since all these processes appear to be disconnected, it seems unlikely that the atmospheric concentration of CO_2 would be constant over geological time. In 1861, John Tyndall further authenticated the theory of the greenhouse effect. By measuring the absorption and infrared emission of the various gases present in the air, he demonstrated that nitrogen or oxygen are essentially transparent to infrared rays and that the greenhouse effect of our planet is caused primarily by gases in very small quantities, in large part by water vapor, but also carbon dioxide, methane, nitrous oxide and ozone. Tyndall then suggested that all of the climate changes discovered by geologists, including ice ages, could be explained by changes in the levels of atmospheric greenhouse gases.

But it was the Swedish chemist Svante Arrhenius (1896) who managed to calculate the effect of carbon dioxide on the climate, in an attempt to explain the ice ages. Based on geological data on moraine positions during glacial periods, he estimated a cooling of 4 or 5 °C and calculated that this could be explained by a reduction of about 40% in the atmospheric concentration of CO_2. The measurement of pCO_2 from this glacial period, carried out on air bubbles from ice cores taken from Antarctica in the years 1980–1990, confirmed his calculations: the atmospheric concentration of CO_2 was indeed 30% lower during the ice age. This scientific prediction, nearly a century before it could be confirmed through observation, is a good illustration of the essential role of greenhouse gases in the deployment of the Quaternary cycles. Arrhenius also considered that future global warming would be linked to anthropogenic CO_2 emissions. He calculated an increase in global temperatures of about 5 °C for a doubling of CO_2, a figure surprisingly close to the most recent estimates of about 3.5 °C, another prediction likely to become true in the not too distant future.

All these arguments were underscored by the American geologist Chamberlin, who highlighted the succession of at least five glacial stages in the United States. According to Chamberlin, there was a sort of oscillation between the climate and the geochemistry of the Earth: a decrease in CO_2 leading to cooling, with the effect of reducing carbon sinks on Earth by reducing the burial of organic matter as well as the erosion of silicates. This would then lead to a gradual increase in atmospheric CO_2 until it switches over to the opposite situation. Chamberlin therefore attempted to formulate an 'internal oscillation' to explain the succession of ice ages, without calling on a '*Deus ex machina*' such as the astronomical forcing.

It is interesting to note that these two opposing theories of the glacial periods have existed since the middle of the nineteenth century and are still largely valid today: it remains to be understood how they relate and complement each other.

Astronomical Parameters and Insolation

Before proceeding further, it is useful to review the various astronomical parameters that influence the energy received at the top of the Earth's atmosphere, which we call 'insolation'.

Eccentricity

According to Kepler's first law, the Earth's orbit is an ellipse. This is characterized by a major parameter, the semi-major axis, often denoted a, by a shape or flattening parameter, the eccentricity, often denoted e, and also by three parameters defining the position of this ellipse in space, two of which define the orbital plane (the inclination i with respect to a reference plane and the longitude of the ascending node Ω defined by the intersection of these two planes), and another to define the absolute position of the perihelion (the longitude π). In fact, as soon as the system is made up of three material bodies (the Sun with two planets) or more, the movement is no longer strictly an ellipse, and there is no analytical solution to the problem of celestial mechanics at N bodies, for $N > 2$. It is therefore appropriate to calculate the perturbations or the approximate numerical resolutions. The notion of terrestrial orbit nevertheless still makes sense because the perturbations are secondary. It is therefore useful to reason in terms of elliptical orbit, which deforms and moves over time.

The perturbations induced by the other planets do not modify the semi-major axis of the ellipse a, only the terrestrial trajectory, i.e. the eccentricity e and the orientation

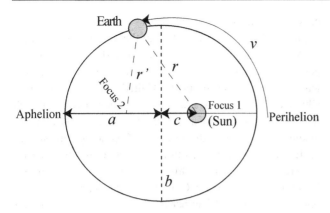

Fig. 28.1 An ellipse can be defined as the locus of the points whose sum of distances to the two foci is constant: $r + r' = 2a$. The eccentricity is defined as the ratio $e = c/a$. The semi-minor axis b is therefore given by the theorem of Pythagoras: $b = a(1 - e^2)^{1/2}$

parameters of the ellipse (i, Ω, π). So a will be a constant (at least over hundreds of millions of years). Moreover, the orientation of the ellipse in space does not have a direct consequence on the solar radiation received by the Earth. The only orbital parameter which is liable to modify the insolation is therefore the eccentricity e.

The eccentricity e is defined by the ratio between the distance from the focus to the center of the ellipse c, and the semi-major axis a, as shown in Fig. 28.1. Although today it is 0.0167 (i.e. a flattening of 1.67%), it has varied between values of almost-zero and 0.06, with periodicities appearing to be around 100,000 and 400,000 years.

Eccentricity is the only parameter capable of modifying the average annual energy received by the Earth. As the semi-major axis is constant, the average distance between the Earth and the Sun depends on its eccentricity. The second law of Kepler (conservation of angular momentum) is written as:

$$r^2 \frac{dv}{dt} = \frac{2\pi a^2 \sqrt{1 - e^2}}{T}$$

The 'solar constant' S_0, defined as the average energy received by the Earth, is deduced by integrating it into a complete orbit:

$$S_0 = \frac{S}{T} \int_0^T \frac{a^2}{r^2} dt = S \int_0^{2\pi} \frac{dv}{2\pi \sqrt{1 - e^2}} = \frac{S}{\sqrt{1 - e^2}}$$

The average annual energy received by the Earth, S_0, depends on the energy S (assumed here to be truly constant) received at the distance a from the Sun and on the eccentricity e. A higher eccentricity leads to a more flattened orbit, on average closer to the Sun, and thus to greater overall solar energy being received by the Earth. However, the variations remain very small, since for $e = 0.06$, the maximum is calculated as $S_0 = 1.0018\,S$, i.e. an increase of only 0.18%.

These tiny variations have almost no effect on the climate. They are of the same order of magnitude as the observed variations in solar flux S over the 11-year cycle, which generate temperature variations of the order of 0.1 °C. On the other hand, as we will see a little later, it is through the modulation of the effects of precession that eccentricity plays an essential role in climate.

It is relevant to note that the solar system is chaotic. This means that the calculation of the orbital parameters in general, and in particular that of eccentricity, is only possible for instants not too far from the current period. In fact, errors increase exponentially with time, and it is not possible to arrive at an estimation beyond a certain time. For the eccentricity, this timeframe is only around 20–30 million years, which is very short compared to the age of the Earth. Beyond that, although variations in eccentricity remain similar in nature (with identical periodicities of approximately 100,000 and 400,000 years), it becomes impossible to say whether the eccentricity was minimal (close to zero) or maximum (close to 0.06) 600 million years ago. In other words, the phase of the oscillations becomes theoretically unreliable over the long term (Laskar et al. 2004).

Obliquity

In addition to the parameters of the Earth's orbit, the position of the rotational axis of the Earth relative to the orbital and ecliptic planes must also be taken into account. This position is given by two axial parameters; the obliquity, denoted ε, representing the inclination of this axis relative to the ecliptic; and the precession of the equinoxes, which indicates its absolute position compared to the stars. The position of the Earth's axis is modified by the differential attraction of the Moon (and, to a lesser extent, the Sun) at the equatorial bulge. In fact, our planet is slightly flattened, because of the Earth's rotation, and the gravitational pull of the Moon towards the Earth is therefore not exactly symmetrical. The equatorial bulge, at an incline relative to the lunar orbit, is subject to attraction forces which create a torque on the Earth's axis and modify its orientation. Contrary to orbital parameters, such as eccentricity, which depend only on point mechanics, the axial parameters (obliquity and precession) depend on the shape of the Earth, which introduces new sources of error and uncertainty. The timeframe beyond which the calculation of the axial parameters becomes impossible is therefore probably shorter than for the eccentricity. The calculations become more complicated further back than a few million years if the shape of the Earth changes slightly under the influence of glaciations due to the enormous volumes of ice accumulating on the continents of the northern hemisphere at the glacial maxima. It has been suggested that this could have consequences for the calculation of axial parameters further back than only a few million years. Similarly, internal convection in the Earth's mantle potentially induces changes in the distribution of masses which are difficult to take into account in these astronomical calculations. The phase of the obliquity, like that of the precession, is therefore subject to caution when extrapolating calculations for the distant past or future beyond about ten million years. Moreover, the periodicity of these axial parameters depends on the Earth-Moon distance, but the lunar recession is rather poorly constrained in the distant geological past, and the frequencies of the obliquity and precession also become uncertain.

The obliquity today is 23° 27′, which defines the latitude of the polar circles (67° 33′ north and south) and the tropics (23° 27′ north and south). This value oscillates between extremes of around 21.9° and 24.5°, with a periodicity of around 41,000 years. It is clear that any change in obliquity will have consequences for the climate by altering the size of the polar and tropical areas. Thus, on the island of Taiwan, in the county of Chiayi (Jia-Yi), for almost a century, there has been a monument marking the Tropic of Cancer. However, the current decline in obliquity, at a rate of 0.46 arc-seconds per year, means that there has been a displacement of the tropics of 14.4 m per year, i.e. 4 cm per day and thus more than 1 km since the monument was first erected. The Taiwanese have therefore regularly built new monuments to follow the southward movement of the Tropic of Cancer.

Although the average global incident radiation has not changed, its geographical distribution depends on the obliquity. To be precise, if one calculates the average annual solar incident radiation, it is found that this depends mainly on the obliquity and, to a lesser extent, on the eccentricity as mentioned above. An increase in obliquity ε results in an increase in insolation at high latitudes and a decrease in the tropics. At the poles (north and south) and at the equator, incident is calculated as follows:

$$W_{Year}(\text{pole}) = \frac{S}{\pi\sqrt{1-e^2}} \sin \varepsilon;$$
$$W_{Year}(\text{equator}) = \frac{2S}{\pi^2\sqrt{1-e^2}} E(\sin \varepsilon)$$

where $E(x) = E(\pi/2, x)$ is the secondary complete elliptic integral. For a eccentricity e of zero, when ε goes from 21.9° to 24.5°, changes of around 1%, i.e. 18 W/m² are obtained for W_{Year} (pole) and changes of around 0.4%, i.e. 5 W/m² are obtained for W_{Year} (equator). It should be noted that for very large obliquities (for $\pi \sin \varepsilon > 2 E(\sin \varepsilon)$, in other words, for $\varepsilon > \varepsilon c = 53,896°$), the poles receive a higher annual energy average than the equator. This is currently the case on Uranus and Pluto.

Moreover, the phenomenon of the seasons is directly linked to the obliquity, and it is all the more marked when the obliquity is large. For example, for a circular orbit ($e = 0$), we obtain the following expression of daily insolation at the poles during solstices:

$$W_{summer}(\text{pole}) = S\sin\varepsilon; W_{winter}(\text{pole}) = 0.$$

If insolation at the winter solstice remains zero, the summer solstice will vary between W_{summer}(pole) = 0.373 S and W_{summer}(pole) = 0.415 S, when the Quaternary ε goes from 21.9° to 24.5°, i.e. an increase of around 4%, that is to say more than 50 W/m². This is far from negligible.

In addition, it is essential to notice that, contrary to the precession which we detail below, the effect of the obliquity on the insolation is symmetrical relative to the equator. This is an important aspect of Milankovitch's theory: contrary to the theories of Croll and Adhemar, which involve winter insolation (mainly dependent on the precession), Milankovitch's theory is based on summer insolation. which strongly depends on the obliquity. On a planet with symmetrical topography, this would imply the presence of ice caps oscillating largely in phase in both hemispheres. Of course, the distribution of continents on Earth is not symmetrical and several other factors will also affect how climates are distributed on Earth.

Precession of the Equinoxes and Climate Precession

In addition to the alternation of day and night and the phenomenon of seasons which have been known since the dawn of time, astronomers have observed since ancient times a slow drift of the polar axis relative to the celestial sphere. This discovery is generally attributed to Hipparchus (130 B.C.) who estimated the drift at approximately 1° per century (that is to say, a periodicity of about 360 centuries). This estimate is remarkable since we now know that the precession does have a periodicity of 25,765 years (1.397° per century). It is also likely that the Egyptians and the Mesopotamian astronomers were already well aware of this phenomenon because, centuries earlier, they identified the position of the celestial pole as well as the constellations of the zodiac. changed the orientation of certain temples to 'follow' this movement. As Nicolas Copernic already noted, this is the 'third movement' of the Earth, the first two corresponding to the day and year. It was therefore logical for Adhémar to focus on the consequences for the climate of this 'third movement'.

Nevertheless, a distinction should be made between the precession of the equinoxes and the climate precession. Indeed, the 'absolute' position of the axis of the Earth,

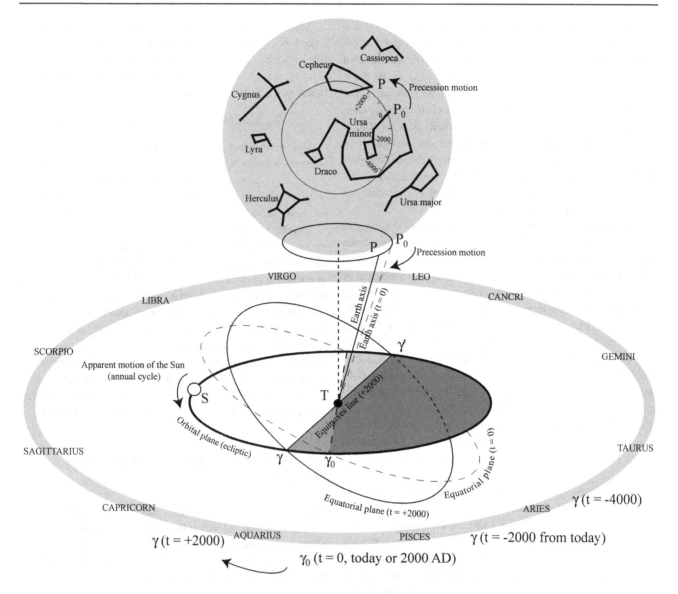

Fig. 28.2 The precession of the equinoxes corresponds to the rotational motion of the Earth's axis (T). The pole thus moves relative to the stars (from P_0 today, pointing towards the polar star, to P in the future in the constellation of Cepheus). The same applies to the equinox line $\gamma\gamma'$ (intersection of the orbital plane and the equatorial plane). The current point γ_0 is in the Pisces constellation and is drifting towards the constellation of Aquarius

relative to the stars, has, in itself, no effect on the climate. However, when this axis is oriented differently, so too is the equatorial plane of the Earth. The equinox line is defined as the intersection of the equatorial plane and the ecliptic (orbital) plane. The precession of the equinoxes (from which the origin of the name) corresponds to a drift of the equinoxes relative to the constellations, as shown in Fig. 28.2.

Thus, while the Sun is now in the constellation of Pisces on the day of the spring equinox (thus defining what astronomers call the vernal point γ), it was in Aries of the time of the Greeks and in Taurus at the time of the Egyptians. The symbol used (γ) since antiquity comes from the zodiacal symbol of the ram. This succession (Taurus, Aries, Pisces) certainly has close symbolic links, via astrology, with the history of religions (the pre-Hebraic bull, the ram or the lamb for the Jews, and the fish for the Christians). The vernal point, and therefore the position of the seasons on the Earth's orbit, moves with the precession of the equinoxes. Knowing that the orbit is elliptical makes it easy to understand that the seasons will be situated at different distances from the Sun, depending on whether the vernal point is closer or further from the perihelion or the aphelion. The effect of the precession on climate is therefore measured by the relative position of the vernal point and the perihelion. The latter also moves, as mentioned in the paragraph concerning eccentricity (this is the orbital parameter π)

according to a movement called 'precession of the perihelion'. The combination of the precession of the equinoxes and the precession of the perihelion thus makes it possible to define the relative position of the seasons and the principal axes of the ellipse. The climate precession, denoted by $\tilde{\omega}$ ('curvilinear pi'), is defined as the angle between the vernal point and the perihelion. If the vernal point carries out a complete cycle in approximately 25,700 years, the perihelion does the same in about 112,000 years. As these two movements occur in opposite directions, we can deduce an average periodicity of 21,000 years for climate precession (1/25.7 + 1/112 ∼ 1/21).

However, there is a small additional complication. When the orbit is circular ($e = 0$), there is no longer a perihelion. The angle $\tilde{\omega}$ is then not defined. Moreover, it is clear that the effect of changes in precession $\tilde{\omega}$ on the climate will be greater as the eccentricity increases, since the distance between the Earth and the Sun will be greater between its maximum $a(1 + e)$ (aphelion) and its minimum $a(1 - e)$ (perihelion).

This effect will be zero when $e = 0$. For all these reasons, it is appropriate to introduce the 'climate precession parameter' $e \sin \tilde{\omega}$, which cancels out when ϖ is not defined (for $e = 0$) and which increases with e. In fact, it is mathematically useful to replace the pair of parameters (e, $\tilde{\omega}$), defined only if e is not zero, with the pair ($e \cos \tilde{\omega}$, $e \sin \tilde{\omega}$), which is always well defined, in other words, a polar-Cartesian coordinate change. The effect of the precession is thus modulated by the eccentricity, as can be seen in the following insolation formula. This results in a duplication of frequencies (more precisely, a multiplication, since e has itself multiple periodicities). If e varies with a single periodicity of 100 000 years, as for example the function $|e0 \cos(t/200)|$, and $\tilde{\omega}$ has a cycle of 21 000 years, we can deduce:

$$e \sin \tilde{\omega} = |e_0 \cos(t/200)| \sin(t/21)$$

hence the periodicities of 19,000 and 23,000 years (1/21 + 1/200 ∼ 1/19 and 1/21 − 1/200 ∼ 1/23), which have been detected in oceanic paleoclimate records and form a strong argument in favor of Milankovitch's theory.

Calculations of Insolation, Calendar Problems

Knowing the three astronomical parameters e, ε, $\tilde{\omega}$, it is easy (with the application of some trigonometry) to determine the radiation received by the Earth, or insolation, for each location (latitude ϕ) and for each season. It is common practice to use the daily insolation, by giving an orbital position with respect to the spring equinox (i.e. the moment in the year) identified by a longitude λ (for example, $\lambda = 90°$ at the

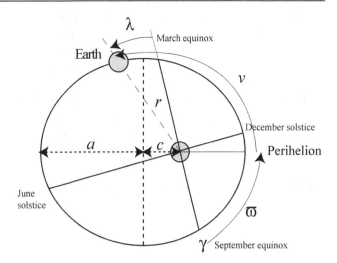

Fig. 28.3 Definitions of the longitude λ, of the climate precession $\tilde{\omega}$, and the anomaly v, with respect to the seasons, the perihelion and the vernal point γ

summer solstice, or $\lambda = 270°$ at the winter solstice), as shown in Fig. 28.3. It is then assumed that this longitude λ is 'fixed' during the day, as the astronomical parameters are. The only movement that is taken into account and which is averaged is therefore the rotation of the Earth on itself in a day.

By formulating:

$$s = \text{Max}(0; 1 - \sin^2 \phi - \sin^2 \delta) = \text{Max}(0; 1 - \sin^2 \phi - \sin^2 \varepsilon \sin^2 \lambda)$$
$$p = \sin \phi \sin \delta = \sin \phi \sin \varepsilon \sin \lambda$$

The following expression of daily insolation can be used:

− if $s = p = 0$, $W_D = 0$
− otherwise $\left(\sqrt{s + p^2} \neq 0 \right)$,

$$W_D = S \left(\frac{1 - e \cos(\lambda - \tilde{\omega})}{1 - e^2} \right)^2 \left(\frac{p \arccos\left(\frac{-p}{\sqrt{s+p^2}} \right) + \sqrt{s}}{\pi} \right)$$

In this formulation, three factors can be identified:

1. The solar constant (S);
2. A term for the Earth-Sun distance that depends on the time of year (λ), climate precession ($\tilde{\omega}$) and eccentricity (e);
3. A geometric term that depends only on the time of year (λ), the obliquity (ε) and the latitude of the location (ϕ).

In particular, the geometrical term remains unchanged if ϕ is changed to $-\phi$ and λ is changed to $\lambda + \pi$ simultaneously (if the hemisphere and half-year are changed). Similarly, the distance term remains unchanged if $\tilde{\omega}$ is changed to

$\tilde{\omega} + x$ and λ is changed to $\lambda + x$ simultaneously (if the precession and the moment of the year are shifted by the same quantity). We note also that the integral over the year (between $\lambda = 0$ and 2π) of the insolation does not depend on the precession $\tilde{\omega}$.

In these formulas, it is important to note that the longitude λ is the angle that represents the position of the Earth in its seasonal cycle, but that λ is not quite proportional to the time that elapses during the year. In fact, when the Earth is close to the perihelion, its velocity is greater and λ changes more rapidly than when the Earth is at the aphelion.

To be precise, the equation of the ellipse (in polar coordinates, centered on the Sun) is:

$$r = \frac{a(1-e^2)}{1+e\cos v}$$

where v is the position relative to the perihelion ($v = \lambda - \tilde{\omega} + \pi$).

The second Kepler equation $\left(r^2 \frac{dv}{dt} = \frac{2\pi a^2 \sqrt{1-e^2}}{T} \right)$ can be written:

$$a^2(1-e^2)^2 \int \frac{dv}{1+(e\cos v)^2} = \frac{2\pi a^2 \sqrt{1-e^2}}{T} = \int dt$$

This fits into:

$$E - e\sin E = \frac{2\pi}{T} t \quad \text{with} \quad \tan\frac{E}{2} = \sqrt{\frac{1-e}{1+e}} \tan\frac{v}{2}$$

which is called the 'Kepler equation', where the new angle E is called the eccentric anomaly. With these equations, the time $t_2 - t_1$ necessary to pass from orbital position λ_1 to λ_2 is deduced.

The duration of the seasons thus changes with the climate precession $\tilde{\omega}$. This poses a problem to define the calendar. In particular, our Gregorian calendar is partly adjusted for shorter winters and longer summers (CDD (Cooling Degree Days): 90.25 days, MAM: 92 days, JJA: 92 days, SON: 91 days), partially in line with the true duration of the seasons (winter: 89.0 days, spring: 92.8 days, summer: 93.6 days, autumn: 89.8 days). While it is important for paleoclimate data to be based on the astronomical calendar (and therefore on seasons defined by the solstices and equinoxes), models need above all a temporal seasonal axis and have to take into account variable durations for the seasons. Most often, these are defined by a fixed time interval (either a quarter of a year or based on the current schedule) using the March equinox as a reference point. This results in a significant lag with the astronomical seasons, up to about two weeks, especially in September, the month furthest from the reference point. The alternative is to rely on the real astronomical seasons which do not have the same number of days, leading to diagnoses more complicated to implement (Joussaume and Braconnot 1997).

Which Astronomical Forcing Should Be Applied to the Climate?

In addition to the definition of the seasons, another critical issue is the clarification of the concept of 'summer' insolation, which is used as a forcing term for Milankovitch's theory of the evolution of ice cover in the northern hemisphere. Should the value of this insolation be taken on a given day (for example, the summer solstice)? Or should an average for the whole season be taken? Or possibly over another orbital interval $[\lambda_1, \lambda_2]$? Or over a time interval? Or should every day of all the seasons be taken and applied to an explicit coupled climate-ice cap physical model? While this latter solution is the most relevant one, it is, in practice, difficult to implement and it is still useful to understand the bases behind the theory by formulating simpler versions.

Milankovitch calculated an average of the isolation for half the year, centered on the June solstice which he called 'calorific insolation'. The usual practice since the 1970s is to choose a given day, often the summer solstice. The typical forcing used is thus the daily insolation at 65°N on the June solstice. Nevertheless, insolations averaged over durations greater than one day may be useful.

For example, it was suggested (Huybers 2006) that a much more relevant astronomical forcing would be an integral of insolation above a critical threshold, since this forcing will melt or not, the ice or snow cover, and corresponding to temperatures above or below the zero degree Celsius threshold. An insolation integral above a threshold is a reasonably good fit with what glaciologists use as a climate forcing, i.e. a temperature integral called Positive Degree Days (PDD).

In practice, the longer the integration period chosen, the more important the role of obliquity in the result. If one integrates over the whole year, precession is eliminated, leaving only the obliquity. In any case, it is essential to be aware of the non-uniform motion of the Earth in its orbit. Indeed, the integration of insolation W_D must be done according to the time variable, even if the result concerns an orbital interval $[\lambda_1, \lambda_2]$. The calculation leads to elliptic integrals that can then easily be evaluated numerically.

The Successes and Difficulties of Milankovitch's Theory

From Hypothesis to Evidence

Milankovitch's theory was not accepted by the majority of geologists for a long time. In fact, stratigraphic studies by Penk and Brückner had made it possible to define only four successive glacial episodes in the Alps (Günz-Mindel-Riss-Würm), and not a succession of regular events. It was not until the middle of the twentieth century that thanks to studies on marine sediments that the number of glacial cycles identified increased considerably (see Chap. 20, volume 1). In particular, in the 1950s, C. Emiliani carried out the first isotopic measurements on marine carbonates, showing more than a dozen glacial-interglacial successions with a clear cyclicity, which put an end to the traditional denomination (Günz-Mindel-Riss-Würm). From then on, astronomical theory became accepted.

Emiliani thus defined the 'isotopic stages' which are still used today to designate glacial and interglacial periods, odd numbers for interglacial periods and even numbers for glacial periods. It is interesting to note that Stage 3 appears to be an exception, since it is now unanimously considered to be part of the last glacial period (which includes Stages 4, 3 and 2). This apparent inconsistency stems from the simple fact that astronomical theory predicts a dominant periodicity associated with variations in obliquity, i.e. cycles of 41,000 years.

The very rare chronological information available from this time suggests a stage 3 around 30–50 ka BP, preceded by many other older cycles of greater amplitude, but with uncertain dating. It was therefore logical to begin the numbering of past interglacials from stage 3, in accordance with the idea of a dominant cyclicity linked to the obliquity. It was not until the 1960s and 1970s, thanks to the Pa–Th datings, that the main cyclicity appeared to be around 100,000 years, an observation which seemed strange in the context of astronomical theory (Broecker and van Donk 1970). This famous 'problem of the 100,000 years' is still a major obstacle, as will be explained later.

Advances in dating techniques using radioisotopes have resulted in a gradual refinement of the chronology, in particular through the use of coral reefs and magnetic reversals that can be identified in both marine sediments and volcanic flows. This enabled a more precise time frame to be proposed and the paleoclimate periodicities to be precisely defined. The paper by Hays et al. (1976) identified cycles of 23,000 years, 41,000 years, and 100,000 years, which correspond well to astronomical frequencies (Berger 1978). This demonstrated unambiguously the astronomical imprint on the climate and the value of Milankovitch's theory that it is astronomy that drives the Quaternary climate cycles. Nevertheless, this paper also highlights the main problem. The 100,000 year cycle is the dominant cycle, but according to the theory, it should hardly appear. It seems that there is a link between the dominant cycle of 100,000 years and eccentricity, but the causes are unknown. In other words, this 'historical' paper demonstrates both that the astronomical theory is necessary, but that, alone, it does not explain the observations.

A Quasi-linear System for Precession and Obliquity

Although the astronomical periodicities are indeed present in paleoclimate records, a simple relation connecting the two has yet to be verified. Various analytical techniques have been used to this end. It is thus possible to assess the consistency, in other words, the correlation in the spectral domain (or frequency domain) between the astronomical forcing and the paleoclimatic, geochemical and micropaleontological data. The results are quite significant for the periodicities of 23,000 and 41,000 years, i.e. the variations of obliquity and precession. There is therefore a close link between astronomical changes and climate for these two frequencies, which can be interpreted in terms of a 'quasi-linear' model. Another way to be sure of this is to observe the amplitude modulation of the astronomical forcing and that of the climate for these periodicities. For example, for obliquity, we note that the greater the amplitude of the variation in the obliquity, the greater the climate response around the corresponding periodicities (i.e. 41,000 years), as illustrated in Fig. 28.4.

This close relationship between astronomical forcing and 'climate' thus seems sufficiently well established to be used

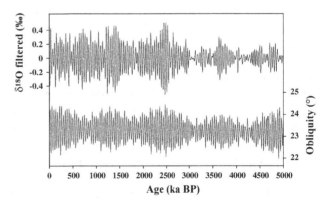

Fig. 28.4 Isotope recording (ODP 659, Tiedemann et al. 1994) filtered around 41 ka (above) and variations in obliquity (Laskar et al. 2004) (below). It is clear that when the obliquity variations increase, so too do the associated climate variations

to date past recordings. Indeed, a persistent difficulty in paleoclimate reconstructions is the establishment of a reliable chronology. As such, absolute dating obtained from the radioactive decay of certain isotopes (^{14}C, ^{40}K/^{39}Ar, ^{40}Ar/^{39}Ar, U/Th) is often rare and difficult to obtain. Moreover, they are imprecise. For example, a 1% error in the measured age translates to an error of 10,000 years for samples of one million years. The error only grows as one goes back in time. It is therefore extremely tempting to use astronomical theory to identify the cycles measured in a paleoclimate recording by referring to the cycles of the astronomical forcing. This link seems particularly relevant for precession or obliquity, at least for the Quaternary. In the case where correct identification of all the cycles is possible, which turns out to be quite frequently, this methodology has the enormous advantage of having an associated error which does not increase disproportionately with time. This is because if one does not 'skip' a cycle, then only the phase relationship between the forcing and climate is cast into doubt, which limits the error to a few thousand years, even if we go back tens of millions of years into the geological past. In this case, isotopic stratigraphy shows how extraordinarily efficient it is, and the establishment of an astronomical chronology for a substantial part of the Earth's history is currently underway. However, the chaotic nature of celestial mechanics mentioned above imposes limits on the calculation of certain astronomical parameters. Conversely, it is possible that we may be able to place geological constraints on the evolution of the parameters of the solar system (Pälike et al. 2004).

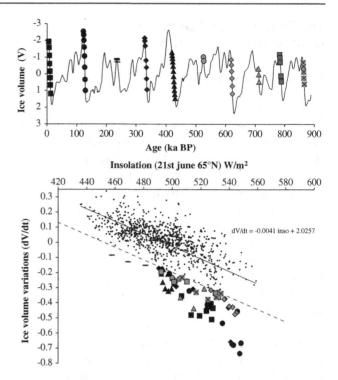

Fig. 28.5 On top: SPECMAP isotopic recording (Imbrie et al. 1984) (normalized) interpreted like a record of ice cap volume. Bottom, this volume of ice is derived according to the time (taking the difference between two successive points), then the derivative is represented as a function of summer insolation at 65°N. Good correlation between the two, for the majority of points, is interpreted as a proof of Milankovitch's theory. The points which deviate substantially from this correlation are marked by symbols which are also shown in the top figure. It is these specific points that are the terminations

The Difficulty of the 100,000-Year Cycles

What is true for the 23,000 and 41,000 year cycles is not true for the major climate cycles that occur more or less every 100,000 years. The first two are highly asymmetric a marked by a much shorter deglaciation phase than the 100,000-year cycle, and as was already noted in the 1970s, the latter did not fit very well into the framework of asronomical theory (Broecker and van Donk 1970). These deglaciations are consequently called 'terminations'. This is not only a visual impression, and it is possible to define these mathematically, by observing that they all correspond systematically to an accelerated decrease in the volume of the ice caps, as illustrated in Fig. 28.5. As such, the terminations are therefore, from the start, outside the scope of Milankovitch's theory.

Moreover, we can no longer observe any real link (in consistency or amplitude modulation) between the variations of eccentricity and the major climate cycles. The very notion of the 100,000-year cycle is problematic because during the Quaternary these have only existed for about a million years and so we only have about ten of these cycles. As a result, statistics have struggled to attribute a specific periodicity to them. It seems that the periodicity of 100,000 years is merely an average between cycles with each having significantly different durations (see Paillard 2001, Table 1). Some authors even suggest that these so-called '100,000-year cycles' are really a double or triple obliquity period (i.e. $2 \times 41 = 82$ ka, or $3 \times 41 = 123$ ka) (Huybers and Wunsch 2005).

To put it simply, as noted above, variations of eccentricity only have a negligible role on the energy received by the Earth, so, it is necessary in any case to imagine relatively complex processes to achieve a climate response in this frequency band where the forcing is almost non-existent.

The simplest way of doing this is to assume the existence of thresholds in the climate system. For example, the system would not function in exactly the same way during the large terminations as during the rest of the cycle. This strategy can be assessed by very simple conceptual models.

A Few Simple Models

Milankovitch's accomplishment was to calculate the orbital parameters of the Earth and the associated insolation occurring over hundreds of thousands of years. Neither a model of evolution of the ice caps nor a climate model was available at this time. He simply linked the minimums in the astronomical forcing with episodes of glaciation. He explained that due to a very high probably inertia in the caps, there would be a gap of several thousand years between the astronomical forcing and the evolution of the caps, even to the point of 'smoothing' the astronomical forcing if the caps do not have the time to react. Indeed, the volume of the caps is not directly related to the insolation, but rather it is the variation in the volume of ice which is a function of insolation. In other words, an extremely simple model of evolution of the caps consists of integrating the astronomical forcing over time. As the important physical object is the cap, it is essential to look at the dynamics of the cap, not only the forcing.

Surprisingly, no scientist seems to have focused on this task, following either in the footsteps of Adhémar or Croll, or those of Milankovitch. It was therefore a journalist (Calder 1974) who would publish a scientific article describing, for the first time, a model of ice cap evolution in the Quaternary, more than a century after the ice ages were observed. This model was both very simple and informative. It is based on a simple integration of the forcing [see Paillard (2001, 2010, 2015)]:

$$dV/dt = -k(i - i_0).$$

where V is the volume of the cap and $i(t)$ is the astronomical forcing, i.e. summer insolation at the high latitudes of the northern hemisphere. Above a fixed isolation value i_0, the volume of ice V decreases proportionally to $i-i_0$ and, below this value, V increases. This model works relatively well provided that an asymmetry between the melting and accumulation episodes is introduced, with a different k coefficient in the two cases, $k = k_F$ and k_A depending on whether $i > i_0$ (melting) or not. In addition, V is imposed as a positive value. Although this model fails to closely reproduce the observations, it nevertheless has some remarkable characteristics. In particular, it correctly predicts the position of the large terminations, where many other more sophisticated models fail, see Fig. 28.6. We will return to this point.

This model is quite unstable and small changes in the parameters (i_0 or k_A/k_F) lead to significantly different results. This is easily understood, since the volume of ice is, ultimately only the integral of the insolation. Small changes in the threshold or the coefficient quickly lead to very different volumes.

A much more robust model was formulated by Imbrie and Imbrie in 1980 (see Paillard 2001):

$$dV/dt = (-i - V)/\tau$$

where, in a way, the threshold i_0 has been replaced by the volume of ice V. This time, the insolation i is taken as normalized (by subtracting its average value and dividing it by the standard deviation) and so it is possible for the volume of ice to also be negative. As before, the coefficient, or time constant, τ, will have two different values depending on whether the cap melts ($-i-V < 0$) or grows ($-i-V > 0$). This model reproduces quite well the evolution of the cycles linked to precession and obliquity, but fails to reproduce the 100,000-year cycles. In particular, it produces a strong 400,000 year cycle, clearly present in the eccentricity, but not in the paleoclimate data on the volume of the ice caps, see Fig. 28.7. This model thus illustrates the 'Stage 11 problem'. About 430,000 years ago, the eccentricity was very low and therefore the variations in the precession parameter $e \sin \tilde{\omega}$ were minimal. This was reflected in very little variation in the Imbrie model. Conversely, the paleoclimate data show that this period corresponds to a major transition between a very intense glacial stage (stage 12) and a very marked interglacial

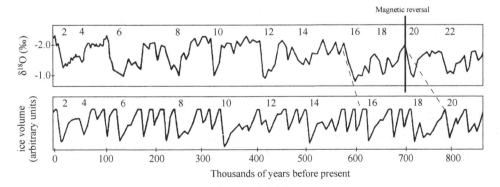

Fig. 28.6 Comparison between the Calder model (bottom) and the V28-238 core (top) (see Paillard 2015). The glacial-interglacial transitions are predicted for the correct dates by the model, while the isotopic data in 1974 were displaced in time, with a too-early date for the Brunhes-Matuyama magnetic reversal, now fixed at around 772 ka (good match between cycles indicated by the dotted lines). In hindsight, the prediction made by the Calder model was quite remarkable

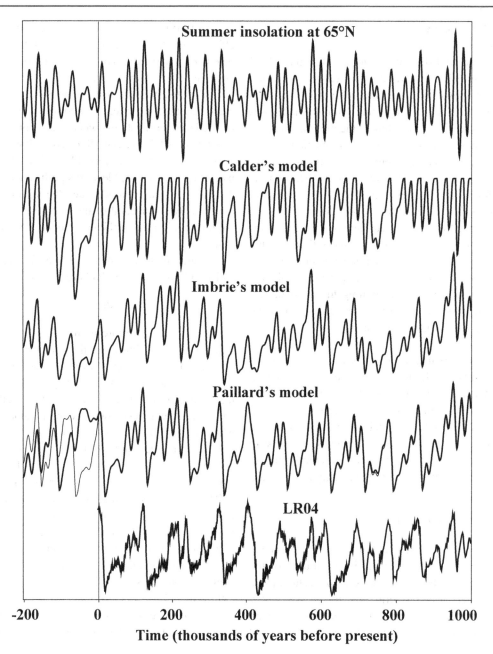

Fig. 28.7 Comparison between the various simple models discussed, over the last million years and the next 200,000 years. From top to bottom: daily summer insolation (65°N, June solstice) (Laskar et al. 2004), results from the Calder model; Imbrie model; Paillard model 1998 and marine isotopic data LR04 (Lisiecki and Raymo 2005). Note that in Paillard's threshold model, there are two possible solutions for the future climate cycles, depending on whether the threshold for entry into glaciation has already been crossed or not (see Paillard 2001)

(stage 11). How can small variations in insolation lead to the greatest of transitions? A natural solution is to view the great cycles at 100,000 years, not as a linear oscillation around an equilibrium point whose amplitude is inevitably linked to the amplitude of the forcing, but rather as a relaxation oscillation between two different climate modes between which the system can switch as soon as certain thresholds are crossed. This is what is proposed in the Paillard model (1998):

$$dV/dt = (V_R - V)/\tau_R - F/\tau_F$$

This time, the volume of ice is 'relaxed' towards different V_R values: the 'climate mode' R is changed as a function of certain threshold overruns on the astronomical forcing i and on the volume of ice V. In particular, an essential point emerging from the study of this model is that in order to predict deglaciations at the right position, they must be linked to the glacial maxima: the switch between glacial-interglacial regimes needs to be triggered by a threshold overrun of the ice cap volume. This relationship is what allows Calder's model to finally work 'well', because Calder is more successful at predicting glacial maxima than transitions. In other words, for still unknown physical reasons, deglaciations are facilitated by the occurrence of a glacial maximum, which triggers a relaxation oscillation, propelling the system into the opposite state.

Although this astronomical forcing is at the root of the climate variations in the Quaternary, the understanding of the climate mechanisms in play is limited. In Milankovich's theory there is a very simple hypothesis which, in a way, equates the concept of 'climate' with the expansion of the ice caps in the northern hemisphere. Milankovitch's theory, strictly speaking, is not a climate theory, but rather, an ice cap theory, a point emphasized by the conceptual models above. The action of insolation on the ice caps of the northern hemisphere accounts for certain Quaternary phenomena, but does not explain all the observations, in particular the existence of the 100,000-year cycles, punctuated by exceptional deglaciations (or terminations). This theory therefore needs to be further completed to make it a true climate theory.

Recent Advances

The Vital Role of Atmospheric CO_2

Shortly after the discovery of the role of astronomical periodicities in the climate (Hays et al. 1976), analysis of air bubbles in Antarctic ice cores demonstrated that the last glacial period was also characterized by a significantly lower atmospheric concentration of CO_2, in line with the predictions of Arrhenius. Since the work on the Vostok ice cores (Petit et al. 1999) and Dôme C ice cores (Monnin et al. 2001), it is now well established that the glacial-interglacial cycles also correspond to cycles in atmospheric greenhouse gas composition, and in particular of CO_2, which varies between about 280 ppm (cm^3/m^3 of air) during the interglacial period and 180 ppm during the glacial period. These measurements make it possible to demonstrate that the two traditional theories, astronomical and geochemical, are not mutually exclusive, but that both are necessary. This was largely confirmed by numerous numerical experiments in the simulation of the glacial climate: in order to explain the paleoclimate observations, it is essential to take into account the 30% decrease in partial CO_2 pressure. In addition, during the terminations and in particular during the last deglaciation, it is well established that the concentration of CO_2 increased several thousand years before the rise of the sea level associated with the melting of the ice caps, or, in other words, the actual deglaciation, as illustrated in Fig. 28.8.

Although it is relatively easy to apply Milankovitch's theory to most of the past cycles, considerable difficulties arise for the deglaciations as is highlighted in Fig. 28.5. It is therefore for these specific moments, when the astronomical theory alone is insufficient, that other mechanisms need to be explored.

In other words, glacial-interglacial changes are not limited to changes in the expansion of the ice caps that could subsequently influence the rest of the climate system. They are, on the contrary, a combination of changes for the caps, but also for the biogeochemical cycles and the climate as a whole. Milankovitch's theory only accounts for part of this reality, the other part most likely involves the carbon cycle coupled with climate variations.

Unfortunately, our understanding of the carbon cycle during the Last Glacial Maximum is very patchy. To make a first approximation, it is reasonable to consider the ocean + atmosphere + terrestrial biosphere system as isolated, i.e. with no significant exchanges of geological carbon (via volcanoes or rivers). The problem is therefore to reduce the atmospheric reservoir by about 200 GtC (billion tons of carbon) while increasing the others by the same amount. However, the terrestrial biosphere was considerably reduced during the glacial period (between 300 GtC and 700 GtC), making the problem all the more difficult, since all this atmospheric and biospheric carbon needs to be trapped in the ocean. Many hypotheses have been put forward to try to explain this low level of pCO_2 during the Last Glacial Maximum, but no consensus has yet emerged. A complex combination of multiple factors (physical and biogeochemical) is one possibility which would explain a glacial-interglacial difference of 100 ppm, but the high level of similarity between the climate recordings around the Antarctic and the pCO_2 records argue for a relatively simple mechanism which would link the Southern Ocean and its climate with the atmospheric concentration of CO_2.

Towards a Consolidation of Astronomical and Geochemical Theories?

Nevertheless, very recent progress makes a forthcoming solution possible. The conceptual models mentioned above suggest viewing the 'glacial' and 'interglacial' states as distinct states able to account for a relaxation oscillation between two (or more) different modes of operation. In this context, it is interesting to mention the hypothesis of a glacial ocean with very cold and above all salty bottom waters. This hypothesis is largely supported by measurements of interstitial fluids in marine sediment cores (Adkins et al. 2002), which directly estimate the salinity of the ocean floor in the past. The glacial ocean was therefore likely to have been profoundly different from the current ocean, with strong stratification between the upper and lower halves of

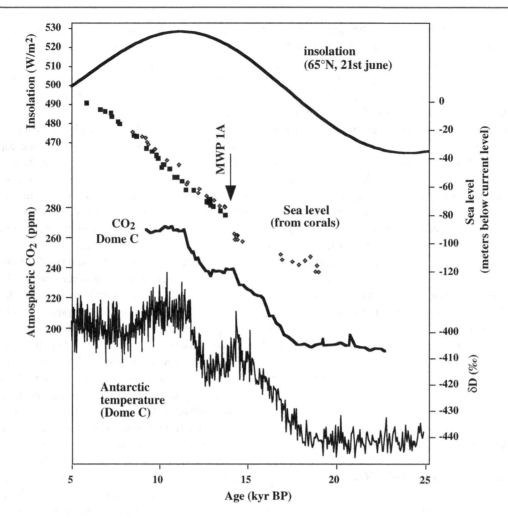

Fig. 28.8 The last deglaciation. From top to bottom: daily insolation (65°N summer solstice) (Laskar et al. 2004); sea level (Bard et al. 1996); atmospheric CO_2 and temperature at Dome C in Antarctica (Monnin et al. 2001). At about 15 kyr BP (15,000 years before the present day), while the sea level remains close to its ice age level (about −100 m), atmospheric CO_2 has already increased by about 60 ppm above its ice age value, more than half of the full transition

the water column. These highly saline bottom waters are likely to form around Antarctica, due to the salt deposits that occur during the formation of sea ice. Indeed, thanks to the continental shelf, these very salty waters are able to flow along the topography (Ohshima et al. 2013) and reach the abyss of the Southern Ocean. The consequences for the carbon cycle of this ocean configuration are considerable. With a mechanism of this type, it is possible to store a lot of carbon at the bottom of the ocean and to explain the low level of atmospheric CO_2. This mechanism was recently confirmed in a relatively simple model coupling climate and carbon cycle (Bouttes et al. 2011). This makes it possible to formulate a scenario that accounts for the glacial-interglacial cycles, both in terms of changes in the expanse of the ice sheets, but also in terms of atmospheric CO_2 and hence global climate (Paillard and Parrenin 2004), as shown in Fig. 28.9.

In this model, terminations are explicitly induced by an increase in atmospheric CO_2, itself caused by the previous glacial maximum, which destabilizes the stratification of the deep ocean. Since this model is based on a bi-modal system, it correctly reproduces certain characteristics already present in even simpler models (Paillard 1998), such as the possibility of switching between dominant periodicities, from 23,000-year cycles before the Quaternary glaciations became established, to 41,000-year cycles 3 million years ago, and then to 100,000-year cycles in the last million years.

Pre-quaternary Astronomical Cycles

Periodic variations of insolation have existed throughout the history of our planet. Although their effects on the climate have been particularly marked for about a million years

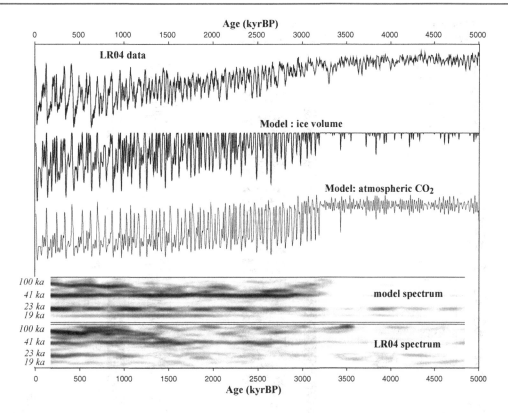

Fig. 28.9 Results of the Paillard and Parrenin model (2004) which includes a mechanism coupling the evolution of ice caps and changes in the carbon cycle, forced by summer insolation at 65°N. By adding a slow drift to the critical threshold parameter, it is possible to account for periodicity switches with the emergence of the 41,000-year cycles about 3 million years ago, and then the 100,000-year cycles for the last million years. From top to bottom: ice volume data (Lisiecki and Raymo 2005), the model results in terms of ice caps and CO_2, and then the decomposition into periodicities (spectrum) for the model and the data

through the glacial-interglacial alternations, it is normal to expect possibly more subtle types of changes in climate when the Earth was less ice-covered or even not at all. Climate variations of astronomical origin may well involve other components of the Earth system besides the ice caps, such as monsoons, biological productivity, or other aspects of our planet. These astronomical cycles are also often incorrectly called 'Milankovitch cycles', although, for the most part, they have nothing to do with changes in the expansion of the ice caps, which were, more often than not, non-existent throughout the history of the Earth. Milankovitch's theory is a theory of the evolution of the ice caps. It is not a climate theory and therefore does not apply to all of the changes in climate due to astronomical causes. Theoretically, there is no reason to select summer insolation in the northern hemisphere as a dominant forcing parameter for components of the system other than the ice caps in the northern hemisphere. According to measured indicators (sedimentology, isotopes of oxygen or carbon, color, etc.), depending on the sites and the geological periods under consideration, the cycles line up with different astronomical periodicities corresponding to variations in precession, obliquity or eccentricity. These alternating sedimentary layers may be from very different origins, and they are sometimes rather poorly understood. There are examples of this in more or less all the epochs of the history of the Earth.

A prime example concerns deposits of organic matter in the Mediterranean Sea, which occur regularly in the form of clearly identifiable layers of black silt called sapropeles. These layers rich in organic matter are explained either by an increase in biological production at the surface or by a change in the circulation of the deep waters of the Mediterranean, which would have been poorly oxygenated during these events, just as the Black Sea is today. In fact, these sapropel events correspond to rainfall episodes in Saharan Africa, or even over the Mediterranean basin as a whole, which brought significantly large supplies of freshwater via the Nile and the rains and disrupted the formation of deep waters and therefore the oxygenation levels of the Mediterranean. Whatever the mechanism, these sedimentary levels appear, with a few exceptions, to be governed by precessional variations since the Miocene, about 14 million years ago, until the last event designated S1, at the beginning of our interglacial period, about 7000 years ago. As shown in Fig. 28.10, this cyclicity is sufficiently well-marked to be used not only as a dating method but also to calibrate the

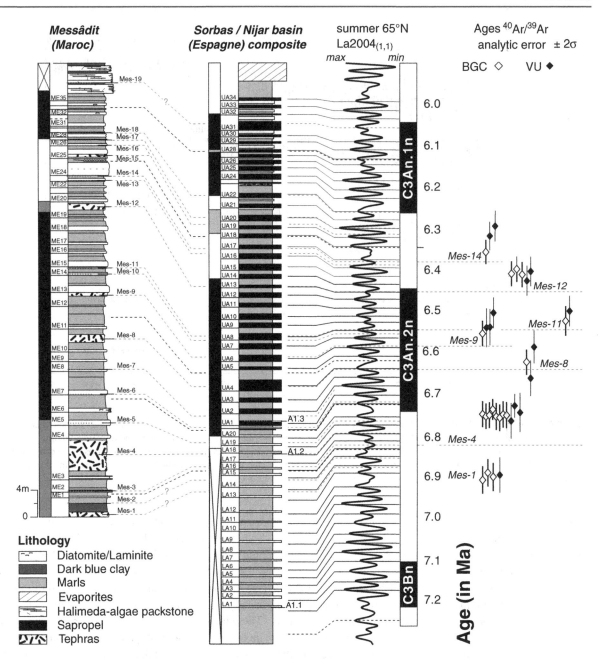

Fig. 28.10 According to Kuiper et al. 2008. Astronomical calibration of the Messinian (between 7.2 and 5.3 Ma) where the alternating marl/sapropel series can be correlated with the astronomical forcing (in agreement with many other stratigraphic markers). This astronomical calibration is then used to better constrain the $^{40}Ar/^{39}Ar$ method (measured here in tephras), since the disintegration constant of ^{40}K is only known with to an error of about 3 or 4% (($5.463 \pm 0.214) \times 10^{-10}$ an^{-1})

$^{40}Ar/^{39}Ar$ radiometric method (Kuiper et al. 2008). As mentioned above, this way of establishing age scales, or cyclostratigraphy, makes it possible to achieve a level of precision far superior to the usual radiometric methods, the uncertainties of which increase as we go back in time.

The chronology is thus often the first point of interest when identifying astronomical cycles in old sedimentary series. The most famous example of cyclicity found in geological recordings concerns the alternating marl-limestone series. As early as the end of the nineteenth century, Gilbert suggested that these sedimentary successions from the Cretaceous, which he studied in the limestone formations of the Green River in Colorado, were probably caused by astronomical changes. Extrapolating from the limited outcrops that he had at his disposal, he assumed that they were linked to the precession cycles, which led him to

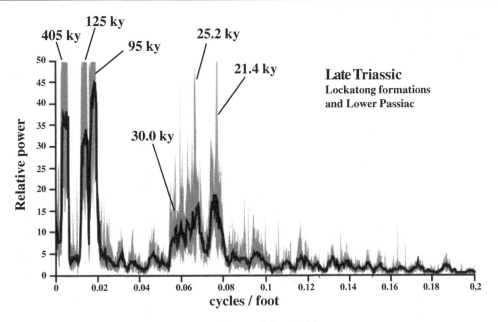

Fig. 28.11 Spectrum of the Triassic lake levels in New Jersey (Olsen and Kent 1996)

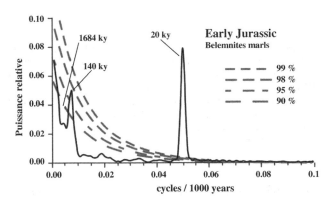

Fig. 28.12 Spectrum of the percentage of carbonate, Jurassic, England (Weedon et al. 1999)

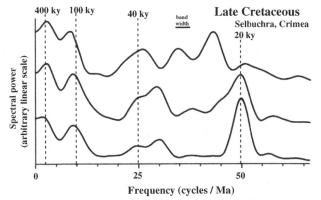

Fig. 28.13 Spectra of sedimentary reflectance, Cretaceous, Crimea (Gale et al. 1999) with various chronological hypotheses

suggest that the Upper Cretaceous lasted between 20 and 40 million years. This is in line with recent estimates (34 million years). Numerous studies have since confirmed the existence of these Triassic, Jurassic and Cretaceous cycles, as illustrated in the following Figs. 28.11, 28.12 and 28.13. In particular, it can be seen that although the precession is often dominant, the cycles linked to the obliquity or to the eccentricity also play very important roles.

The influence of astronomical cycles has also been demonstrated in the Paleozoic, although there are fewer studies on this subject. This is due to the difficulty in finding chronological benchmarks that are sufficiently precise and reliable to be able to unambiguously attribute the cyclicity found to astronomical parameters. For example, during the Carboniferous, alternating marine sediment and coalbed layers, called cyclothems, probably correspond to changes in sea level in the delta regions, where abundant vegetation during low sea levels (regressions) was followed by sedimentary deposits during high sea levels (transgressions). This is ultimately to be expected, as the Carboniferous, like the Quaternary, corresponds to a 'glacial' period with consequential ice caps which are likely to fluctuate in line with astronomical changes. Similarly in the Devonian, sedimentary cyclicities are observed and these are usually interpreted in terms of the evolution of the large ice caps present at that time. It is also possible to find much older cycles, such as the Archean, for example, more than 2 billion years ago (Hofmann et al. 2004).

The further back in time we go, the more important it is to take possible changes in astronomical periodicities into account. In particular, the movements of the Earth's axis (precession and obliquity) will be strongly affected by the Earth-Moon distance which increases over time due to tidal dissipation. Thus, the periodicity of the precession of the equinoxes, between 25,700 years ago and today, was noticeably faster before. The same applies to the cycles of climate precession and obliquity, which are directly related to it, and which are today 19, 23 and 41 ka. Considering the current rate of lunar recession, we obtain respective periodicities of 16, 18.7 and 29 ka for 500 million years ago. These values are too quick and are inconsistent with geological observations, which underscores the need for a slower lunar recession in the past, due to the current isostatic rebound but also to changes in sea level, in the topography of the ocean floors, and even ocean stratification. Generally speaking, even more than finding a given periodicity, the aim is to find a coherence between several periodicities which corresponds well to the astronomical forcing. For example, the presence of three periodicities in a 1:5:20 ratio is often interpreted as the precession (~20 ka) and the eccentricity (~100 ka and ~400 ka) periodicities, even though the frequencies have varied in the past, and even if the chronological information does not allow the recorded periodicities to be determined with confidence.

If one had to choose a particularly stable periodicity in the past, the 405 ka cycle associated with variations in eccentricity would probably be the best choice (Laskar et al. 2004). As stated in paragraph IIa, the solar system is chaotic and it is impossible to calculate precisely its evolution further back than a few tens of millions of years. However, if the perturbations associated with the internal planets (from Mercury to Mars) become quickly unpredictable, the outer planets (in particular, Jupiter and Saturn) have much more regular long-term movements. This produces high stability for the periodicity at 405 ka, which can therefore be used as a chronological reference over several hundred million years. According to Laskar et al. (2004), the spread of solutions over the last 250 million years for this periodicity is less than one cycle (<400 ka). This therefore, creates the opportunity to construct a very precise absolute chronology not only for the Cenozoic (since 65 Ma) but also for the whole of the Mesozoic (between 65 Ma and 250 Ma). These cycles can be systematically numbered from the present time to the distant past, thus offering a new way of stratigraphic tracking on the geological scale. For example, in Fig. 28.14, the clear presence of these cycles is observed during the Oligocene period. Moreover, the Eocene-Oligocene (Oi-1 event) or Oligocene-Miocene (Mi-1 event) transitions correspond to particularly 'cold' climate periods according to the oxygen isotopes. These extreme values can be related to the minima of the astronomical forcing linked to the obliquity (minima of the amplitude modulation).

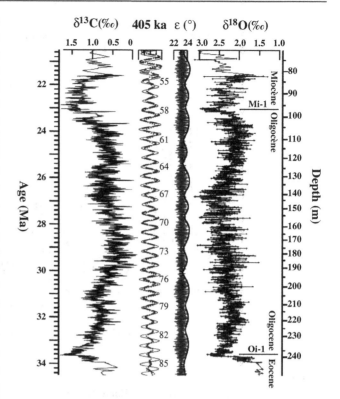

Fig. 28.14 According to Pälike et al. 2006. Stable ^{13}C and ^{18}O benthic isotopes of the ODP 1218 site, obliquity (ε) with its amplitude modulation, and filtering of the eccentricity and the isotopic signals at 405 ka. The 405 ka periodicity is clearly visible in the isotopic signals, particularly ^{13}C, which makes it possible to define an astronomical chronology (numbering the 405 ka cycles from the current one). The minimum amplitude in variation of the obliquity very often correspond to maxima for the ^{18}O

These 405 ka eccentricity cycles probably play an important role in the evolution of climate in the Quaternary. Although changes in ice cap volume are largely dominated by the 100 ka cycle, this is not necessarily the case for other indicators. In particular, we find a clear signature in this frequency band in the ^{13}C of benthic foraminifera with cycles between 400 and 500 ka (Wang et al. 2004). These changes in the global carbon cycle are probably the cause of the different climate phases of the last million years: the beginning of the glaciations (Pliocene-Pleistocene transition around 2.6 Ma), as well as the transition between the 41 ka cycles and the 100 ka cycles (Mid-Pleistocene transition around 0.7 Ma) corresponding to the amplitude modulation of this 400 ka eccentricity cycle (Paillard 2017).

Although cycles were probably present throughout the Earth's history, it is important to avoid systematically ascribing an astronomical origin to all periodicity observed in sedimentary records. Indeed, it is quite conceivable that certain components of the Earth system could generate more or less periodic internal oscillations at frequencies that have little to do with celestial dynamics. Thus, in the Quaternary, there is a so-called 'sub-Milankovitch' variability corresponding to the Dansgaard-Oeschger and Heinrich events. It is likely that cyclicities of this type have also existed at other times in Earth's history. For example, the 470 m thick Triassic strata, observed in Latemar in the Italian Dolomites, have a well-marked periodic structure showing about 600 cycles. There is controversy between those who support the hypotheses of astronomical cycles, who see precession cycles and therefore a total recording time of around 10 million years, and the proponents of much quicker environmental variations, using radiometric dating to estimate a total duration of around a million years only, hence an order of magnitude faster. Although the debate is not yet settled, it highlights that it is not enough to identify the cycles in geological records to systematically find an astronomical signature. The dynamics of the Earth system are indeed likely to reveal many more surprises.

Conclusions

The dynamics of the glacial-interglacial cycles of the Quaternary are far from fully understood and this is even more true for the pre-Quaternary astronomical cycles. Moreover, the climate during glacial periods can sometimes change very abruptly during the Dansgaard-Oeschger or Heinrich events, for reasons unrelated to astronomical forcing or to atmospheric CO_2 variations. It is noteworthy that the last deglaciation was punctuated by such rapid events. It is therefore quite possible that they also play a decisive role in the dynamics of large cycles, and even more so if one considers that they are based on tipping points that allow a shift from a glacial state to an interglacial state.

The conceptual models presented here are far from sufficient to account for the physical and biogeochemical interactions involved in these climate changes. The most sophisticated climate models that are used to simulate the twenty-first century, or 'general circulation models', are unfortunately unable to simulate these changes because the time scales involved are far too long. In addition, the ice cap changes and the dynamics of the main biogeochemical cycles must also be taken into account, which these models are not yet able to do correctly. For all these reasons, simpler models of the 'Earth system' (or models of intermediate complexity) are used to address this type of question. Recent advances, both in terms of paleoclimate reconstructions but also in terms of modeling (Bouttes et al. 2011), provide hope that a new theory of the glacial-interglacial cycles of the Quaternary that takes into account the various aspects of terrestrial environmental changes (ice caps, carbon, vegetation, climate) can be developed in the years to come. An overview of this type would provide a much better understanding of the great climate changes experienced so far by *homo sapiens*.

References

Adkins, J., McIntyre, K., & Schrag, D. (2002). The salinity, temperature and $\delta^{18}O$ of the glacial deep Ocean. *Science, 298*, 1769–1773.

Arrhenius, S. (1896). On the influence of carbonic acid in the air upon the temperature of the ground. *Philosophical Magazine and Journal of Science, 41*, 237–276.

Bard, E. (2004). Greenhouse effect and ice ages: Historical perspective. *C. R. Geoscience, 336*, 603–638.

Bard, E., Hamelin, B., Arnold, M., Montaggioni, L., Cabioch, G., Faure, G., et al. (1996). Deglacial sea-level record from tahiti corals and the timing of global meltwater discharge. *Nature, 382*, 241–244.

Berger, A. (1978). Long-term variations of daily insolation and quaternary climatic change. *Journal of the Atmospheric Sciences, 35*, 2362–2367.

Bouttes, N., Paillard, D., Roche, D. M., Brovkin, V., & Bopp, L. (2011). Last glacial maximum CO_2 and $\delta^{13}C$ successfully reconciled. *Geophysical Reseach Letters, 38*, 1–5.

Broecker, W., & van Donk, J. (1970). insolation changes, ice volumes and the O^{18} record in deep-sea cores. *Reviews of Geophysics and Space Physics, 8*, 169–197.

Calder, N. (1974). Arithmetic of ice ages. *Nature, 252*, 216–218.

Gale, A. S., Young, J. R., Shackleton, N. J., Crowhurst, S. J., & Wray, D. S. (1999). Orbital tuning of cenomanian marly chalk successions: Towards a milankovitch time-scale for the late cretaceous. *Philosophical Transactions of the Royal Society A, 357*, 1815–1829.

Hays, J., Imbrie, J., & Shackleton, N. J. (1976). Variations in the earth's orbit: Pacemakers of the ice ages. *Science, 194*, 1121–1132.

Hofmann, A., Dirks, P. H. G. M., & Jelsma, H. A. (2004). Shallowing-upward carbonate cycles in the belingwe greenstone belt, zimbabwe: A record of archean sea-level oscillations. *Journal of Sedimentary Research, 74*, 64–81.

Huybers, P. (2006). Early pleistocene glacial cycles and the integrated summer insolation forcing. *Science, 313*, 508–511.

Huybers, P., & Wunsch, C. (2005). Obliquity pacing of the late pleistocene glacial terminations. *Nature, 434*, 491–494.

Imbrie, J., Hays, J., Martinson, D., McIntyre, A., Mix, A., Morley, J. J., Pisias, N., Prell, W., Shackleton, N., Berger, A., Kukla, G., & Saltzman, B. (1984). The orbital theory of pleistocene climate: Support from a revised chronology of the marine $d^{18}O$ record. In A. dans Berger (Ed.), *Milankovitch and climate* (pp. 269–305), Dordrecht: Kluwer Academic Publishers (Nato ASI Ser. C).

Joussaume, S., Braconnot, P. (1997). Sensitivity of paleoclimate simulation results to season definitions. *Journal of Geophysical Research D, 102*, 1943–1956.

Kuiper, K., Deino, A., Hilgen, F. J., Krijgsman, W., Renne, P. R., & Wijbrans, J. R. (2008). Synchronizing rock clocks of earth history. *Science, 320*, 500–504.

Laskar, J., Robutel, P., Joutel, F., Gastineau, M., Correia, A. C. M., & Levrard, B. (2004). A long-term numerical solution for the insolation quantities of the earth. *Astronomy & Astrophysics, 428*, 261–285.

Lisiecki, L. E., & Raymo, M. E. (2005). A pliocene-pleistocene stack of 57 globally distributed benthic $\delta^{18}O$ records. *Paleoceanography, 20*, PA1003. https://doi.org/10.1029/2004pa001071.

Milankovitch, M. (1941). *Kanon der Erdbestrahlung und seine Andwendung auf das Eiszeiten-problem.* (p. 633).

Monnin, E., Indermühle, A., Dällenbach, A., Flückiger, J., Stauffer, B., Stocker, T., et al. (2001). Atmospheric CO_2 concentrations over the last glacial termination. *Science, 291*, 112–114.

Ohshima, et al. (2013). Antarctic bottom water production by intense sea-ice formation in the Cape Darnley polynya. *Nature Geoscience, 6*, 235–240.

Olsen, P. E., & Kent, D. V. (1996). Milankovitch climate forcing in the tropics of pangaea during the late triassic. *Palaeogeography Palaeoclimatology Paleoecology, 122*, 1–26.

Paillard, D. (1998). The timing of pleistocene glaciations from a simple multiple-state climate model. *Nature, 391*, 378–381.

Paillard, D. (2001). Glacial cycles: Toward a new paradigm. *Reviews of Geophysics, 39*, 325–346.

Paillard, D., & Parrenin, F. (2004). The antarctic ice-sheet and the triggering of deglaciations. *Earth Planet Science Letters, 227*, 263–271.

Paillard, D. (2010). Climate and the orbital parameters of the earth. *C. R. Geoscience, 342*, 273–285.

Paillard, D. (2015). Quaternary glaciations: from observations to theories. *Quaternary Science Reviews, 107*, 11–24.

Paillard, D. (2017). The plio-pleistocene climatic evolution as a consequence of orbital forcing on the carbon cycle. *Climate of the Past, 13*, 1259–1267.

Pälike, H., Laskar, J., & Shackleton, N. (2004). Geologic constraints on the chaotic diffusion of the solar system. *Geology, 32*, 929.

Pälike, H., Norris, R. D., Herrle, J. O., Wilson, P. A., Coxall, H. K., Lear, C. H., Shackleton, N. J., Tripati, A. K., & Wade, B. S. (2006). The Heartbeat of the oligocene climate system. *Science, 314*, 1894–1898.

Petit, J.-R., Jouzel, J., Raynaud, D., Barkov, N., Barnola, J.-M., Basile, I., et al. (1999). Climate and atmospheric history of the past 420,000 years from the Vostok ice core, Antarctica. *Nature, 399*, 429–436.

Tiedemann, R., Sarnthein, M., & Shackleton, N. (1994). Astronomic timescale for the pliocene atlantic $\delta^{18}O$ and dust records of ocean drilling program site 659. *Paleoceanography, 9*, 619–638.

Wang, P., Tian, J., Cheng, X., Liu, C., & Xu, J. (2004). Major pleistocene stages in a carbon perspective: The South China sea record and its global comparison. *Paleoceanography, 19*, PA4005. https://doi.org/10.1029/2003pa000991.

Weedon, G. P., Jenkyns, H. C., Coe, A. L., & Hesselbo, S. P. (1999). Astronomical calibration of the jurassic time-scale from cyclostratigraphy in British mudrock formations. *Philosophical Transactions of the Royal Society A, 357*, 1787–1813.

Rapid Climate Variability: Description and Mechanisms

Masa Kageyama, Didier M. Roche, Nathalie Combourieu Nebout, and Jorge Alvarez-Solas

The previous chapters focused on climate variations stemming from factors external to the climate system: tectonics, orogeny, variations in insolation. Yet the study of glacial records and marine sediments over the past three decades has revealed major changes within the climate system over much shorter time scales than previously envisaged. These abrupt reorganisations of the climate system, which cannot be explained by forcings external to the system, are true climate 'surprises'. They have been the subject of many studies, both to describe the expression of these events on a global scale and to model these events and their impacts. This work is still going on to better characterize and understand this type of climate variability, called millennial variability, in contrast to the time scales associated with the Milankovitch forcings (see Chap. 7), or abrupt (rapid) variability, because the transitions between climate states take place on even shorter time scales, ranging from ten to one hundred years. This chapter presents a synthesis of current knowledge on rapid climate variability.

M. Kageyama (✉) · D. M. Roche
Laboratoire des Sciences du Climat et de l'Environnement, LSCE/IPSL, CEA-CNRS-UVSQ, Université Paris-Saclay, 91191 Gif-sur-Yvette, France
e-mail: masa.kageyama@lsce.ipsl.fr

N. C. Nebout
UMR 7194 CNRS/UPVD/MNHN, HNHP-Histoire Naturelle de l'Homme Préhistorique, Département Homme et Environnement, Muséum National d'Histoire naturelle, Paris, France

N. C. Nebout
Institut de Paléontologie Humaine, 1 rue René Panhard, 75013 Paris, France

J. Alvarez-Solas
Departamento de Física de la Tierra y Astrofísica, Facultad de Ciencias Físicas, Universidad Complutense de Madrid, 28040 Madrid, Spain

J. Alvarez-Solas
Instituto de Geociencias, Consejo Superior de Investigaciones Científicas-Universidad Complutense de Madrid, 28040 Madrid, Spain

© Springer Nature Switzerland AG 2021
G. Ramstein et al. (eds.), *Paleoclimatology*, Frontiers in Earth Sciences,
https://doi.org/10.1007/978-3-030-24982-3_29

Rapid Climate Changes During Glacial Periods: Heinrich and Dansgaard-Oeschger Events

The Discovery

Abrupt climate changes were first discovered in the context of the last glacial period (see the review by Hemming 2004). In 1977, Ruddiman showed that during Marine Isotopic Stages 4, 3 and 2, large quantities of coarse detrital material transported by icebergs detaching from the Northern Hemisphere ice sheets were deposited at the mid-latitudes of the North Atlantic between 40 and 65° N in a band now known as the Ruddiman Belt, while the main deposition area for Stage 5 is closer to Greenland and Newfoundland. The concentration of detrital material measured in marine sediments is thus linked with the size and expansion of continental ice sheets over the Milankovitch timescale. In 1988, H. Heinrich showed that six major events of coarse detrital material deposition occurred in the Ruddiman band during the last glacial period at intervals of about 10,000 years, and therefore at time scales shorter than those of Milankovitch. These six events are often accompanied by a major change in the composition of assemblages of planktonic foraminifera, with a predominance of the left-coiling *Neogloboquadrina pachyderma* polar species, an indicator of a particularly cold environment. Heinrich initially based his interpretation of these results on the fact that the period of 10,000 years is about half a precession cycle. He speculated that the coarse detrital material was transported either by icebergs (in orbital conditions favoring a 'cold' period) or by the melting of the ice sheets ('warm' situation). This hypothesis did not stand up to more precise dating and analyses of marine records in the North Atlantic but it shows that, at the time of this discovery, the variations in orbital insolation were considered the main contributors to the evolution of the climate system. At that time, sudden changes to the system were not envisaged. The discovery however initiated much research on these major deposits of coarse detrital material, which were named Heinrich events (HE) by Broecker et al. in 1992. Heinrich events are defined by these authors in terms of both their characteristic low concentration of planktonic foraminifera and high concentration of detrital material. They further demonstrate that these events are related to armadas of icebergs that broke off from the ice sheets rather than to the melting of these ice sheets. The layers of detrital material may be several meters thick in the Labrador Sea. This thickness is less, although still significant, close to the European coasts. Subsequent studies of the composition and properties of this ice-rafted detrital material show that its origin is mostly from the Canadian shield and has therefore been transported by icebergs which detached from the North American ice sheet.

During the same period, the analysis of the $\delta^{18}O$ isotopic signal from ice cores taken from the summit of the Greenland ice cap (Dansgaard et al. 1993) shows the contrast between interglacial periods, and particularly between the remarkably stable Holocene, and the last glacial period, characterized by high amplitude oscillations. This amplitude can be as much as half of the glacial-interglacial difference in Greenland. The warming events were named 'Dansgaard-Oeschger events' (D/O), because these two authors had already detected rapid $\delta^{18}O$ fluctuations in the ice in a core from Camp Century. At the time, the record appeared dubious due to large simultaneous variations in the CO_2 content in the air bubbles trapped in the ice. We now know that the latter are artefacts caused by the presence of carbonate dust attacked by the sulfuric acid present in the ice. However, the rapid fluctuations of $\delta^{18}O$ are indeed significant.

D/O warming appears to have occurred particularly quickly, over a few decades, solely when ice sheets developed on the continents of the northern hemisphere. The relatively warm period following the D/O event is called the 'interstadial period' and is characterized by gradual cooling. It ends with a rapid return to the coldest levels recorded, known as 'stadial periods'. This return completes a D/O cycle, which lasts a total of about 1500 years. There are therefore more D/O events than Heinrich events.

The existence of abrupt variations in both the surface climate in Greenland and in the ocean conditions over time scales far shorter than the Milankovich cycles contributed to the emerging idea that the climate system could have multiple equilibrium points and that it could be abruptly reorganized as it transfers from one equilibrium to another. This idea was reinforced by the correlations between marine and glacial records, first demonstrated by Bond et al. (1993). These authors suggest a perspective of glacial variability that integrates the Heinrich and Dansgaard-Oeschger events. The glacial millennial variability is organized in cycles, later called "Bond cycles" (Fig. 29.1). Each cycle begins at the end of a Heinrich event, by the first Dansgaard-Oeschger event, which has a large amplitude. This is followed by a few other Dansgaard-Oeschger cycles, of diminishing amplitude. The last cycle ends with a massive discharge of icebergs from the Laurentide sheet, in other words, by a Heinrich event. Each cycle lasts approximately 7000–10,000 years.

Since these discoveries in the early 1990s, millennial variability has become a subject of intense research. Numerous records were analyzed with as fine a resolution as possible and they showed that variability at shorter time scales than those of Milankovitch was not limited to the North Atlantic and adjacent regions. The challenge is then to be able to synchronize these different records to better characterize this type of variability, as well as to better understand the connections between climate signals recorded all over the globe. In the following sections, we focus on the signature of the Heinrich and D/O events, on their impact on climate and on the efforts to understand these climate instabilities through modeling.

The Regional Impacts of the Heinrich and Dansgaard-Oeschger Events: North Atlantic and Adjacent Regions

Atlantic Ocean

The Heinrich Events

The melting of massive armadas of icebergs profoundly modified the ocean surface conditions in the North Atlantic. The changes are recorded in ocean sediment cores, particularly in $\delta^{18}O$ signals from planktonic foraminifera. Indeed, excursions towards lighter $\delta^{18}O$ are measured in the calcite of these foraminifera, indicating either less saline surface waters or warmer temperatures. The fauna of fossil foraminifera in sediments reveal very cold conditions during Heinrich events (see, for example, Hemming 2004; Cortijo et al. 2005). The melting of icebergs is responsible for a huge inflow of freshwater, desalinating the surface of the ocean and introducing a highly negative $\delta^{18}O$ signal, typical of ice contained in the ice sheets.

Other paleoceanographic records reveal significant reorganizations during the Heinrich events. In particular, the study of $\delta^{13}C$ in sediment cores proves that the bottom waters of the Atlantic Ocean were less well ventilated (Elliot 2002), demonstrating a reorganization of ocean circulation. These studies were corroborated by analyses of the magnetic properties of sediment cores collected in the North Atlantic (see Kissel 2005 for a compilation), which show that the melting of icebergs was accompanied by a slowing of deep currents and of the thermohaline circulation.

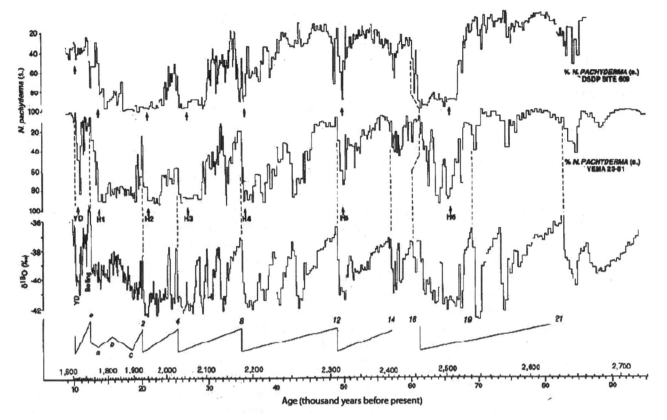

Fig. 29.1 Bond cycles (from Bond et al. 1993). The top two curves represent the abundance of *N. Pachyderma* (s.) in sediment records from the DSDP-609 and VEMA23-51 cores from the North Atlantic at 50 and 54° N, respectively. The Heinrich events and the Younger Dryas are indicated by arrows under the second curve. The third curve shows the $\delta^{18}O$ record in an ice core from the GRIP project in Greenland. These curves show how events recorded in marine cores and those recorded in ice cores correspond with each other. The fourth curve is a schematic representation of the Dansgaard-Oeschger and Heinrich events organized into cycles now called 'Bond cycles'. The figures over the age axis indicate the depth in the GRIP core

The isotopic study (Sr/Nd) of the grains brought by the icebergs to the North Atlantic during the Heinrich events made it possible to determine where these icebergs came from. Grousset et al. (1993) first demonstrated that within the Heinrich levels, although most of the detrital material brought to the Ruddiman belt came from the Laurentide ice sheet, there were some detrital elements from European and Icelandic ice sheets. Subsequent, more detailed studies of the Heinrich events complemented this panorama by showing that during each Heinrich event a level of detrital material from the European ice sheets preceded the level coming from the North American ice sheet. This almost systematic phase relationship led these authors to question the possibility of a causal relationship: could the icebergs from the European ice sheet be triggering the discharge of icebergs from the North American ice sheet? Another interpretation of these observations is to imagine that the European ice sheets were smaller and oscillated faster than the larger Laurentide ice sheet, causing them to precede it during the Heinrich events, with no obvious causality between the two.

The Heinrich events continue to attract a lot of attention from researchers, and many connections between these iceberg armada events and other records sensitive to temperature, precipitation and other climatic factors have been highlighted. This has led, in some publications, to a shortcut being made between the iceberg armada event and its supposed climate impact. It should be noted that, strictly speaking, a Heinrich event can only be defined by the presence of coarse detrital elements in sedimentary cores taken from the Ruddiman band. This implies in particular that the presence of detrital elements in an ocean core from the Fennoscandian ice sheet margin does not necessarily indicate a Heinrich event, but could be due to the melting of icebergs from European ice sheets. In addition, it is not generally possible to define Heinrich events from marine records outside the Ruddiman band and, even less so, from continental records. As a result, the term 'cold Heinrich event' is often inaccurately used because it applies only to the Ruddiman band, and is simplistic, because the drop in temperature is only one aspect of the climate changes associated with a Heinrich event. One exception is pollen records from marine cores containing detrital material. These provide simultaneous records from the adjacent continent (pollen) and from the iceberg armadas originating from the North

American ice sheet. They can therefore safely be used to reconstruct continental climate changes associated with Heinrich events (Sánchez-Goñi et al. 2002; Combourieu-Nebout et al. 2002).

Dansgaard-Oeschger Events

When the cores are studied at century-scale resolution, records of calcite $\delta^{18}O$ from foraminifera from the ocean surface indicate that abrupt changes also occurred during Dansgaard-Oeschger events. The variations recorded by marine indicators are very similar to those obtained from Greenland ice data. It is therefore tempting to directly link an observed variation in one set of records to one observed in the other and to assume that they are responses to a common cause. While this is certainly partially true (the $\delta^{18}O$ of polar ice is above all a response to temperature and the $\delta^{18}O$ of the foraminiferal calcite also contains a temperature signal), especially in the case of marine records close to the ice sheets, it should be borne in mind that it is paleoclimate indicators and not climate variables, such as temperature, that are being measured. Directly correlating every small variation in the two records is therefore hasty and inordinate. Despite these difficulties, marine records reveal that during Dansgaard-Oeschger events, the surface ocean and the deep ocean undergo changes of great amplitude in the surface temperature and/or the $\delta^{18}O$ of the sea water (see Rasmussen et al. 1996; Shackleton et al. 2000). The reconstruction of ocean surface temperatures from independent indicators shows that the temperature signals in Greenland cores are matched in the neighboring North Atlantic.

Associated with these changes in temperature and hydrological conditions are signals of relatively small amplitude in oceanic $\delta^{13}C$ records (Elliot 2002). Since $\delta^{13}C$ is an indicator of the ventilation of ocean water bodies, it would appear that at least for the North Atlantic and the Arctic, transitions between stadial and interstadial periods are not associated with large anomalies in ventilation, and therefore with any drastic change in thermohaline circulation in the Atlantic. This observation shows the different mechanisms operating during the Dansgaard-Oeschger events and Heinrich events, the latter having very marked anomalies in ocean $\delta^{13}C$ records.

Neighbouring Continents

The evolution of continental paleo-conditions is recorded in numerous environments, such as in caves (e.g. in concretions or speleothems), lakes (lake sediments), peat bogs, loess or marine cores. For continental regions adjacent to the Ruddiman belt, these records show a clear correlation between the evolution of the ocean and continent during periods of strong glacial variability. Pollen data, for example, show that during this period, rapid climate changes, whether D/O oscillations or Heinrich events, had repercussions on vegetation.

In Western Europe, D/O events resulted in periodic changes in forest cover (Fig. 29.2), with warm events corresponding to the expansion of oak forests, also implying wet conditions (e.g. Sánchez-Goñi et al. 2002; Combourieu-Nebout et al. 2002). On the other side of the Atlantic, in Florida, over the same periods, the oak is associated with large quantities of herbaceous plants (*Ambrosia, Poaceae*), indicating drier phases (Grimm et al. 2006). During the Heinrich events, the vegetation cover of western Europe became steppe-like, indicating a cold and very dry climate, especially in the Mediterranean region (Fig. 29.2). These cold conditions prevailing over Europe are associated with high $\delta^{13}C$ values recorded in several stalagmites in the south of France. They show up in a slowing down, or even a curtailment of growth as the cold conditions prevent the infiltration necessary for the formation of stalagmites (Genty et al. 2005).

At the same time, vegetation in Florida was characterized by an abundance of pine trees and the regression of herbaceous plants and oak trees which is interpreted as a consequence of a warmer and more humid climate (Grimm et al. 2006). The strong contrast between the responses by climate and vegetation to abrupt climate events on the two sides are examined with the cases of Europe and Florida. In Florida, this response may seem to be counter-intuitive, since the climate becomes warmer and more humid during the Heinrich events, although they are responsible for a major cooling in the North Atlantic. However, as records obtained for Florida have a relatively poor temporal resolution, it is possible that this contrasting response with the European one is due to problems of synchronization in the reconstructions. Another possibility is that the climates of Europe and Florida do not have the same sensitivity to freshwater incursions into the North Atlantic and the associated cooling. Europe, indeed, has a regime of prevailing westerly winds and is therefore under the direct influence of the cooling of the North Atlantic, unlike Florida on the other side of the Atlantic. It is thus crucial to be able to explain the complexity of these different signals. Modeling, as will be shown later in this chapter, can provide a coherent framework for reconstructions of climate changes in regions geographically distant from each other.

Millennial-Scale Variability in Other Regions of the World

After the discovery of abrupt climate changes in Greenland, the North Atlantic and adjacent regions, rapid changes in the characteristics of the climate system were discovered in

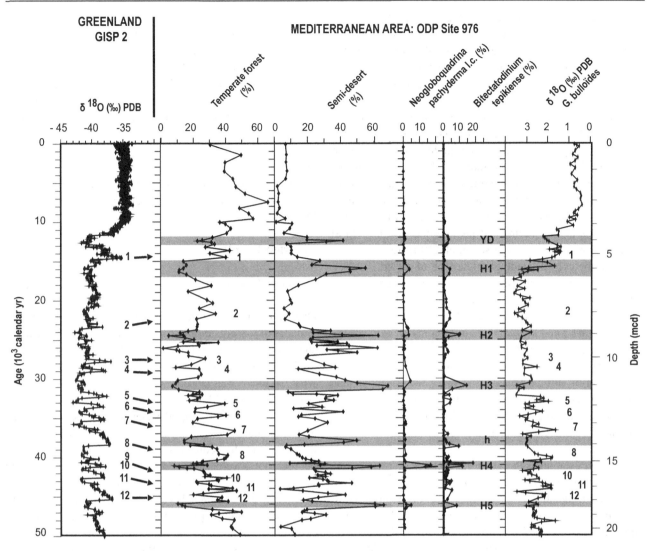

Fig. 29.2 Comparison of the $\delta^{18}O$ curves of the GISP2 ice core from Greenland with the paleoenvironmental reconstructions obtained for the ODP 976 (Alboran Sea) site. The curves are presented as a function of calendar age (on the left) and as a function of depth in the ODP 976 core, in m (on the right). D/O events are indicated by the numbers 1–12 on the GISP2 curve, with the corresponding numbers on the Mediterranean curves. The Heinrich events and the Younger Dryas are indicated by the gray bands. Adapted from Combourieu-Nebout et al. (2002)

other parts of the world, in some cases far from the North Atlantic. Abrupt changes in the δD of the ice at Vostok in Antarctica were reported as early as 1994 by the team of M. Bender, who established a correlation with the Greenland records, through the oxygen isotopes in the air preserved in the ice. In 1995, abrupt changes in ventilation were observed in the Santa Barbara Basin off California. In 1998, abrupt changes in biological productivity in the Arabian Sea, strongly related to monsoon intensity, were shown over time scales similar to Dansgaard-Oeschger cycles. In 2000, significant variations in the hydrological cycle in the low latitudes of the North Atlantic (Cariaco basin) were shown to correspond with events observed in the North Atlantic. Voelker et al. (2002) list 183 sites where variations of this type are recorded. Since this review, new records have been produced, confirming the global nature of these abrupt climate variations, and suggesting the involvement of a mechanism operating on a global scale.

Rapid climate variations of great amplitude have thus been found at many points on the globe. The most difficult part of interpreting these records is to synchronize them. The ideal would be to have an absolute dating for each record, on the same scale of the abrupt events, i.e., on a decadal scale. For most paleoclimate indicators, this is impossible to obtain. A dating method such as that based on ^{14}C does not offer sufficient precision over the glacial period to be able to link abrupt events recorded in two different places, in other words, to establish the phase differences between the various records. In addition, knowing that the residence time of ^{14}C in the ocean (reservoir age variations) varies during abrupt events makes absolute dating based on this single variable even more difficult. However, these variations are interesting because they give us information on changes in ocean ventilation and ocean circulation (Chap. 21).

However, it is possible to synchronize the different records with each other using parameters known to be the same or to vary in a similar manner in different locations on the planet. This allows us to study the chronology of events recorded in different places and to propose hypotheses about the mechanisms behind glacial abrupt climate variability. Specific events, such as volcanic eruptions, whose falling ash appears in ice, marine and lake sediments, can help to synchronize records (Rasmussen et al. 2014). Other parameters can be used depending on the type of record.

One example is the use of variations in the atmospheric concentration of methane to place Greenland and Antarctic ice records on the same time scale. Methane is well mixed in the atmosphere and so variations in its concentration are the same everywhere on the globe. Members of the EPICA community (2006) used this method to compare the expression of abrupt events at the two poles. Comparisons show an anti-correlation between isotopic signals in Greenland and Antarctica. In addition, the strength of the warming in Antarctica is related to the length of the cooling in Greenland.

Another example is the use of magnetic properties to synchronize marine records (see Chap. 7). At the scale of the ocean basin (e.g., the North Atlantic, Elliot 2002), variations in rock magnetism can be used to put marine records on a common timescale. At the interhemispheric scale, paleomagnetic intensity can be used, especially for episodes of sudden and intense variation, such as around the Laschamp event.

Mechanisms

The climate reconstructions described above demonstrate the existence of rapid climate variability of great amplitude that involve all the components of the climate system. Understanding all aspects of this type of variability is still a challenge for the scientific community. An approach to test our understanding of these phenomena is to build models that include processes that are thought to contribute in a critical way to rapid variability. Well-informed choices from the possible forcings and processes can highlight the importance of each one for a particular type of event or for a level of variability (see Chap. 25, volume 2). Models have thus been constructed to explain the iceberg armadas (the Heinrich events themselves, strictly speaking) or to estimate the inflow of freshwater to the North Atlantic that could explain the observations or even to establish the connection between the inflow of freshwater to the North Atlantic and the climate consequences. Each study focuses on one aspect of the rapid climate variability. For the time being, no study has tried to include all of the factors of rapid climate variability in a single model, which would require a representation of the ice sheets, climate, ocean, chemical composition of the atmosphere, conditions of land surfaces, etc. constituting an almost complete model of the Earth system, which does not currently exist. The following sections present some of the suggested mechanisms behind Heinrich events and how they impact on global climate as well as on the Dansgaard-Oeschger cycles.

Heinrich Events

Ice-Sheet Instabilities

One of the first explanations of the cycles of spectacular armadas of icebergs discharged from the North American ice sheet was proposed by MacAyeal in 1993. He constructed a model of the part of the Laurentide ice sheet positioned on what is today the Hudson Bay, that is, on a sediment-covered bedrock. When the base of the ice sheet reaches its melting point, causing liquid water to be present there, the ice slides much more easily than it would if the base of the ice sheet and the sediments were frozen or if it was on a bedrock not covered with sediment. The ice basal layer can warm up through the input of geothermal energy when the ice sheet is thick enough, because the temperature at the base of an ice column is dependent on the pressure exerted by this column and because of the ice insulating its base from the cold conditions at its top. Based on these properties, MacAyeal proposes the following cycle: (1) the ice sheet grows due to the accumulation of snow on the surface (assumed to be constant over time in this model) and the base of the ice is frozen; (2) once the ice sheet is thick enough, its base melts and a layer of liquid water forms at the base-sediment-ice interface; (3) the ice sheet then quickly slides towards the ocean and its elevation decreases; (4) the elevation of the ice sheet decreases sufficiently so that the basal layer freezes again, thus slowing down the movement of the ice, and returning to step 1. MacAyeal calls this model 'binge/purge'. In this model, the time between two "purge" events (i.e. between two episodes of iceberg break-up) is determined by the characteristic time necessaryfor the ice sheet to ticken and hence depends on the configuration of the ice sheet (size, distance to margins, characteristics of the underlying surface) and on its surface mass balance. It should be stressed that this mechanism works even for a constant climate. MacAyeal demonstrates an inherent oscillation in the ice sheets and shows that for the Laurentide

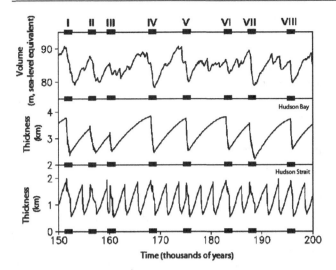

Fig. 29.3 Modeling of the instabilities of the atmosphere-ocean-North American ice sheet system in the context of a glaciation (Calov 2002). The climate model used is the CLIMBER-2 model, the model for the North American cap is the SICOPOLIS model in which the possibility of the ice cap sliding when the base is not frozen is incorporated. The model simulates a series of instabilities of the North American cap. **a** Ice volume of the ice cap. **b** Altitude of the ice cap above Hudson Bay (85.5° W, 61.5° N). **c** Altitude of the ice cap near the mouth of Hudson Bay (67° W, 60.75° N). The periods indicated in bold indicate periods when the ice cap releases armadas of icebergs, in other words, Heinrich events simulated by the model

ice sheet the length of the cycle is about 7000 years, which is of the same order of magnitude as the observed period. One criticism of this explanation is precisely that the ice sheets can oscillate independently, whereas a certain consistency between the break-up events in North America and Scandinavia has been observed (Grousset et al. 1993). However, the MacAyeal mechanism was reproduced in a more complex model (Calov 2002), which included the ocean, atmosphere, vegetation and ice sheets of the northern hemisphere, the characteristics of the bedrock at the base of the Laurentide ice sheet being imposed and spatially variable (Fig. 29.3). This latter study shows that although instabilities of the Laurentide ice sheet exist under constant forcing, the rest of the climate system can play a synchronization role for the events.

Hulbe et al. (2004), with MacAyeal as co-author, proposed an alternative to the "binge-purge" mechanism. Their explanation relies on the speculated presence of a massive ice shelf on the Labrador Sea. A break-up of this floating part of the Laurentide ice sheet would indeed be very efficient source of the icebergs defining the Heinrich events in the North Atlantic. A drawback of this theory, however, consists in the difficulty of identifying the ultimate reason of the ice-shelf break-up during cold surface conditions such as those observed during Heinrich events. Alvarez-Solas et al. (2013) postulated that such an ice-shelf break-up could be triggered by warmer subsurface waters. By using the hybrid ice-sheet/ice-shelf thermomechanical model GRISLI (Ritz et al. 2001), they simulated the effects that such an ice-shelf removal has on the inner part of the Laurentide ice sheet. Under this theory, not only the floating parts of the ice sheet contribute to the generation of the iceberg armadas but also the subsequent acceleration of the ice streams terminating in the ocean. For this mechanism to work, the ice shelves need to be destabilized from beneath and thus it assumes the existence of an external (to the ice sheet) forcing of Heinrich events. At the same time, Marcott et al. (2011) showed through the analysis of Mg/Ca in benthic foraminifera that the North Atlantic Ocean indeed experienced marked warmings of its subsurface waters during Heinrich events. Later on, in 2015, in a study entitled "Icebergs not the trigger for North Atlantic cold events", Barker et al. analysed several marine sediment records. They demonstrated that the melting of icebergs into the North Atlantic could not be the cause of the observed surface cooling because the icebergs simply arrive too late to these sites. They concluded that although the freshwater from icebergs could provide a positive feedback for lengthening stadial conditions, it does not trigger northern stadial events. Recently, Bassis et al. (2017) further expanded on the idea of the Heinrich events being triggered by the ocean by simulating the response of the Laurentide grounding line and the associated ice-streams acceleration.

Therefore, recent advances both in modelling and paleo-record analyses suggest that Heinrich events are a consequence of oceanic circulation changes rather than their primary cause. In this theory, a shift into a cold surface stadial condition in the Northern Hemisphere caused by a weakening of the Atlantic oceanic circulation would be accompanied by a warming of the subsurface waters of the Labrador Sea, facilitating the occurrence of a Heinrich event. An important fact here is that Heinrich events appear at the middle of the stadial phases. Subsequently, the massive presence of icebergs in the North Atlantic will amplify the decrease of the oceanic circulation intensity or even halt it, therefore enhancing stadial conditions. This interpretation of the chain of events involved in Heinrich events remains fully compatible with the observed excursions towards the lighter $\delta^{18}O$ observed in planktonic foraminifera. Nonetheless, this theory does not explicitly explain yet why Heinrich events do not appear for every stadial. The likeliness of this theory does not remove any value to the modeling exercises based on mimicking Heinrich events by means of freshwater flux injections into the North Atlantic. These have been very helpful in understanding the mechanisms by which a weakened or suppressed Atlantic circulation can have impacts on the rest of the climate system.

Evaluation of the Freshwater Influx Associated with a Heinrich Event: Example of a Joint Model-Data Approach

The thickness and location of the detritic levels found in the North Atlantic open the question of the volume of icebergs released in this region, especially as the records of sea level change during the deglaciation do not show the Heinrich 1 event having any impact. This would lead us to assume that the volume of icebergs is low (a maximum of 3–5 m of sea level). However, estimates achieved using other methods (Roche et al. 2004) have quite different results. Yet, in order to reproduce the evolution of these events in climate models, we must know the influx of freshwater represented by the corresponding armada of icebergs. It is therefore necessary to determine not only the volume of icebergs released into the ocean, but also the timeframe over which they were released. The classic ^{14}C dating method presents a major problem related to the reorganization of the ocean. Indeed, the postulated shutdown of the thermohaline circulation is causing a change in the distribution of carbon within the ocean reservoir, the largest in the Earth system. Basically, the stopping of the thermohaline circulation would make the surface waters appear younger, and the bottom waters appear older during the event, and it would make the surface waters appear older, and the bottom waters younger after the event, making it difficult to evaluate the duration of the event.

As the results from conventional methods assessing the duration and volume of icebergs emitted during these events have a very high degree of uncertainty, a more detailed estimation using different methods is necessary. An approach combining model and data was undertaken by Roche et al. (2004), based on the simulation of $\delta^{18}O$ of water in the ocean. In fact, there is a large number of $\delta^{18}O$ records of foraminiferal calcite from marine sedimentary cores from the North Atlantic, which constrain tightly the evolution of this indicator during the Heinrich 1 and 4 events. The basis of this new evaluation method is to consider that the geographical distribution of the $\delta^{18}O$ anomaly recorded in planktonic foraminifera provides information about the duration and volume of icebergs emitted. If the thermohaline circulation slows down, the slight anomaly in $\delta^{18}O$ created by the melting of the armadas of icebergs will tend to remain longer at the surface (and vice versa). The maximum recorded anomaly should also be related to the maximum of the iceberg flux. To assess this relationship, Roche et al. (2004) performed a large set of simulations by varying the duration and influx of additional freshwater for all of the feasible values (determined from the data). Then, the simulated $\delta^{18}O$ anomaly was compared with the anomaly measured in the marine sediment cores. The simulations that best represent the distribution described by the data were selected to provide a new estimate of the volume of icebergs discharged and the duration of these events: for the Heinrich 4 event (about 45 ka BP) the most likely duration is 300 ± 100 years and the volume of icebergs discharged is equivalent to about 3 m of sea level. These results were later confirmed by a method based on a sediment/iceberg model (Roberts et al. 2014).

Interpretation of the Isotopic Signal Measured in the North Atlantic

We have seen that the $\delta^{18}O$ records from Greenland cores and from ocean sediment cores appear to have very similar signals during periods of rapid climate variability. These similarities should not lead to the conclusion that both signals have the same climatic cause. Indeed, $\delta^{18}O$ records from foraminiferal calcite are complex, sensitive to both temperature and hydrological changes. Hydrological changes have a particularly strong impact. To separate out the influences of the different causes of $\delta^{18}O$ variations in calcite, one solution is to simulate this indicator within a climate model to analyze the importance of the various processes at work. Roche and Paillard (2005) performed this type of simulation for the series of Dansgaard-Oeschger events surrounding the Heinrich 4 event (about 40 ka BP). The result is shown in Fig. 29.4. The first outcome is the accuracy of the model in reproducing the variations measured in the ocean record. The advantage of using a climate model is that not only the $\delta^{18}O$ of calcite can be simulated but also temperature and salinity, prognostic variables for the climate model. The 'temperature' and 'hydrological cycle' components can then be extracted from the $\delta^{18}O$ signal of the calcite. The surprising result from Fig. 29.4 is that, in the model, the strongest changes in both temperature and $\delta^{18}O$ of the water over the entire simulated period are shown to be during the Heinrich 4 event. In the $\delta^{18}O$ record in calcite, the greatest variation is associated with the Dansgaard-Oeschger events. The reason for this difference is that since the variations in temperature and in the $\delta^{18}O$ of water have, from the point of view of the calcite $\delta^{18}O$, opposing signatures, their effects partially cancel each other out in the $\delta^{18}O$ of the calcite, hiding the strongest signal in the marine sedimentary data. This example shows how difficult it is to interpret isotopic paleoclimate indicators in terms of climate and how an integrated data-model approach can provide a better understanding of climate dynamics in the past.

Transmission of the Signal to the Continents

The pollen records extracted from the marine cores provide information on the evolution of the vegetation contemporaneous with the evolution of the oceanic conditions. The two

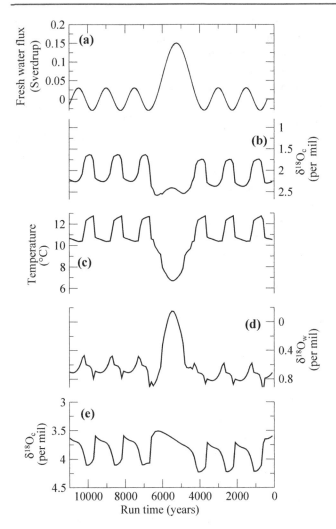

Fig. 29.4 Analysis of glacial variability (D/O and Heinrich events) in the North Atlantic Ocean, in the intermediate complexity model CLIMBER2-ISO, an atmosphere-ocean-vegetation coupled model including calculations of ^{18}O concentration, which allows direct comparison of the model results with measurements from marine sediments (foraminiferal calcite). Adapted from Roche and Paillard (2005). **a** Scenario of freshwater influx in Sverdrup (10^6 m^3/s). All model outputs are given at 41.25° N in the Atlantic for easy comparison with the MD95-2042 core. **b** $\delta^{18}O$ of the calcite in the ocean model's surface layer, i.e. the first 50 m (expressed in parts per thousand versus SMOW). **c** Evolution of the temperature at the surface simulated by the model (in °C). **d** Results of the simulation in $\delta^{18}O$ of the water (in parts per thousand versus SMOW). **e**) $\delta^{18}O$ of simulated calcite at a depth of 3000 m. The results for the temperature and the $\delta^{18}O$ of the water are not included for this depth because their variations are in the same direction and each contribute half of the $\delta^{18}O$ signal of the calcite

Atlantic, Mediterranean and the west coast of Europe (Sánchez-Goñi et al. 2002; Combourieu-Nebout et al. 2002, cf. Figure 29.2). The question is then to understand these relationships. The climate in a region such as Western Europe depends not only on the ocean surface conditions in the North-East Atlantic and the Mediterranean, but also on much more distant marine and continental conditions which impact on atmospheric and oceanic circulation.

To continue with the example of the impact a Heinrich event has on the climate and on the vegetation of Western Europe, we have seen that the signature of this event in terms of ocean surface temperature is firstly, a cold anomaly of several degrees Celsius at the mid-latitudes of the North Atlantic. We know, moreover, that at these latitudes the atmospheric circulation is dominated by westerly winds. A simple transmission mechanism of the oceanic signal would be its advection, i.e. its transport, by the prevailing atmospheric circulation, towards the 'leeward' continents, in this case, Western Europe. However, it is difficult to estimate how far the signal can travel and the atmospheric circulation itself can be modified by the changes in sea surface temperatures. The initial response may also be reinforced or negated by positive or negative feedbacks, such as, for example, snow cover or clouds (Chap. 1). Another question is: what can be said a priori about changes in precipitation associated with a cold event in the North Atlantic? Two processes need to be taken into account. A first constraint is that a colder atmosphere contains less water vapor (Clausius-Clapeyron relation) which makes it less conducive to the formation of precipitation. Second, in the mid-latitudes, over the oceans and the western side of the continents, most precipitation comes from weather systems that form over the oceans and pushed towards the western coasts of the continents by the prevailing winds. These perturbations are the result of atmospheric instabilities related to the meridional temperature gradient. A shift of the areas of higher gradient is important because it causes a displacement of the precipitation zones. A negative anomaly of ocean surface temperatures at the mid-latitudes of the North Atlantic implies a shift of the high meridional gradient zone of ocean surface temperatures towards the south. This should favor a southward migration of the prevailing wind belt and the precipitation associated with weather systems. If these weather systems are stronger, this can offset the direct influence of temperature on precipitation. This shows how important it is to have information from multiple locations in order to interpret a given record. For example, in order to interpret reconstructions of precipitations in Western Europe, it is important to know the temperatures in the region, which, in general, is relatively easy to infer from the same core, but it is also important to know the meridional temperature gradient over the North Atlantic and this requires having data from several cores located far from, but yet

types of reconstructions, marine and continental, are synchronized because they come from the same sediment core. Thus, by looking for pollen associations that appeared at the same time as the levels of detritus, the vegetation present during Heinrich events can be reconstructed. More generally, it is possible to study the relationship between ocean conditions and vegetation, as has been done for the Eastern synchronized with, the original core. This shows how important it is to synchronize the different cores. This can be complex. Given the application of the comparison of results from different cores, it is important to understand how this synchronization is done.

At this stage of interpretation, climate models (Chap. 25) can also be useful to better understand the possible links between changes in climate recorded at different locations. We now continue our analysis of the impact of a Heinrich event on climate and vegetation in Western Europe. In a first series of numerical experiments we hypothesized that for the climate of Western Europe, the main forcing linked to a Heinrich event is a cooling by 4 °C of the North Atlantic at the mid-latitudes. We used the LMDZ atmospheric general circulation model (developed at the Laboratoire de Météorologie Dynamique—Dynamic Meteorology Laboratory, Paris) to estimate the impact of such an anomaly in the context of glaciation, in this case, the Last Glacial Maximum. We therefore carried out two experiments by forcing the atmospheric model with glacial boundary conditions (see http://pmip1.lsce.ipsl.fr): ice sheets as reconstructed for the LGM, concentrations of greenhouse gases as measured for this period from ice cores, orbital parametres as they were 21,000 years ago. In the first experiment, we used the surface temperatures of the oceans as reconstructed by the CLIMAP project (Chap. 21). The only difference in the second experiment is the surface temperatures of the oceans in the North Atlantic. In this experiment, we decrease these temperatures by 4 °C between 40 and 50 °N. The second experiment is a sensitivity experiment to the ocean surface temperatures of the North Atlantic at the mid-latitudes.

Kageyama et al. (2005) show the results of this experiment for climate in France and the Iberian Peninsula. Figure 29.5 summarizes their results. The model does not simulate a propagation of the cooling imposed in the North Atlantic very far inland over the European continent. The place on the European continent where this cooling is most important in terms of temperature of the coldest month is in the northwest of the Iberian Peninsula. It brings about an increase of only 1 °C in this region, which is low compared to the 4 °C imposed in the North Atlantic. On the other hand, the precipitation anomaly simulated by the model in response to the cooling imposed in the North Atlantic is much greater: it reaches −200 mm/year (a shortfall of 200 mm/year, a drop of about 30%) over the Iberian Peninsula. Examining the results of the model for the North Atlantic and Europe, we see that the band of strong westerly winds is shifted southward, contributing to the decrease of precipitation over Europe. The slight precipitation increase over northwestern Africa can also be partly attributed to this atmospheric circulation change.

Kageyama et al. (2005) also show the impact of this change in climate on vegetation, as simulated by the dynamic ORCHIDEE vegetation model. The climate changes simulated by the climate model for a cold event in the North Atlantic, as weak as the impacts may seem on the land masses, result in a significant decrease in vegetation cover, of both trees and herbaceous plants. This result suggests that during glaciations vegetation in Europe and the Mediterranean is, as indicated by pollen records, extremely sensitive to changes in climate.

It may be noted that there is an area on the Mediterranean side of the Iberian peninsula where precipitation increases in the sensitivity experiment on colder ocean surface temperatures at the mid-latitudes of the North Atlantic. This anomaly, in opposition to the existing reconstructions for this zone (Combourieu-Nebout et al. 2002; Kageyama et al. 2005), shows the limitations of this experiment. The hypothesis that the factor responsible for the changes in climate in western Europe was the differences in surface temperatures of the North Atlantic Ocean at mid-latitudes is not sufficient to explain all the climate anomalies reconstructed around the Mediterranean region. A first explanation could be that the differences in surface temperature of the Mediterranean Sea were not taken into account. A second explanation could simply be that the model is not able to simulate the climate correctly for this region. A third explanation could be an error in the interpretation of the records. In a case like this, it is very instructive for both modelers and palynologists to compare their data. It is also for this reason that the results of the models should be analyzed not only from the point of view of surface climate variables but also from the point of view of the circulation and physics of the atmosphere, was this makes it possible to identify the mechanisms responsible for the simulated changes in climate and to improve the model and future experiments.

These experiments allow the influence of the various mechanisms suggested above to be quantified: propagation of the ocean temperature anomaly to the adjacent continents by the mean circulation, the shift of the prevailing winds following the migration of the zones with a strong meridian temperature gradient, the impact of atmospheric instabilities and average temperature on precipitation. In this way, a model can help interpret reconstructions and, in particular, can confirm or rebut scenarios established using several records located far away from each other.

It is important to be aware of the limitations of experiments of this type: regions where the anomaly is imposed, lack of vegetation or ocean feedback, assumptions in the model design and experimentation. These aspects can be evaluated by additional sensitivity experiments (see Chap. 25).

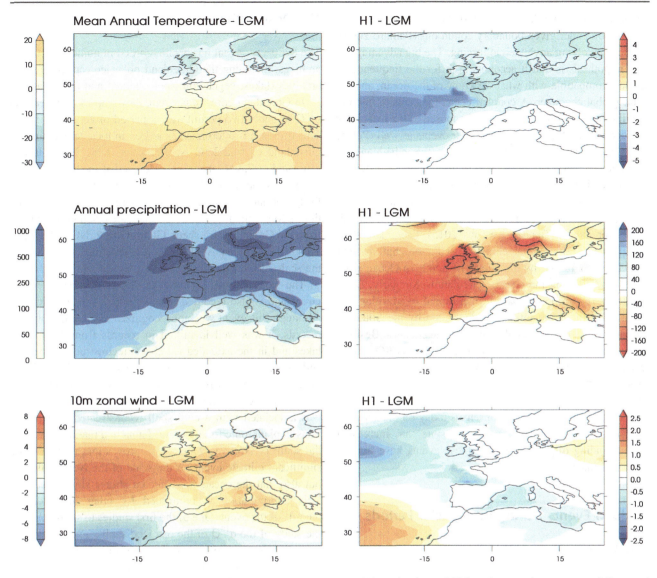

Fig. 29.5 Simulated climate variables for the LGM and the Heinrich 1 event (H1). Left column: LGM results: annual temperatures (°C), annual precipitation (mm/year), surface zonal wind (m/s). Right column: H1—LGM anomalies (simulations described in Kageyama et al. 2005)

Dansgaard-Oeschger Cycles

The Dansgaard-Oeschger cycles modify the climate on a global scale (see Voelker (2002) for a review). Therefore, the underlying mechanism must either have a global origin as well, or else must modify a major element of the climate system, which has a global impact. We will now consider three possible mechanisms that fulfil these requirements.

One of the simplest ways to obtain a global response is a variation in a forcing external to the climate system. It could be a mechanism operating on long time scales (Milankovitch forcing) or on short ones (volcanic changes, for example). It has therefore been proposed that variations in solar forcing may be responsible for Dansgaard-Oeschger events (Braun et al. 2005). However, the relationship between the small variations in the solar cycle at the relevant highlighted times and the observed climate variations remain questionable.

The component of the climate system most likely to be the cause of the Dansgaard-Oeschger events is undoubtedly the ocean, and more particularly ocean circulation. It has two main characteristics of interest here: it reacts on a global scale, on time scales compatible with the Dansgaard-Oeschger cycles, and it can respond to a relatively local forcing. Studies such as Ganopolski and Rahmstorf (2001) have shown that a semi-periodic freshwater forcing can, under certain conditions, reproduce oscillations that resemble the cycles observed in the paleoclimate data. However, this does not provide any indication of the origin of this freshwater forcing and therefore gives no new explanation as to the mechanism at work.

To compensate for this lack of a proven periodic climate forcing, some authors propose mechanisms with oscillations internal to the climate system. One of the most interesting explanations today relies on an internal oscillation of oceanic circulation under given boundary conditions (Paillard 2004), without involving any external forcing. The big advantage of this hypothesis is that it accounts both for the frequency of cycles, compatible with what we know of the ocean, and for their global character. A coupled ocean-atmosphere internal oscillation has more recently been proposed as the reason for the occurence of Dansgaard-Oeschger events as well (Peltier and Vettoretti 2014). This study has the advantage of showing such an internal oscillation of the climate system in a more realistic General Circulation Model.

Nevertheless, opposing theories coexist nowadays when trying to explain the triggering mechanisms of Dansgaard-Oeschger events. Banderas et al. (2015) also relate their occurence to an internal climatic oscillation, but in their case, it is controlled by the effects of Southern-Ocean wind stress and CO_2 on the oceanic circulation. Similarly, Dima et al. (2018) proposed that the meridional overturning circulation changes are ultimately controlled by the Southern Ocean with links to gradual global climate changes. Besides, it has recently been proposed that Dansgaard-Oeschger events owe their existence to the interactions between the North Atlantic ice shelves, sea ice and oceanic circulation (Boers et al. 2018).

To conclude, there is no consensus today on the cause of the Dansgaard-Oeschger events, despite an increasing understanding of the effects that these events have had on global climate.

Sudden Events During Interglacial Periods

The Discovery

Although the climate oscillation called 'the 8200-year event' left traces in the Greenland cores, the first publications presenting the isotopic records from the ice in Greenland from GRIP cores (Dansgaard et al. 1993) described the climate of the Holocene as stable, as the weak oscillation occurring around 8200 years before today was considered to be background noise. Even the discovery of a second record with exactly the same oscillation did not allow the correct recognition of this climate event. It was not until four years later that the notion of 'the 8200-year event' was introduced in an article by a team led by R. Alley. Efforts to place this climate evolution event on a hemispheric scale have indicated that this event had in fact been identified since 1973 from other records.

The Observations

The 8200-year event has been recorded in numerous climate archives, including the Greenland polar ice cap, continental archives (lakes, vegetation, speleothems) and the ocean.

In Greenland cores, this event shows up as a decrease in $\delta^{18}O$ of about two permil over a period of about 100 years. This decrease is interpreted as a drop in local temperature in Greenland of 6 ± 2 °C (see Thomas et al. 2007). It was estimated at 7.4 °C using an independent method based on oxygen isotopes ($\delta^{15}N$) (Kobashi et al. 2007). This was a fairly dramatic event, at least over Greenland. This oscillation is found in all the cores of adequate resolution obtained from the Greenland ice sheet: GRIP, GISP2, NGRIP, Dye-3. This is therefore an event with a regional impact, at least, on the scale of Greenland.

The cold period equating to the 8200-year event in Greenland is also recorded in other paleoclimate indicators. Although this event was not recognized as such in polar ice records, it was well identified in records from Norwegian lakes and in the extended ice sheets reconstructed from the end moraines. Indeed, the cold and dry period 'Finse[1]' was identified (Dahl and Nesje 1994), and then connected to the 8200-year event in the ice cores. Simultaneously, this event was associated with a cold spell reconstructed from variations in planktonic foraminiferal fauna in a marine core from the south of Norway. This record proves that it is indeed a significant climate event on the scale of the North Atlantic, but also that the ocean was affected almost simultaneously. This interpretation is reinforced by other more distant records of the event. One of the first studies to show this was the study of the $\delta^{18}O$ of the ostracod shells from a Bavarian lake (Ammersee, Germany), which recorded changes in the $\delta^{18}O$ of the precipitations associated with cooling (Chap. 15, von Grafenstein 1998).

In the oceans, records marking the 8200-year event are rare, as identification requires a temporal resolution of much less than a century. Of these records, those from the northern seas show a decrease of about 3 °C in ocean surface temperature. This study proves that this event was the largest during the entire Holocene in this region. Finally, these authors show that the oceanic change affected not only the surface but also the deep ocean. These last changes have been confirmed recently in a core from southern Greenland, which shows changes in the formation of deep water bodies in the North Atlantic.

On the continent, the 8200-year event is expressed as a cooling especially evident in Europe. Studies of the changes

---

[1]Named after the Finsevatn, the Norwegian Hardanger Lake where the record originated.

in the level of European lakes indicate that during this event the climate was wetter between 50 and 43° N, while north and south of this zone, it was drier. Pollen records rarely show this event, perhaps because these climate variations were not significant enough to modify the vegetation in a major way, or because it did not affect the growing period of the main plants characteristic of these paleoenvironments, as was shown in Northern Europe and North America.

The Mechanisms

The scale of changes observed as well as the multiplicity of their impacts (precipitation, temperature, oceanic circulation), very quickly pointed to the Atlantic thermohaline circulation as a major actor in this abrupt change in climate. The experience of Heinrich events (see above) suggested that here too, a sudden influx of fresh water to the ocean could have altered the thermohaline circulation and the climate around the North Atlantic. However, 8200 years ago, only a small part of the Laurentide ice sheet could have been involved, with the disappearance at this time of the dome covering what is currently the Hudson Bay. Traces of paleo-shorelines of lakes also indicate the disappearance of two large pro-glacial lakes (the Ojibway and Agassiz lakes) around this period, which could have contributed an influx of fresh water to the ocean. However, the first ^{14}C datings did not enable a precise chronology of these events. More precise dating, as well as an evaluation of the volume released, make it possible to pinpoint the massive draining of Lake Agassiz as a cause of this cold event (Clarke et al. 2004). These data indicate a draining of about 100,000 km^3 of water in about a year, thus causing a massive influx to the ocean.

Modeling and the Global and Hemispheric Consequences

Nevertheless, can the consequences for the climate of this extremely abrupt but yet short-lived influx be replicated in coupled climate models? Although the impact of this huge influx to the ocean was quickly identified, the first simulation of the 8200-year event using boundary conditions consistent with the time period was achieved in the early 2000s (Renssen et al. 2001). These authors were able to show that by forcing an ocean general circulation model coupled with a simplified atmospheric model using an idealized water flux of 0.75 Sv for twenty years, they produced a slowing of the thermohaline circulation in the Atlantic for about three centuries, with a temperature drop of 1–5 °C on the continents bordering the North Atlantic. This result fits well with the data, although the freshwater forcing is relatively larger than is accepted today. Even more importantly, it revealed disparities in the seasonal climate response, with a greater anomaly in summer than in winter over northern Europe and a north-south bipolarity, with cooling in the north and warming at 60° S. Subsequent studies confirmed that a freshwater influx, even one lasting only one year, was sufficient to obtain a climate response consistent with the one obtained from the analysis of paleoclimate indicators. The climate impact was not immediate, as can be seen in Fig. 29.6. For example, in Greenland, the maximum point was reached after about thirty years.

Other authors have also highlighted the role of climate 'noise' in the duration of the ocean's response to the freshwater forcing. Indeed, they show that by adding noise (with an average value of zero) to the freshwater forcing, the duration of the response by the model can be changed from a few decades to two or three centuries. This highlights the importance of the initial climate state to the response to a given forcing, as the response can have a more or less global effect as was the case for the 8200-year event, or have a local impact on a decadal scale. The 8200-year event was also reproduced in a coupled general circulation model incorporating water isotopes (Legrande et al. 2006), demonstrating that the results were consistent not only with temperature estimates but also with the paleoclimate indicators themselves, directly simulated in the model.

These authors were able to correctly predict the variations in $\delta^{18}O$ associated with the 8200-year event from the calcite of marine sediments before the publication of the first data showing the 8200-year event in marine sediment cores.

It is important to note that although the 8200-year event is considered an abrupt event during the interglacial period, this classification is not quite accurate. In fact, it is a final manifestation of the glacial climate through the draining of a periglacial lake, the result of a melting ice sheet. To date, no abrupt temperature changes involving glacial ice sheet behavior have been observed during the Holocene.

Outlook

Global Connections

The work of the last thirty years has made it possible to demonstrate a variability in the climate system on millennial time scales, characterized by transitions occurring, in some cases, over only a few decades. This variability was first observed in North Atlantic marine sediment cores for Heinrich events, and in Greenland ice cores for Dansgaard-Oeschger events. Gradually, climate records have been analyzed at increasingly fine temporal resolutions and variations with temporal characteristics similar to those in the North Atlantic and Greenland have been discovered in areas

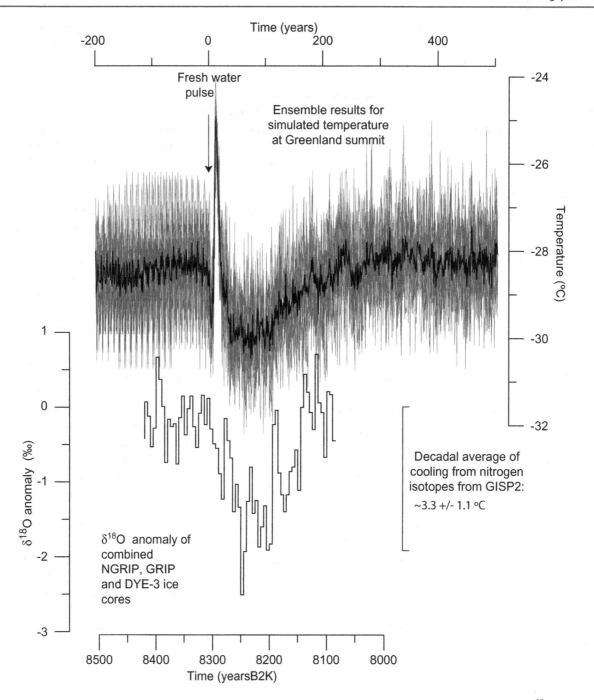

Fig. 29.6 Simulated surface air temperatures for the summit of Greenland for a set of ten simulations (gray). The average is drawn in black. These results are compared with the reconstructed $\delta^{18}O$ profile from measurements from Dye-3, GRIP and NGRIP ice cores. The same scale is used for the time axes, the $\delta^{18}O$ curve is positioned in relation to the model results, by aligning the first decrease in $\delta^{18}O$ with the first decrease in simulated temperature. The temperature axis is selected so that the decrease of 3.3 °C reconstructed from the nitrogen isotopes in the GISP2 core corresponds to the minimum anomaly in $\delta^{18}O$. Adapted from Wiersma et al. (2011)

very far from the original regions, as far away as the South Pacific and Antarctica, and including China and the northern Indian Ocean, influenced by monsoons. The interpretation of these data is based on the assumption that the time scales associated with each record can be accurately synchronized. This crucial synchronization is still a challenge despite advances made in methods of relative synchronization (such as methane concentration in ice cores or paleo-magnetic properties in marine cores) for the scientific community seeking to characterize rapid variability.

A comprehensive description on a global scale is very important for our understanding of the climate and of the variations that may have been experienced contemporaneously in parts of the globe very distant from each other. Finding the explanation for these connections is also a challenge faced by modellers, because attempting to model these events means testing our knowledge of climate as it is interpreted in the models which are also used to predict future climates. In terms of modeling, the first experiments testing the sensitivity of the ocean-atmosphere system to freshwater discharges in the North Atlantic in a glacial context have been carried out in the years 2000.

These experiments, if they are to be carried out with a resolution sufficiently fine to enable comparison between model results and reconstructions, require a calculation time at the forefront of the capability of the most powerful modern computers. For example, Kageyama et al. (2009), using an ocean-atmosphere general circulation model, show that the link between the cooling in the North Atlantic and the decrease in the Indian monsoon can be explained by a decrease in the summer temperature gradient in the upper half of the troposphere between the Indian Ocean and the Tibetan plateau. However, these results are obtained for conditions corresponding to the LGM and not to Marine Isotopic Stage 3, and further experiments will be required to better understand the signal transmission mechanisms between the North Atlantic and Asia. Furthermore, although these experiments reproduce certain climate events contemporaneous with the Heinrich events (southward migration of the Intertropical Convergence Zone, reduction of the Indian monsoon, cooling and drying in Western Europe), they do not manage to reproduce others, such as variations in the Southeast Asian monsoon. The modeling and understanding of the millennial variability within the climate system thus remains in many respects a challenge for modellers.

We have, until now, attempted to describe climate variability on the millennial scale mainly in terms of the "physical" Earth system, that is, including the ocean, the atmosphere and the cryosphere. However, the records contain more information than this simple description. Take, for example, the case of methane concentrations. We have shown that they are useful to synchronize the records from Greenland and Antarctica but this important greenhouse gas emitted mainly from the wetlands in the tropics and high latitudes shows significant variations. The role of biogeochemical phenomena in millennial climate variability also needs to be better understood and modeled.

Abrupt Event Interactions—Large Climate Transitions

Although abrupt events are the result of internal variability in the climate system, they are nonetheless sensitive to its large longer term variations. In particular, these events are more numerous and of greater amplitude in glacial periods than in interglacial periods. This suggests a different expression of millennial variability depending on the size of the ice sheets in the northern hemisphere. The high-resolution records of the most recent climate transitions (the deglaciation between LGM and the Holocene as well as the entry into the last glaciation) may provide information on the conditions favoring events of larger amplitude. However, these transitions are punctuated by abrupt climate events. This raises the question of the role that these events might play in the transition itself, knowing that a rapid event can have, in the case of Dansgaard-Oeschger events in the Greenland cores, an amplitude equivalent to half the difference between glacial and interglacial states.

Let us consider the case of the last deglaciation (Fig. 29.7). A few thousand years after the LGM came the Heinrich H1 event. Climate conditions returned to an almost glacial level. This event was followed by a warm phase, with the transition between these two events showing up as abrupt in many records, particularly around the North Atlantic. This is the Bølling-Allerød phase, whose climate is almost interglacial. However, this period was followed by the cold period of the Younger Dryas, which is sometimes considered to be the most recent Heinrich event (H0), which is wrong because there is no corresponding layer of detritic elements in the Ruddiman belt. It is after this last cold phase that the climate of the current interglacial, the Holocene, became definitively established (apart from the 8200-year event).

This shows that the last deglaciation was not a smooth transition. On the contrary, it is a series of abrupt events, as if the climate system 'hesitated' between two equilibria, one glacial and the other interglacial. The role of abrupt events during this climate transition is therefore important, but they still need to be understood and modeled.

Entry into the last glaciation at the end of the Eemian is also characterized by the appearance of abrupt events, ~110 ka before today, in a context where glacial ice sheets had already developed over Canada. One might think that these cold events would promote entry into glaciation, but this assumption ignores the fact that cooler air at high latitudes also contains less water and is therefore less able to supply the water needed to build the ice sheets up at a significant rate. These compensating factors need to be assessed. Here again, the influence of abrupt events on the evolution of the ice sheets has yet to be assessed, in comparison with other mechanisms and feedbacks, such as the slower changes in state of the ocean, cryosphere, vegetation, atmospheric concentrations of greenhouse gases, as well as external changes such as changes in insolation. Models of the Earth system can lead to a better understanding of the reconstructed signals by conducting sensitivity experiments for each of these factors.

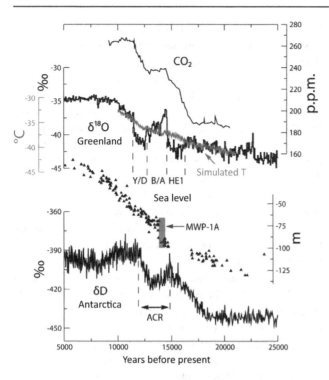

Fig. 29.7 An overview of the last deglaciation. From top to bottom: atmospheric CO_2 as measured in air bubbles from Antarctic ice cores; $\delta^{18}O$ measured in Greenland ice (GRIP members 1993) as a first-order indicator of local temperature, superimposed with the simulated average annual temperature in Greenland from the iLOVECLIM model; the scaling between model and data implies that the sudden increase of $\delta^{18}O$ at the HE1/B/A limit is 10 °C. The periods corresponding to the Heinrich 1 event (HE1), Bolling-Allerod (B/A) and the Younger Dryas (Y/D) are indicated below. It may be noted that there is no abrupt event in the temperature simulated by the climate model. Sea-level (Peltier and Fairbanks 2006) (black triangles) in meters relative to the current sea level, as reconstructed from corals with the abrupt event "Meltwater Pulse-1A" (MWP-1A) indicated, representing a rise of about 20 m in ~300 years. Isotopic abundance of deuterium in Antarctic ice (Peltier and Fairbanks 2006) as a first-order indicator of local temperature with the Antarctic Cold Reversal (ACR) indicated, a cold spell interrupting the warming of the deglaciation

References

Alvarez-Solas, J., Robinson, A., Montoya, M., & Ritz, C. (2013). Iceberg discharges of the last glacial period driven by oceanic circulation changes. *Proceedings of the National Academy of Sciences, 110*(41), 16350–16354.

Banderas, R., Alvarez-Solas, J., Robinson, A., & Montoya, M. (2015). An interhemispheric mechanism for glacial abrupt climate change. *Climate Dynamics, 44*(9–10), 2897–2908.

Barker, S., Chen, J., Gong, X., Jonkers, L., Knorr, G., & Thornalley, D. (2015). Icebergs not the trigger for North Atlantic cold events. *Nature, 520*(7547), 333.

Bassis, J. N., Petersen, S. V., & Mac Cathles, L. (2017). Heinrich events triggered by ocean forcing and modulated by isostatic adjustment. *Nature, 542*(7641), 332.

Boers, N., Ghil, M., & Rousseau, D. D. (2018). Ocean circulation, ice shelf, and sea ice interactions explain Dansgaard-Oeschger cycles. *Proceedings of the National Academy of Sciences, 115*(47), E11005–E11014.

Bond, G., et al. (1993). Correlations between climate records from North Atlantic sediments and greenland ice. *Nature, 365*, 143–147.

Braun, H., Christl, M., Rahmstorf, S., Ganopolski, A., Mangini, A., Kubatzki, C., et al. (2005). Possible solar origin of the 1470-year glacial climate cycle demonstrated in a coupled model. *Nature, 438*(7065), 208.

Broecker, W., et al. (1992). Origin of the northern Atlantic's Heinrich events. *Climate Dynamics, 6*, 265–273.

Calov, R., et al. (2002). Large-scale instabilities of the Laurentide Ice sheet simulated in a fully coupled climate-system model. *Geophysical Research Letters, 29*, 2216. https://doi.org/10.1029/2002gl016078.

Clarke, G. K. C., et al. (2004). Paleohydraulics of the last outburst flood from glacial lake agassiz and the 8200 BP cold event. *Quaternary Science Reviews, 23*, 389–407. https://doi.org/10.1016/j.quascirev.2003.06.004.

Combourieu-Nebout, N., et al. (2002). Enhanced aridity and atmospheric high-pressure stability over the western mediterranean during the North Atlantic cold events of the past 50 k.y. *Geology, 30*, 863–866.

Cortijo, E., et al. (2005). Heinrich events: Hydrological impact. *Comptes Rendus Geoscience, 337*, 897–907. https://doi.org/10.1016/j.crte.2005.04.011.

Dahl, S. O., & Nesje, A. (1994). Holocene glacier fluctuations at hardangerjøkulen, Central-Southern Norway: A high-resolution composite chronology from lacustrine and terrestrial deposits. *The Holocene, 4*, 269–277.

Dansgaard, W., et al. (1993). Evidence for general instability of past climate from a 250-kyr ice-core record. *Nature, 364*, 218–220.

Dima, M., Lohman, G., & Knorr, G. (2018). North Atlantic versus global control on dansgaard-oeschger events. *Geophysical Research Letters*.

Elliot, M., et al. (2002). Changes in North Atlantic deep-water formation associated with the dansgaard-oeschger temperature oscillations (60–10 ka), *Quaternary Science Reviews, 21*, 1153–1165.

EPICA community members. (2006). One-to-one coupling of glacial climate variability in Greenland and Antarctica. *Nature, 444*, 195–198.

Ganopolski, A., & Rahmstorf, S. (2001). Rapid changes of glacial climate simulated in a coupled climate model. *Nature, 409*, 153–158. https://doi.org/10.1038/35051500.

Genty, D., et al. (2005). Rapid climatic changes of the last 90 kyr recorded on the European Continent. *Comptes Rendus Geoscience, 337*, 970–982.

Greenland Ice-core Project (GRIP) Members. (1993). Climate instability during the last interglacial period recorded in the GRIP ice core. *Nature, 364*, 203–207. https://doi.org/10.1038/364203a0.

Grimm, E. C., et al. (2006). Evidence for warm wet heinrich events in Florida. *Quaternary Science Reviews, 25*, 2197–2211. https://doi.org/10.1016/j.quascirev.2006.04.008.

Grousset, F. E., et al. (1993). Patterns of ice-rafted detritus in the Glacial North Atlantic. *Paleoceanography, 8*, 175–192.

Heinrich, H. (1988). Origin and consequences of cyclic ice rafting in the Northeast Atlantic ocean during the past 130000 years. *Quaternary Research, 29*, 142–152.

Hemming, S. R. (2004). Heinrich events: Massive late pleistocene detritus layers of the North Atlantic and their global climate imprint. *Review of Geophysics, 42*, RG1005.

Hulbe, C. L., MacAyeal, D. R., Denton, G. H., Kleman, J., & Lowell, T. V. (2004). Catastrophic ice shelf breakup as the source of Heinrich event icebergs. *Paleoceanography, 19*(1).

Kageyama, M., et al. (2005). Le Dernier Maximum glaciaire et l'événement de Heinrich 1 en termes de climat et de végétation autour de la mer d'Alboran: une comparaison préliminaire entre modèles et données. *Compte Rendus Geoscience, 337*, 983–992.

Kageyama, M., et al. (2009). Glacial climate sensitivity to different States of the Atlantic meridional overturning circulation: Results from the IPSL model. *Climate of the Past, 5*, 551–570.

Kissel, C. (2005). Magnetic signature of rapid climatic variations in Glacial North Atlantic, a review. *Comptes Rendus Geoscience, 337*, 908–918. https://doi.org/10.1016/j.crte.2005.04.009.

Kobashi, T., Severinghaus, J. P., Brook, E. J., Barnola, J. M., & Grachev, A. M. (2007). Precise timing and characterization of abrupt climate change 8200 years ago from air trapped in polar ice. *Quaternary Science Reviews, 26*(9–10), 1212–1222.

Legrande, A., et al. (2006). Consistent simulations of multiple proxy responses to an abrupt climate change event. *Proceedings of the National Academy of Sciences of the United States of America, 103*, 837–842. https://doi.org/10.1073/pnas.0510095103.

MacAyeal, D. R. (1993). Binge/purge oscillations of the Laurentide ice sheet as a cause of the North Atlantic's heinrich events. *Paleoceanography, 8*, 775–784.

Marcott, S. A., Clark, P. U., Padman, L., Klinkhammer, G. P., Springer, S. R., Liu, Z., et al. (2011). Ice-shelf collapse from subsurface warming as a trigger for Heinrich events. *Proceedings of the National Academy of Sciences, 108*(33), 13415–13419.

Paillard, D. (2004). Modelling rapid events within the climate system. *Comptes Rendus Geoscience, 336*, 733–740. https://doi.org/10.1016/j.crte.2003.12.019.

Peltier, W. R., & Fairbanks, R. G. (2006). Global glacial ice volume and Last Glacial Maximum duration from an extended Barbados sea level record. *Quaternary Science Reviews, 25*, 3322–3337.

Peltier, W. R., & Vettoretti, G. (2014). Dansgaard-Oeschger oscillations predicted in a comprehensive model of glacial climate: A "kicked" salt oscillator in the Atlantic. *Geophysical Research Letters, 41*(20), 7306–7313.

Rasmussen, T. L., et al. (1996). Rapid changes in surface and deep water conditions at the faeroe margin during the last 58,000 years. *Paleoceanography, 11*, 757–771. https://doi.org/10.1029/96PA02618.

Rasmussen, S. O., et al. (2014). A stratigraphic framework for abrupt climatic changes during the Last Glacial period based on three synchronized Greenland ice-core records: refining and extending the INTIMATE event stratigraphy. *Quaternary Science Reviews, 106*, 14–28. https://doi.org/10.1016/j.quascirev.2014.09.007.

Renssen, H., et al. (2001). The 8.2 Kyr BP event simulated by a global atmosphere-sea-ice-ocean model. *Geophysical Research Letters, 28*, 1567–1570. https://doi.org/10.1029/2000gl012602.

Ritz, C., Rommelaere, V., & Dumas, C. (2001). Modeling the evolution of Antarctic ice sheet over the last 420,000 years: Implications for altitude changes in the Vostok region. *Journal of Geophysical Research: Atmospheres, 106*(D23), 31943–31964.

Roberts, W. H. G., et al. (2014). A new constraint on the size of Heinrich Events from an iceberg/sediment model. *Earth and Planetary Science Letters, 386*, 1–9.

Roche, D. M., & Paillard, D. (2005). Modelling the oxygen-18 and rapid glacial climatic events: A data-model comparison. *Comptes Rendus Geoscience, 337*, 928–934. https://doi.org/10.1016/j.crte.2005.03.019.

Roche, D. M., et al. (2004). Constraints on the duration and freshwater release of Heinrich event 4 through isotope modelling. *Nature, 432*, 379–382. https://doi.org/10.1038/nature03059.

Ruddiman, W. (1977). Late quaternary deposition of ice-rafted sand in the subpolar North Atlantic (lat 40° to 65° N), *GSA Bulletin*, 1813–1827.

Sánchez-Goñi, M. F., et al. (2002). Synchroneity between marine and terrestrial responses to millennial scale climatic variability during the last glacial period in the Mediterranean region. *Climate Dynamics, 19*, 95–105. https://doi.org/10.1007/s00382-001-0212-x.

Shackleton, N. J., et al. (2000). Phase relationships between millennial-scale events 64,000–24,000. *Paleoceanography, 15*(6), 565–569.

Thomas, E. R., et al. (2007). The 8.2 Ka event from greenland ice cores. *Quaternary Science Reviews, 26*, 70–81. https://doi.org/10.1016/j.quascirev.2006.07.017.

Voelker, A. H. L., et al. (2002). Global distribution of centennial-scale records for marine isotope stage (MIS) 3: A database, *Quaternary Science Reviews, 21*, 1185–1212.

von Grafenstein, U., et al. (1998). A Mid-european decadal isotope-climate record from 15,500 to 5000 years BP, *Science, 284*, 1654–1657.

Wiersma, A. P., et al. (2011). Fingerprinting the 8.2 Ka event climate response in a coupled climate model. *Journal of Quaternary Science, 26*, 118–127. https://doi.org/10.1002/jqs.1439.

An Introduction to the Holocene and Anthropic Disturbance

Pascale Braconnot and Pascal Yiou

The Major Trends of the Holocene

The Different Radiative Disturbances

The Holocene started about 10,000 years ago at the end of the last glaciation. The last thousand years of this period is marked by the growing impact of human activity (changes in land use and atmospheric composition). At first sight, the numerous data show us variations that are less spectacular than the great upheavals engendered by deglaciation. Nevertheless, the general natural trend, driven by changes in solar radiation at the top of the atmosphere, is characterized by radical changes in the monsoon and the El Niño phenomenon in the tropics. In the mid-latitudes of the northern hemisphere, the changes are seen in the characteristics of the main modes of variability. Several abrupt events also punctuate the unfolding of the process.

Slow variations in solar radiation at the top of the atmosphere, caused by variations in orbital parameters, are the driving force behind the evolution of the main climate characteristics on this 10,000-year scale. Variation in the obliquity, from 24.23° at the beginning of the Holocene to 23.44° at the present time, has brought about an increase in the average annual solar radiation at high latitudes of 1.5 W/m^2 over this period. During the same period, solar radiation in the low latitudes dropped by 1.10 W/m^2. Added to this effect of obliquity is the precession of the equinoxes. At the beginning of the Holocene, the summer solstice was located at the perihelion of the ecliptic. It is now at the aphelion. Thus, the solar radiation received at the top of the atmosphere (insolation) in June was 48 W/m^2 higher at 60° N at the beginning of the Holocene than it is today (Fig. 30.1), and 5 W/m^2 higher about 3000 years ago. Also, the amplitude of the seasonal cycle of insolation at the beginning of the Holocene was greater in the northern hemisphere but less in the southern hemisphere. This variation in amplitude is not symmetrical on either side of the equator, with greater variations in the northern hemisphere and in the tropics (Fig. 30.1). The precession also alters the length of the seasons. According to Kepler's laws, the boreal summer, defined as the time between the spring and autumn equinoxes, lasted 172 days 9500 years and 176 days 6000 years ago with 180 days currently. In the northern hemisphere, summer insolation was therefore more intense over a shorter period.

Greenhouse gases, volcanism and the solar constant are other factors that have influenced the climate of the Holocene through their impact on the radiative balance (Fig. 30.2). These are also the dominant factors of the last 2000 years, ever since insolation has reached approximate current values. Indeed, the combined effect of these greenhouse gases has resulted in a reduction in the radiative balance of the planet of around 0.5 W/m^2 from the beginning of the Holocene to the beginning of the industrial era. This slight perturbation of the radiative balance is due to a 7 ppm increase in atmospheric carbon content at the beginning of the Holocene, followed by a decrease of 20 ppm up to the beginning of the industrial era. Methane levels decreased from 730 ppb at the beginning of the Holocene to 580 ppb in the Middle Holocene (6000 years ago), and gradually returned to early Holocene levels in the pre-industrial era. Levels of atmospheric N_2O follow the variations of CO_2 and have varied between 2 and 10 ppb. In more recent times, the evolution of greenhouse gases is dominated by anthropic emissions. The combined effects of human activity correspond to an increase of 1.6 W/m^2 in the radiative balance. This estimate takes into account the dominant effect of the increase in greenhouse gases and the negative contribution of aerosols for the twentieth century (Solomon et al. 2007; IPCC 2013).

P. Braconnot (✉) · P. Yiou
Laboratoire des Sciences du Climat et de l'Environnement LSCE/IPSL, CEA-CNRS-UVSQ, Université Paris-Saclay, 91191 Gif-sur-Yvette, France
e-mail: pascale.braconnot@lsce.ipsl.fr

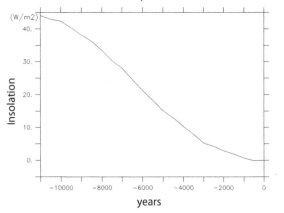

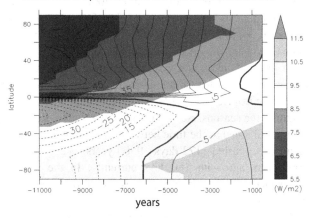

Fig. 30.1 a Difference in insolation at 60°N between the Holocene and the current period. **b** Change in the amplitude of the seasonal cycle during the Holocene, calculated as the difference between the month of maximum insolation and the month of minimum insolation (isolines, W/m^2), and months of maximum difference (shaded, number of the month)

There are many proofs of volcanic activity during the Holocene, but there is no quantified series for the overall period. Data has become more accurate for the last century, although the eruption timing, geographic location and quantities of aerosols emitted are not known with precision. Thus, depending on the sources of information, the amplitude and the date of the Holocene eruptions vary greatly.

The Evolution of Temperature in the Various Records

The recorded variations in temperatures during the Holocene reflect the response of the climate system to the various radiative disturbances listed above, as well as other various feedbacks, such as water vapor and melting snow and ice caps. The climate optimum at the beginning of the Holocene did not occur simultaneously at all latitudes (Fig. 30.3), as insolation depends on latitude and the season and feedbacks produce a delay between the forcing and its maximum summer temperatures 10,000–8000 years ago. Reconstructions for the mid-latitudes of the northern hemisphere are characterized by slow decreases in sea-surface temperature (SST) during the Holocene. In North America, the hottest period was between 7000 and 5000 years ago. Pollen data and macrofossil remains (see Volume 1, Chap. 10 for methodologies) indicate that temperate forests were found further north than they are today and that glaciers had retreated. The early warming at high southern latitudes at the beginning of the Holocene cannot be explained simply as a response to local insolation conditions, and appears to be a manifestation of a large-scale reorganization of atmospheric and oceanic heat transport. Yet most tropical regions show a gradual warming during the Holocene, reflecting an increase in insolation in these regions.

Estimates of the forcings induced by changes in the solar constant and volcanism are more open to controversy because they are more difficult to measure directly. Estimates from satellite data suggest that the differences between periods of solar activity and inactivity are due to fluctuations of about 0.08% in solar irradiance (1 965 W/m^2) over the last twenty years. For older periods, cosmogenic isotopes, the number of sunspots and observations of the aurora borealis are all indicators of solar activity (see Chap. 1, Volume 1). The amplitude of solar variations over the last millennium has recently been revised downwards and current best estimates suggest an increase of between 0.05 and 1.2% in solar irradiance between the Maunder minimum (between 1650 and 1720) and the current period, which corresponds to a radiative disturbance of 0.1 to 0.3 W/m^2.

Very intense volcanic eruptions introduce sulfate aerosols into the stratosphere, which reflect back solar radiation and thus contribute to a cooling of the climate in the year following the eruption.

The growing abundance of chronicles relating the impacts of climate extremes (due to the expansion of humanity) provides a description of the intra-annual climate variability over the last 2000 years. This can be superimposed on the secular fluctuations mentioned above. Figure 30.4 summarizes the current state of knowledge of temperature changes since the year 700 AD. Taking uncertainties into account, the temperature of the northern hemisphere is fairly stable between the year 700 and the end of the nineteenth century, and begins to increase continuously from the twentieth century onwards (Fig. 30.4c). The shape of this curve resembles a hockey stick, hence its iconic name. Excluding the twentieth century, for which there is an abundance of meteorological recordings, the value of the curve is that it shows a relatively warm phase between the year 1000 and 1300 (often called the 'Medieval Warm Period') and a relatively cold phase between 1350 and 1850 (called the 'Little Ice Age'). The causes of the inter-annual variations in

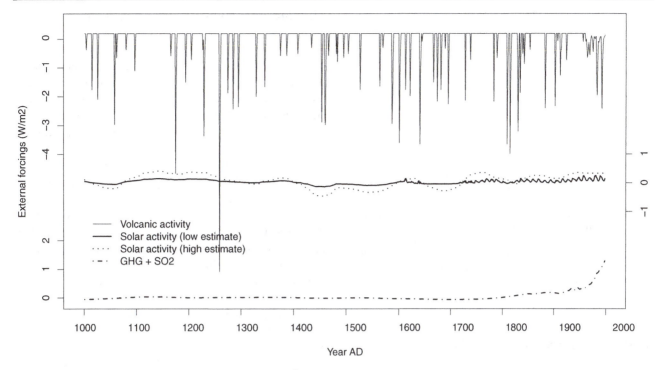

Fig. 30.2 Evolution of the estimates of radiative forcing (W/m^2) since the year 1000 AD. Top panel: volcanic activity; middle panel: solar activity (with low and high estimates from paleoclimate reconstructions); bottom panel: greenhouse gases (GHG) and sulfate (SO$_2$) aerosols (Jansen 2007)

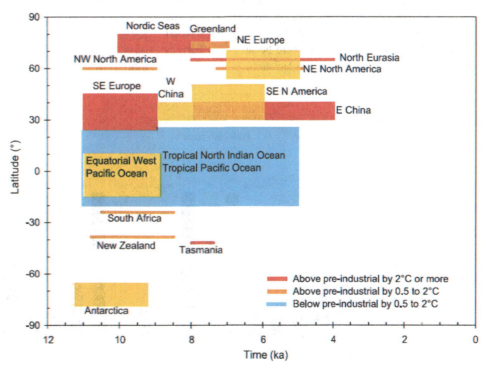

Fig. 30.3 Epochs (in thousands of years) and intensities of the maxima of temperature differences relative to the pre-industrial era, by latitude. The colors indicate the level of variance from pre-industrial levels (according to Jansen 2007)

temperature over the last two millennia are still not fully understood, especially since climate reconstructions are sometimes in disagreement for certain cold periods. Large-scale volcanic eruptions cool the atmosphere by one to two degrees, but these effects don't last longer than two years (Fig. 30.2). On the other hand, solar activity has a more modest effect on the radiative balance but one that is much more durable, as the phases of solar activity (Fig. 30.2) last several decades. In particular, the solar irradiance minimum, the 'Maunder Minimum' occurring in the middle of the seventeenth century corresponds to a marked cooling in the northern hemisphere.

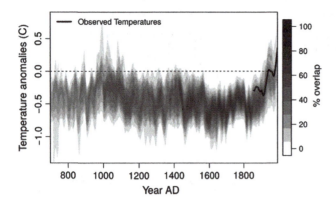

Fig. 30.4 Ensemble of reconstructions of the temperature anomalies of the northern hemisphere since the year 700 (shaded areas) (Jansen 2007) and average measured temperature anomalies (solid line). Anomalies of temperature are defined as differences with respect to a modern climatology (here 1960–1990). The shading accounts for the uncertainties of the reconstructions

Important Events of the Holocene

Several events punctuate the history of the Holocene. Here we will focus on the end of the humid period in Africa, about 4500 years ago, on the Medieval Warm Period, on the Little Ice Age and on the first lasting traces of human activity. The cooling observed about 8 200 years before now is discussed in Chap. 15 of volume 1 and not here because it is considered to be linked to the end of deglaciation.

The End of the Humid Period in Africa

About 4500 years ago, dry tropical Africa experienced variations in precipitation of much greater amplitude than it has for the last century. Lake levels fluctuated by several tens of meters and the surface area of Lake Chad reduced from 350,000 to 5000 km^2. Several 'climate crises', occurring over just a few years, punctuated this period and led to the complete drying up of lakes and the cessation of surface runoff (Gasse 2000). Wind-carried sediment recorded in a marine core off Mauritania suggests that a sharp change in atmospheric circulation occurred off the coast of dry tropical Africa between 5500 and 4000 years BP (deMenocal 2000). This period therefore corresponds with the end of the African Wet Period and the establishment of the current climate conditions. The resulting aridification of the Sahara conditioned the settlement of this area, where the activities of the populations were mainly farming and breeding of animals (Kropelin and Kuper 2006). However, rapid fluctuations do not seem to have impacted on the different regions of the Sahara-Sahel complex simultaneously, and signs of rapid events range are found from 4500 years to 2000 years BP depending on the latitude considered and the specific local conditions (Fig. 30.5). For example, the Dahomey dry interval was established about 4000 years ago in the middle of an area occupied by the Guineo-Congolese forest. The forest, fed by underground galleries, disappeared between 4500 and 3500 years ago in northern Nigeria but only 2000 years ago in the western Sahel. There are still many

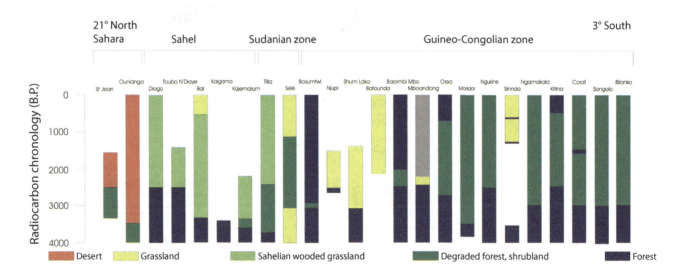

Fig. 30.5 Evolution of vegetation based on pollen data from different sites in Africa covering the transect between the Sahara at 21° N and the Guineo-Congolese zone in the south. This figure shows that the main transitions are not contemporaneous in the different locations (Courtesy of Anne-Marie Lézine)

grey areas concerning this period. Nevertheless, as insolation seasonality varied gradually, with reduced insolation during boreal summer, over the course of the Holocene, the rapid establishment of these dry conditions are the result of non-linear interactions between climates and ecosystems.

Medieval Warm Period

Through the study of historical archives, the British meteorologist Lamb reconstructed in the 1960s, the severity of the winters and the humidity of the summers since the year 1000 (Le Roy Ladurie 1967). He concluded that Europe experienced an escalation of dry summers during the period 1080–1200, unparalleled since then. He named this period the Medieval Warm Period. It is important to note that there are no direct temperature measurements for this time, since the thermometer did not yet exist. More recent studies (Jones and Mann 2004) suggest that Lamb's conclusions cannot be generalized to the entire planet.

From more extensive datasets, several studies have confirmed that many regions of the northern hemisphere experienced warmer conditions during the eleventh and twelfth centuries. However, several regions do not present signs of having experienced this optimum, or else did at a different time in history. For example, it appears that even though the summers were warmer at that time, winters were often very harsh in Western Europe until 1170. These apparent contradictions, probably due to a small number of observations, may be completed by new climate reconstruction programs in the future. Meanwhile, it should not be ruled out that this Medieval Warm Period was purely a regional phenomenon. It is also not a direct analogue to the current warming, as it was not associated with an increase in the atmospheric content of greenhouse gases.

The Little Ice Age

The Medieval Warm Period was interrupted at the beginning of the fifteenth century by the arrival of the Little Ice Age, which took hold until 1850. Again, there is no precise date for the beginning of this period: the Scandinavians felt a cooling at the end of the nineteenth century, whereas in France, the degradation of the weather was felt around 1430 (Bradley 1999; Le Roy Ladurie 1967). During this Little Ice Age in France there was a growth of alpine glaciers, very harsh winters and a succession of terrible summers. For example, the Bossons glacier, near Chamonix, descended 1000 m lower than presently. Though possibly influenced by the artistic trends of the time, the study of old engravings and paintings shows a striking difference with the present period. The strongest cooling was undoubtedly in northern Europe, with lakes being systematically frozen over in winter.

There are also signs of cooling in Equatorial America and New Zealand (Bradley 1999). This relatively cold period (on average, 1 °C cooler than currently in France) was responsible for the loss of crops and spikes in wheat prices, and often for insurrections, as well as an increase in the number of witches burned alive in public squares (Le Roy Ladurie 1967). After 1850, the glaciers began to retreat and the temperature to rise steadily, marking the end of the Little Ice Age.

The Little Ice Age is marked by a few particularly cold decades in Europe between 1650 and 1720. A decrease in the number of sunspots was observed at this time. These decades are called the Maunder Minimum, after the American astronomer Edward W. Maunder (1851–1928), who studied the relationship between sunspots and sun activity. It should be noted that it was the German astronomer Friederich W. G. Spörer (1822–1895) who first noticed the decrease in the number of sunspots between 1650 and 1720. History bestowed the Spörer Minimum between 1420 and 1570 on him. The variation in temperature caused by the direct influence of solar activity on the radiative balance at the Earth's surface (0.5 W/m^2) is less than 0.08 °C for the northern hemisphere (Jansen 2007) and is not sufficient to explain a cooling of between 0.2 and 0.5 °C in the northern hemisphere.

The Little Ice Age was also marked by intense volcanic eruptions that affected temperatures in a visible but not very durable manner (Fig. 30.4). Eruptions having a global effect generally took place in the tropics (Tambora, Krakatoa, Agung, Pinatubo etc.), and injected enough dust into the stratosphere to be homogeneously distributed globally (Robock 2000). Extratropical volcanoes (e.g., Laki in Iceland in 1783, St Helens in the United States in 1980) have had significant effects regionally in the northern hemisphere due to dust transport in the troposphere. However, their impact on a global scale was minor, because the volcanic dust they emitted did not reach the stratosphere.

The Anthropocene Era

The Anthropocene is a recently devised term that refers to the period when the climate and the environment became influenced by human activity. It is accepted that this period started at the beginning of the industrial era (in the middle of the nineteenth century), which also coincides with the first functioning weather service networks. It is possible, however, to argue that man began to impact on the climate from the middle of the Holocene. Land-clearing, the start of farming and slash-and-burn cultivation have emitted enough greenhouse gases into the atmosphere to be detectable in the glacial and sedimentary archives (Ruddiman 2007).

Climate Reconstructions for the Holocene

The Different Archives

The reconstruction of climate conditions during the Holocene poses several major challenges for the scientific community. The low amplitude of the variations in mean temperature, compared to the amplitude observed during the glacial ages, makes it difficult to use the usual indicators to identify a signal. Moreover, the need to take sub-annual time scales into account to describe the significant events means pushing the interpretation of many climate indicators to the limit, or else finding new ones compared to the ones traditionally used in paleoclimatology.

The typology of paleoclimate indicators for the Holocene allows them to be divided into two main categories: natural archives and societal archives. The characteristics of these archival types are summarized in Table 30.1.

This section does not discuss the natural archives treated in volume 1 (ice cores, marine cores, corals, lake sediments etc.). The focus here is on societal archives, because their study requires an expertise not called on previously: that of historians.

Societal Archives

One big difference between the study of climate during the Quaternary and during recent centuries is that detailed and quantitative records made by direct witnesses of the climate events exist for recent times. Societal archives have been mined mainly by historians, most notably by Emmanuel Le Roy Ladurie, who pioneered the study of climate through history on a European scale (Le Roy Ladurie 1967).

The study of climate through history (or of historical climatology) relies on the study of old archives describing chronicles, empirical measures, events (Brazdil 2005). The most important aspect of this study is not necessarily the production of databases that can be used by climatologists, but the verification of sources and the demonstration of their relevance to our understanding of climate.

Indirect Indicators

The climate has an influence on the evolution of the maturation of certain fruits. This is especially the case for grapes (*vitis vinifera*). In France, the harvesting dates for grapes have been recorded in parish or municipal records for several centuries. These records provide valuable indicators of temperature variations, as explained in Chap. 17 of Volume 1.

Similarly, there are series of harvest dates for other crops in France dating back to the Middle Ages, series of the bloom dates of fruit trees (e.g. the Kyoto series in Japan, dating back to 800 AD). These historical phenological indicators are responses to climate variations and provide information on the temperature conditions during the months preceding the harvest. For example, making the reasonable assumption that wheat is harvested when the grain is mature (or shortly afterwards), an understanding of the vegetative cycle of wheat and a knowledge of the date of harvest (which usually takes place at the beginning of summer) gives an idea of the spring temperatures for the region. An important point for the climatologist is to identify the species of wheat grown at that time so as to make an accurate estimate of the phenological cycle.

Societal archives are comprised mainly of chronicles of extreme climate or meteorological events (droughts, heat waves, storms, intense rain etc.) which had destructive impacts on agriculture or buildings. These records tend to contain such records as rogations, i.e. religious processions carried out to ask for divine help when drought or persistent rain threaten crops. The invoked saints are usually thanked when the requested effect occurs. These registers provide indications of the length of these climate episodes (Brazdil 2005; Garnier 2009).

In farming accounts in newspapers and in journals kept by local scholars, we find chronicles describing more or less precisely the influence of certain climate events on society: freezing days, ice jams on rivers, storms etc. This climate information is obtained through accounts of the damage caused and the costs incurred.

One of the reasons for the name 'Little Ice Age' is linked to the progress of the Alpine glaciers into the valleys that contained them from the fifteenth century on. Using dated indicators, such as moraines displaced as the glaciers advanced, it is possible to have an idea of how the

Table 30.1 Types of climate archives used to reconstruct the climate of the Holocene

	Natural archives	Societal archives
Direct climate data	None	Narrative descriptions, instrumented measurements
Indirect organic data	Tree rings, pollens etc.	Vegetation growth
Indirect non-organic data	Ice cores, sediments, glaciers, corals	Precipitations, ice floes, frosts etc.

volume of these glaciers evolved over the centuries. To get a better idea, climatologists also use graphical representations by painters, and later by photographers. It is thus clear that after a spread to maximum size at the beginning of the nineteenth century, the majority of the Alpine glaciers retreated by several kilometers over a century and a half, with an acceleration of this retreat at the end of the twentieth century.

Direct Indicators

Historians classify direct observations of climate and meteorology since the year 1000 into several types of historical phases. These phases qualify the type, abundance and quality of the information.

Before 1300, these were isolated accounts of extreme anomalies and natural disasters (Brazdil 2005; Le Roy Ladurie 1967). These accounts describe in particular the ravages caused by natural disasters: destruction of crops, buildings, floods, increased mortality etc. A detailed study of them, by cross-referencing sources and checking them, makes it possible to track the chronology of extreme events.

From 1300 to 1500, more or less continuous descriptions of the character of summers and winters (and to a lesser extent those of spring and autumn) become available, containing indications of everyday conditions.

From 1500 to 1800, more or less regular descriptions of monthly or daily conditions start to become available. These descriptions can be corroborated by records of processions or rogations organized by the local parishes to end droughts or avert events that might endanger crops (Le Roy Ladurie 1967).

Between 1680 and 1860, the very first instrumental measurements appeared. The barometer was invented by Torricelli and the thermometer by Galileo. The first attempts to establish international meteorological networks were made during this time. In France, the first network of systematic observations dates from the reign of Louis XVI. In 1776, Félix Vicq d'Azyr, secretary of the Academy of Medicine, asked the doctors of the kingdom to record the temperature of the air three times a day, as well as to write a summary of the diseases treated during the month. This initiative was founded on the idea that variations in climate could have an impact on the health of the population. This work was maintained for a few decades, and many doctors in France contributed to the exercise, scrupulously noting temperatures and diseases. Unfortunately, this directive was discontinued, and it is not possible to have continuous data. In addition, the physicians of the day obviously had no knowledge of meteorology or instrumentation, and not all of their measurements were reliable.

From 1860 onwards, meteorology was developed within the framework of national and international networks. Urbain Le Verrier, who discovered the planet Neptune by calculation, was also an influential politician of the Second Empire. Like his astronomer predecessors of the Observatoire de Paris, Cassini, Maraldi and de la Hire, he was also interested in meteorology and, as a politician, in the strategic advantage that could be derived from weather forecasting. Following the disaster of Sevastopol in 1854, when the Allied fleet was destroyed by a storm, Le Verrier claimed that it was possible to predict this event through a network of ad hoc meteorological observations. This marked the birth of centralized meteorological networks via the telegraph, which would for a long time be under the control of the army in most countries in the world.

Statistical Methods for Climate Reconstruction

Methods for obtaining the temperature curves from Fig. 4 are based on statistical regressions between several categories of climate indicators. The general concept behind reconstructing a hemispheric temperature is to use a set of indicators (thickness of tree rings, isotopic concentrations, pollen concentrations, harvest dates etc.) evenly distributed over the hemisphere. This is called a 'multi-proxy' approach because it uses a mix of several types of climate indicators (referred to as proxies from now on).

The strategy of these reconstructions (Jones and Mann 2004) is generally in three steps and requires a temperature dataset with sufficiently good coverage of the globe, hemisphere or well-defined region (e.g., The North Atlantic or the Equatorial Pacific).

The first step is to determine a small number of general statistical characteristics (space or time) for recent observations. The best known temperature reconstructions (Mann et al. 1998) use statistical techniques of decomposition into *principal components* (von Storch and Zwiers 2001), however there are alternatives, depending on the distribution of the proxies and their properties. Thus, the temperature field T, which depends on time t and space x, can be written in the form:

$$T(x,t) \approx \sum_{t=1}^{5} a_k(t) E_k(x).$$

In this equation, the $E_k(t)$ are the spatial modes of the variability of T, and the $a_k(t)$ are the associated time coefficients (von Storch and Zwiers 2001), generally called principal components (denoted PC). In this equation, we only keep 5 modes.

The climate indicators (proxies) are then compared with the evolution of the statistical characteristics of observations over a calibration period, where proxies and observations are available. This calibration period may cover all or part of the twentieth century. There is a technical debate about the selection of this learning period (Jansen 2007) and the number of statistical components to be used, but the result in the end is not very sensitive to this selection. During this period, a linear regression is performed between the proxies and the statistical characteristics observed. This regression attaches weights to proxies in order to maximize their correlation with temperatures. This analysis also makes it possible to eliminate the proxies which have a poor correlation with the temperature signal during the learning period.

A 'verification' period prior to the learning period for which instrumental temperature data are available is then used. This verification period identifies the errors incurred in the regression obtained for the calibration period. For this verification period, it is also possible to determine the error linked to the omission of proxies in the estimation of the temperature. This calculation is essential because it is obvious that the further back in time we go, the fewer climate series are available and the more uncertain the reconstruction of temperatures becomes.

Finally, temperatures can be reconstructed for the last millennium. It should be noted that this reconstruction has a spatial aspect. It is important to bear in mind that this type of method relies on basic assumptions about the temporal stability of the climate modes identified in the temperature data during the calibration period and the temporal stability of the relationship between proxies and temperatures. These two types of stability can be quantified over the verification period, but it is impossible to exclude the possibility of changes over a longer period. Another problem with this type of reconstruction comes from the statistical regression between proxies and main components over the calibration period. Since a regression is generally imperfect, it inevitably leads to an underestimation of the variance when this is used to reconstruct the climate, which can lead to a poor estimation of long-term climate variations. It is possible to partially solve this problem by using proxies that represent different time scales, and are thus sensitive to scales varying from inter-annual to centennial (Jansen 2007).

A major advance in the quality of climate reconstructions lies in the improved understanding of the mechanism linking the 'proxy' to the climate variation, which helps the statistical steps described above to be guided by knowledge of the physics. This research topic is particularly active at the moment.

Several research teams have proposed temperature reconstructions for the last millennium, based on different proxy datasets (Jansen 2007). These reconstructions often have common foundations (often, tree ring data) but the spatial distribution of the proxies used varies considerably from one reconstruction to another.

The evolution of the error bars of these reconstructions shown in Fig. 30.4 shows the discrepancies between the estimates of temperature changes, especially during cold periods. It should be noted in particular that the latter part of the twentieth century emerges significantly from the error bars of the temperature variations for the preceding millennium.

Climate Simulations

Climate simulations provide an understanding of how different forcings affect the climate and quantify the main feedbacks. Moreover, comparison between the model results and the data makes it possible to determine whether the models are capable of representing a climate different from the current one. The models used for Holocene simulations cover the spectrum of models presented in Chap. 4 of this volume which describes the different hypotheses and the protocols to run such simulations. In the case of the Holocene, it is mainly the characteristics of the seasonal cycle that have been analyzed, as changes in insolation, driven by precession, strongly modulate seasonality but have little impact on the annual average of the different climate variables. For the last 2000 years, the emphasis is on understanding the forcing associated with fluctuations in the solar constant and in volcanism, and the identification of associated feedbacks. This is a major step towards gaining perspective on recent centuries strongly disrupted by human activity. Increased understanding of interannual to centennial variability is necessary in order to detect climate change and, where appropriate, to attribute it to human activities.

Holocene Simulations

Major Trends

There are very few simulations that cover the entire Holocene. Those that exist were carried out using models of intermediate complexity (see Chap 4), because it is impossible to simulate a period of 10,000 years within an acceptable time frame using general circulation models, which need an average of one month to achieve 100 years of simulation on supercomputers.

The main objective of simulations of the whole Holocene period is to reproduce the major climate trends caused by variations in orbital parameters and by the concentration of greenhouse gases. They generally do not take into consideration the full set of forcings such as volcanism or the evolution of the solar constant, factors which are not well known for the whole period. The applied models of intermediate complexity take account of atmospheric and ocean circulation in a simplified way, as well as sea ice and vegetation (Crucifix et al. 2002; Renssen et al. 2005). They nevertheless have different levels of complexity.

The first model considers the ocean and the atmosphere in terms of latitudinal sectors, while the second takes a three-dimensional ocean model coupled with a simplified atmospheric model. The simulated annual temperature changes are very low, in accordance with the low amplitude of the change in insolation. Between 60° N and 70° N, these studies show cooling during the Holocene in response to the reduction in summer insolation relatively in line with observations (Fig. 30.3). The increase in greenhouse gas concentrations since the Middle Holocene partly offsets the reduction in temperature by about 0.5 °C. The beginning of the Holocene is also a time when the ocean was still disturbed by the recent deglaciation and when there were larger ice caps than are currently present in North America and Greenland. Simulations (Renssen et al. 2005) indicate that the main effect of the Fennoscandian cap at the beginning of the Holocene was to delay the climate optimum in eastern Canada and Greenland (Fig. 30.3). These studies also suggest an increase in variability during the Holocene, mainly around the northern seas, resulting from feedbacks between atmospheric circulation, ocean and sea ice.

The simulations show no major changes in thermohaline circulation of the ocean during the Holocene, except perhaps a slight increase between 9000 and 8000 years before the present. It also appears that the initial conditions and the long-term variations of the intermediate and deep ocean have had little impact on the characteristics of the climate during the Holocene. Thus, it can be taken that the different periods are more or less in equilibrium with the forcings. Hence, equilibrium simulations for a particular period of the Holocene can be used to correctly identify the main characteristics of the climate. This result is important because the ignorance of oceanic conditions at the global scale at the beginning of the Holocene is a source of uncertainty for numerical simulations. Nevertheless, the equilibrium is not necessarily verified over the last 4000 years because of a 200-year lag between the forcing and the climate response of the reorganization of the boreal forest in the northern hemisphere (Crucifix et al. 2002).

Reference Periods and Analysis of Feedbacks

Simulations from intermediate complexity models describe the major trends and main feedbacks that have shaped the Holocene in the mid and high latitudes of the northern hemisphere but do not allow for a detailed analysis of feedbacks or changes in variability. Some periods, such as the beginning of the Holocene (9500 years ago) and the Middle Holocene (6000 years ago), have been the subject of numerous modeling studies and benefit from international coordination (Joussaume and Taylor 1995; Joussaume et al. 1999; Kutzbach 1988; Braconnot et al. 2007a).

In particular, there is an effort to compile data at the global scale for the Holocene (Prentice and Webb 1998), and this helps, along with simulations, to improve our understanding of the mechanisms and the different feedbacks involved and to assess the ability of climate models used for climate projections for the next century to represent a different climate than the current one (Cane et al. 2006; Braconnot et al. 2012).

Feedbacks from Snow, Vegetation and from Sea Ice in High Latitudes

Feedbacks from ocean, vegetation and snow and ice cover are the main factors that have conditioned the evolution of temperature in the different regions of the northern hemisphere during the Holocene. When forest is replaced by grass, the reflective power of the combination of snow and vegetation is more marked. This results in a positive loop of increasing snow, cooling and withdrawal of the forest. This mechanism in particular describes entry into the last glaciation (de Noblet 1996). The ocean response leads to a phase shift in the seasonal cycle due to its high thermal inertia and evaporation capacity, which limits surface heating. Thus, spring and autumn are more affected than summer and winter by oceanic feedbacks. These effects combine with the feedback from sea ice and snow to reduce seasonal contrasts at high latitudes and to amplify the cooling during the Holocene (Wohlfahrt et al. 2004).

The feedback from snow and sea ice is manifested in a modification of the surface albedo. It is therefore possible to quantify it in terms of energy (W/m^2), using a simple approach to climate sensitivity analysis (Taylor 2007). This allows us to define this effect as:

$$\text{Albedo effect} = (SWncs_{pal} - SWncs_{Ok}) - SWf$$

where $SWf = (1 - \alpha_{Ok})\Delta SWi$ represents the radiative forcing related to the change in insolation and $SWncs$ represents the net solar radiation (SW: short wave) in a clear sky (ncs) for the paleoclimatological (pal) simulation and for the control simulation (Ok). The radiative forcing takes into account the model characteristics via the albedo of the control simulation. Indeed, although the same insolation disturbance SWi is applied to different models, the difference in net incident solar radiation depends on the selected model as a function of the surface characteristics, the representation of clouds etc. which characterize the planetary albedo in the model (Fig. 30.6).

As an example, Fig. 30.6 shows that between February and April, the estimated radiative forcing for two different models ranges from −12 to −4 W/m^2 between 20° N and 50° N. The cooling associated with this insolation deficit allows the snow to remain longer on land.

This snow introduces a local feedback of −4 to −16 W/m^2 on the continent which reinforces the insolation deficit and the cooling. Both simulations produce the same type of feedback, but with different amplitudes and geographical localization. On the other hand, in summer, the

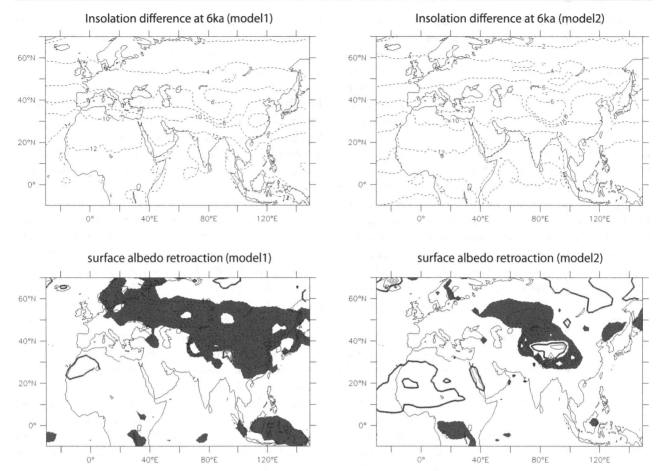

Fig. 30.6 Radiative forcing due to the change in incident radiation at the top of the atmosphere during the Mid Holocene (W/m^2) and feedback resulting from surface changes (mainly change in snow cover) for the months of February to April from two climate models. Adapted from Braconnot et al. (2007b)

lower coverage of sea ice and snow induces a heating surplus of 5 W/m^2 above 30° N. The water vapor feedback adds to this effect, contributing an additional 2.5 W/m^2 to summer warming in these regions. These different feedbacks amplify the direct effect of insolation. The way in which they are represented in the models is the source of the differences between simulations and of the uncertainty on their amplitude.

Monsoons and Ocean and Vegetation Feedbacks

The amplification of monsoon regimes in the northern hemisphere is the main climate characteristic in tropical regions during the first half of the Holocene. The wet belt extending from Africa to India and South Asia recorded strong changes in precipitation regimes, in line with the northward extension of the intertropical convergence zone.

These fluctuations are recorded in both continental and ocean data. Several factors explain the stronger monsoons between the beginning of the Holocene and the Mid Holocene. The first contributory factor is the amplification of the seasonal insolation cycle in the northern hemisphere. The monsoon is the result of differential heating between the hemispheres and between the continents and the oceans. Higher summer insolation increases the contrast in temperature between the land and the ocean. This contrast alone does not explain all the monsoon characteristics. The orography and the warming at high altitudes of the Tibetan plateau constitute a source of energy which contributes to the establishment of this planetary thermal current. As the monsoon becomes established, the release of latent heat at the time of condensation is an additional source of heat enhancing the convergence of moisture in areas of high convective activity.

For the climate of 6000 years ago, Fig. 30.7 shows the changes in temperature, precipitation and large scale circulation linked to the reinforcement of the monsoon in Africa and India in response to increased insolation of the northern hemisphere, as well as the uncertainties arising from differences between the models. Continental warming in July-August reaches about 2.5–3 °C. This heating is

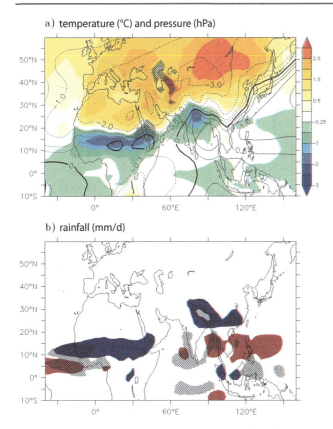

Fig. 30.7 a Temperature change (shaded, °C) and surface pressure (hPa) between the Mid Holocene 6000 years ago and the current period based on the average of the set of simulations from the international PMIP project. **b** As for a, but for ocean surface temperatures (isolines, °C) and precipitation (gray). In figures a and b, the hatching indicates the regions with the most variable results from one model to another

associated with a deepening of the thermal depression which is reinforced where warming is at a maximum. Large-scale circulation leads to increased convergence of ocean winds towards the interior of the continents. Large-scale moisture convergence follows this circulation in the lower layers of the atmosphere, explaining the increase in precipitation in Africa and northern India. These main features are reproduced by all of the models. However, there is great disparity in the magnitude of the simulated changes in many regions such as East Africa, Arabia and northern India.

Several factors amplify or offset the atmosphere's response to insolation. These have been described in several overview articles (e.g. Braconnot et al. 2012) of which the main points are summarized here. The method used to study the ocean and vegetation feedbacks consists mainly of removing the feedbacks one by one.

The Role of the Ocean

The ocean plays a major role in modulating the amplitude and phase of the seasonal cycle. Figure 30.8 summarizes the simulated SST (Sea Surface Temperature) changes for key regions of the different tropical ocean basins (Zhao et al. 2005). Coupled models capture the main features of the SST evolution for the current climate. It should be noted, however, that differences between the simulations range from 0.5 °C to more than 2 °C. The amplitude and phase of the seasonal cycle are relatively well reproduced in most of the models. The results show the greatest amount of disparity for the Indian Ocean as models do not correctly capture the semi-annual SST signal.

Despite the differences for the current climate, the response of the ocean to the change in insolation is very consistent from one model to another. All of the tropical regions are marked by a cooling in the first part of the year, a direct response to the change of insolation. Warming occurs one to two months after the change in insolation hits its peak, depending on the region. The shift in phase between the regions comes from the discrepancy in seasonal insolation between the northern hemisphere and the southern hemisphere and from localized changes in the thermal inertia of the ocean. Although the main changes are relatively similar between the models, dispersion between results can be up to a factor of 2. The phase differences between the models are greater in winter than in summer because the change in insolation is weaker and lasts longer. The tropical ocean is relatively cold in late spring, when the monsoon starts, and accentuates the contrast between land and ocean, promoting moisture advection on the continent. In the Atlantic Ocean, a dipole occurs on both sides of 5° N, with warmer temperatures in the north and colder in the south than is currently the case. It strengthens the low pressure zone between 10 and 20° N, the convergence of humidity in this region and the monsoon influx in West Africa. This differential between the two hemispheres originates in the heating differential linked to the insolation on either side of 5° N. The heating of the surface ocean is reinforced to the north of 5° N by a decrease in evaporation due to a weakening of the trade winds, part of which converges towards the African continent instead of crossing the Atlantic. In addition, enhancing the monsoon influx initiates an Ekman[1] transport in the surface layers of the ocean, contributing to a delay in the warming of the region south of 5° N (Zhao et al. 2005). The system relaxes in autumn when insolation on the equator is sufficient to smooth the temperature gradient. In India and South-East Asia, warming of the western part of the Indian Ocean and of the 'warm pool' promotes the convergence of surface circulation to these warm waters to the detriment of the continent. The Indian monsoon seems

---

[1] The current generated by the surface wind is rotated to the right of the wind in the northern hemisphere under the effect of the Coriolis force and reduces with depth under the effect of friction in the form of a spiral (Ekman's spiral). The transport generated (Ekman transport) over the entire Ekman layer (about 100 m) is perpendicular to the wind.

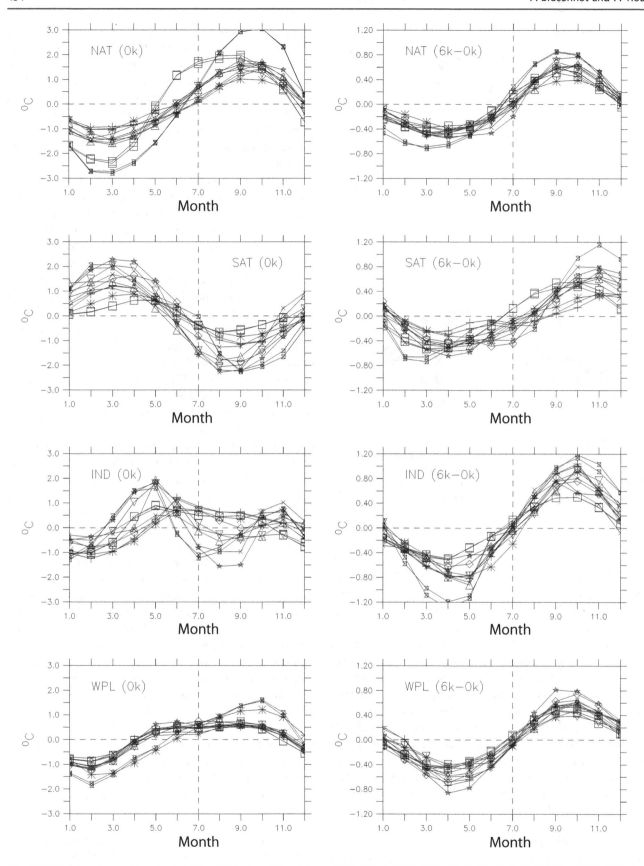

Fig. 30.8 Surface temperature simulated by different models of the PMIP project for the current period (left) and the difference between the Middle Holocene and the current period (right) and four key regions of the tropical ocean, NAT (60° W–20° W; 10° N–20° N), SAT (30° W– 0° W; 10° S–0° S), IND (55° E–75° E; 5° N–15° N) et WPL (110° E– 160° E; 0° N–20° N). Each curve and acronym represents a different model

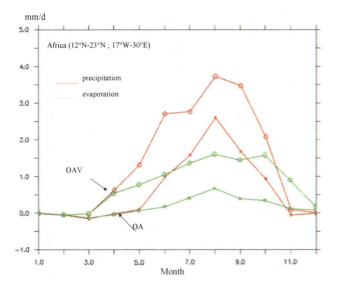

Fig. 30.9 Seasonal evolution of the change in evaporation (mm/d) and in precipitation (mm/d) between the Middle Holocene and the current time as simulated by the IPSL-CM1 model coupled (OAV) and uncoupled (OA) asynchronously with a vegetation model (from Braconnot et al. 1999)

less amplified in the case of an interactive ocean than in older simulations, carried out with atmospheric models forced by the current SST.

The Role of Vegetation

Another factor that affects the characteristics of the monsoon is changing vegetation. Vegetation modifies the surface albedo and the ways in which soil water is used. Depending on the type of vegetation, water losses through interception, evaporation from bare soil and transpiration are distributed differently. All climate models incorporate a more or less complex land surface model to determine the exchanges of heat and water between the surface and the atmosphere. The types of vegetation and the resulting surface characteristics (albedo, roughness, resistance to evaporation) are prescribed from current data. Simulations of the Middle Holocene tend to underestimate the hydrological changes in the Sahelian region (Joussaume et al. 1999). Vegetation characteristics are generally set in line with current ones although pollen data indicate changes in vegetation (Jolly 1998). Change in vegetation, not taken into account in these simulations, is a likely candidate to explain this underestimate.

The first coupled atmosphere-vegetation simulations involved an iterative coupling between climate and vegetation. For this, a vegetation model was used to simulate a vegetation in equilibrium with the climate produced by the atmospheric model and then was reintroduced as a boundary condition to the climate model (Claussen and Gayler 1997; de Noblet 1996). The process is reiterated until climate and vegetation are in equilibrium, i.e. with no major difference between two iterations. New versions of land surface models include the carbon cycle and vegetation dynamics (Krinner 2005), which allows for interactive representation of vegetation changes, taking into account both slow changes in vegetation (replacement of one vegetation type by another) and the interactions of these vegetation changes with the climate. There are still many uncertainties in how these couplings are represented, which results in marked differences in the amplitude of how vegetation reinforces precipitation in monsoon regions (Braconnot et al. 2007b). Nevertheless, some of the broad outlines appear robust.

When vegetation is interactive, the role of local recycling of water becomes more important. Figure 30.9 shows the contributions from advection and local recycling (evaporation) in the Sahelian region, estimated from coupled ocean-atmosphere-vegetation simulations. In this case, the vegetation was coupled asynchronously to the atmosphere model. Simulations with interactive vegetation indicate that the desert in the Sahelian region was replaced by steppe. The albedo was thus decreased in the region, reinforcing the local warming over land at the beginning of the monsoon season, and favoring the advection of humidity and precipitation. During the monsoon season, vegetation is more efficient than bare soil at recycling water, and this also increases precipitation. Finally, at the end of the monsoon season, vegetation absorbs the soil water, which contributes to a lengthening of the rainy season (Texier et al. 2000). All of these effects explain why simulations incorporating interactive vegetation are generally in better agreement with the data for the region of West Africa stretching from 15 to 20° N.

However, there are still great disparities between the results of the different models. These come from small-scale phenomena (clouds, boundary layer, turbulent flows etc.) which need to be better represented in climate models.

Vegetation and the End of the Wet Season in Africa

The role of vegetation and the changing variabilities are two factors that have also been suggested to explain the abrupt end of the wet period in Africa. Early studies using a model of intermediate complexity (Claussen 1999) showed sudden aridification around 5000 years BP in Africa, accompanied by a rapid reduction in precipitation, in line with the data (deMenocal 2000). Rapid vegetation feedback is the source of this abrupt variation, as vegetation can produce multiple equilibria in which green Sahara conditions and desert conditions can coexist.

In the early part of the Holocene, vegetation gradually decreased in response to the reduced summer insolation and the associated reduction in precipitation. When the system reached a threshold where 'green Sahara' conditions could no longer be maintained, the coupled system abruptly produced a desert state accompanied by a sudden drop in precipitation. This sudden change in state is therefore associated with a strong positive feedback from the vegetation. Liu (2007) have another interpretation based on coupled ocean-atmosphere simulations in which the rapid disappearance of vegetation is not accompanied by sudden aridification implying no strong positive feedback induced by vegetation. The mechanism in this case is linked to the internal variability of precipitations. This variability can be considered to be a stochastic forcing that generates slow stochastic variability of soil moisture. This variability interacts with the nonlinearities of the plant system to generate a rapid decrease in vegetation. It is therefore the nonlinear response of the vegetation to the strong internal variability that is responsible for the sudden change of state of the vegetation when a bioclimate threshold is reached. All these studies provide insights into the interactions between long-term insolation variations, climate and vegetation. Nevertheless, we do not yet know the exact reasons and the dominant mechanisms in these African regions. Progress is essential to better represent the semi-arid zones and to obtain hydrological and ecological data at high temporal resolution.

The Recent Climate (Recent Centuries)

Simulating the climate variations of the last millennium is a major scientific challenge for the beginning of the twenty-first century. The first problem is a technological one: these simulations require several months of intensive calculations on the most powerful computers available. This explains why the number of available simulations have gradually increased from one IPCC report to the other (IPCC 2007, 2013). The second problem is how to define an initial state for the ocean and the atmosphere. This problem is avoided by using the output from an ocean model output using pre-industrial conditions and for which no drift is observed. The idea behind this approximation is that the climate will adjust sufficiently quickly to forcing conditions so that the initial conditions of the beginning of the last millennium are not critical. The third problem is purely physical and is the definition of the climate forcings over time. On this time scale, the main forcings are:

- Changes in solar activity. The role of solar activity is not yet clear. Fluctuations in the energy received in the troposphere are theoretically very low. Fluctuations in solar activity also affect the chemistry of the stratosphere. The modulation of reactions with ozone, in particular, can cause heating of the stratosphere and its expansion, thus favoring atmospheric circulation patterns in the troposphere (Shindell 2001). Variations in solar activity in numerical climate simulations can have a purely radiative effect. More sophisticated models include chemical reactions with stratospheric ozone (Shindell 2001) which can produce a warming. This option is more complicated to implement because most of the models used for this exercise do not include stratospheric chemistry, which is very demanding on computing time.
- Volcanic eruptions, particularly the most powerful, which emit sulfurous SO_2 and dust into the stratosphere. The solid particles are quickly pulverized but the sulfurous gas turns into sulfuric acid, which is only washed away after several months. Large volcanic eruptions (e.g. Pinatubo 1991) have the effect of cooling the planet for two to three years (Robock 2000). Volcanism is generally accounted for rather crudely in climate models (Jansen 2007) and is often done by decreasing the solar constant by a few W/m^2 for two years. This representation, although it saves calculation time, cannot account for the dispersion of the volcanic particles across latitude in the stratosphere and therefore for its impact on atmospheric circulation.
- Greenhouse gases added to those already present in the atmosphere. This additional greenhouse effect is noticeable from the twentieth century onwards. These are almost exclusively emissions related to human activity. Sulfate aerosols are also linked to human activity and have a cooling effect on the atmosphere locally.

Estimates of the radiative effects of these forcings during the last millennium are shown in Fig. 30.2.

The details of the atmospheric circulation responses vary widely from one model to another, particularly for GCMs. Simpler models such as EMICs, on the other hand, show similar behaviors because the atmospheric responses have few degrees of freedom, which constrains these models to behave in the same way.

Characteristics of Climate Variability

Extratropical Circulation

The North Atlantic Oscillation (NAO) is the climate regime of the atmospheric circulation dominating the North Atlantic region. This regime was first detected in temperature anomaly structures around the North Atlantic basin. A temperature anomaly, the term used by meteorologists, is the difference between the observed temperature and the average temperature. Sir Gilbert Walker, the British meteorologist, noticed opposing temperature anomalies between Labrador and North Africa on the one hand and the eastern United States and Western Europe on the other. Thus, when it is warmer than usual in the American East (a positive temperature anomaly), it is generally colder than usual in the south of France (a negative anomaly).

This temperature structure is related to fluctuations in the atmospheric circulation in the extra-tropics of the northern hemisphere. Theoretically, this circulation blows from west to east between 30° N and 70° N, and is diverted to the north by the Coriolis force (due to the rotation of the Earth). The North Atlantic has a characteristic zone of low pressure (depression) around Iceland and a zone of high pressure (anticyclone) around the Azores. It is the modulation of the respective influences of these two phenomena that is at the origin of the NAO.

At the level of the North Atlantic basin, the atmospheric flow (of density ρ) is governed by an equilibrium between the pressure forces (P) and the Coriolis force (f perpendicular to the movement). It is the geostrophic equilibrium which is valid on a large spatial scale and over a few weeks:

$$V = \frac{1}{\rho f} \vec{k} \times grad\, P.$$

In this equation, the velocity field V is related to the pressure gradient between low and high latitudes and $\vec{k}$ is a unit vector on the vertical. A strong gradient not only deflects the flow but also accelerates it. The pressure gradient term can be approximated by the pressure difference between two selected locations, for example the Azores and Iceland. This pressure difference is called the North Atlantic Oscillation Index (NAO). A reason for the choice of these locations is that pressure data is available for them since 1825.

So when the pressure gradient between the Azores and Iceland is pronounced (high NAO index), circulation becomes more active and flows towards northern Europe. In this way, moisture is transported to Scandinavia, and during this time, southern Europe is mild and dry. On the other hand, if this gradient is low, the atmospheric flow is towards Southern Europe, causing it to experience wet and cold climate conditions.

During the 1990s, the NAO index was consistently very positive, resulting in a rise in temperatures in Europe (while cold records were being broken in Canada). This consistent index has been interpreted by some as a link with global warming. However, a similar situation with a very strong NAO index was already encountered at the beginning of the twentieth century and is probably not exceptional. Moreover, the index has returned to negative values at the beginning of the 21st century, while the temperature of Western Europe continued to increase.

Thanks to the NAO signature on temperature and precipitations, it has been possible to use paleoclimate data (tree rings or glacial drilling) to extend this 'instrumental' NAO index to the last millennium. This application makes it possible to determine an index based on atmospheric pressure without having access to pressure itself. It is based on large-scale relationships between climate variables and pressure differences, for which the contemporary period can be used to verify. Of course, such reconstructions can yield ambiguities and may have diverged before the nineteenth century, depending on the indices used, because the relationship between the pressure gradient (the NAO index) and variables such as temperature or precipitation may have changed over time.

Looking at the daily weather charts, it can be seen that on some days in winter an anticyclone settles over Scandinavia, causing high pressures that prevents the western winds from reaching France. This kind of weather situation, often referred to as a 'Scandinavian blocking', can last several weeks resulting in generally dry weather. Another meteorological situation that can occur, this time with an abnormally high pressure zone in the middle of the Atlantic is the 'North Atlantic ridge'. The air crossing the Atlantic is diverted to the north of this zone and passes over Greenland, before returning to Europe, causing a cold snap. We can thus count several types of weather with longer or shorter lifetimes. The most common types of weather patterns are the two phases of the NAO, the Scandinavian blocking and the North Atlantic ridge. Figure 30.10 shows the four weather patterns that dominate the atmospheric circulation in winter around the North Atlantic. These four situations are obtained through statistical calculations on pressure data for the last fifty years. The isolines indicate the general direction of the wind.

As part of the heat is transported by atmospheric flow, each of these weather regimes has a different influence on temperature in Europe. In the case of winter in France, cold episodes are more linked to patterns such as the North Atlantic ridge or the negative phase of the NAO and mild spells are more related to the positive phase regimes of the NAO or the Scandinavian blocking. A change in the distribution of these patterns can affect the transport of heat, and thus have an impact on the surface temperature. The effect of this change in atmospheric circulation is comparable to the effect of external forcings (greenhouse gases, volcanism and solar activity) on the variability of temperatures. During the exceptionally hot summer of 2003, these high temperatures were shown to be partly due to the persistence of an anticyclone over Europe, which was in turn related to a precipitation anomaly over the Sahel, which dried up the atmospheric column. It is therefore essential to consider changes in the distribution of atmospheric circulation regimes in order to interpret climate change.

This representation facilitates the interpretation of relationships between meteorology (fluctuations on very short time scales, such as weather regimes) and climatology (centennial climate fluctuations). This concept was applied to a very different climate, such as that of the Last Glacial Maximum 21,000 years before the present. Using numerical simulations of the climate of this period, Kageyama et al. (1999) found that the notion of weather regimes was not

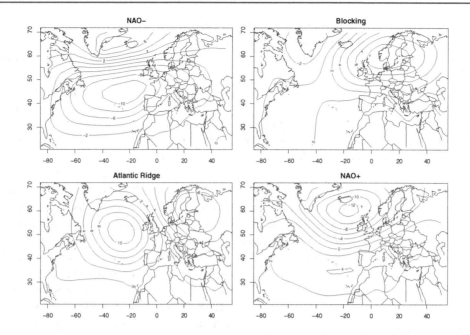

Fig. 30.10 Four common weather regimes around the North Atlantic. These regimes are calculated from sea-level pressure data from 1948 to 2009. General wind direction follows the isobars (iso-pressure lines) from west to east, turning in a clockwise direction around the high pressure zones. From top left clockwise, the weather regimes are: the negative phase of the North Atlantic Oscillation (NAO−), Scandinavian Blocking (Blocking), the positive phase of the North Atlantic Oscillation (NAO+) and the North Atlantic Ridge (Ridge)

applicable at that time, and that the types of stationary circulation observed today were not present then.

The Equatorial Pacific (ENSO)

El Niño events are a major manifestation of the coupled ocean-atmosphere variability of tropical regions originating in the Pacific Ocean (Philander 1990). The normal conditions of the Pacific Ocean are characterized by the 'warm pool' (using the language of climatologists) in the western part of the basin and by the development of an upwelling in the eastern part, brought about by the divergence of the southeast trade wind and, closer to the coast, by the meridian component of the northerly wind along the coast of South America, which favors the development of a coastal upwelling.

The El Niño years are characterized by abnormally warm waters between the date line and the South American coast, and are accompanied by a weakening of the temperature gradient across the Pacific basin (Fig. 30.11).

These oceanic structures are strongly linked to the atmosphere and to the southern oscillation and modulate the intensity of the trade winds, large-scale tropical circulation and precipitation. El Niño occurs every three to seven years and alternates with La Niña periods, which, on the contrary introduces cold conditions in the eastern equatorial Pacific. The impact of El Niño events is not limited to equatorial regions and spreads to mid-latitudes via an atmospheric wave train. Nevertheless, the intensity and frequency of these events, as well as the links between ENSO (El Niño-Southern Oscillation) variability and the average climate state, are still poorly understood.

The Holocene is a period of particular interest to understand these different aspects. Variations in the frequency and intensity of El Niño events for this period were recorded in corals, tropical glaciers and lake sediments. The different studies are in agreement that El Niño events were less intense during most of the Holocene and that the recent period is unusual in this context (Cobb et al. 2013; Carre et al. 2014; Emile-Geay et al. 2016). Although all modeling studies agree on this reduction in activity in the Mid

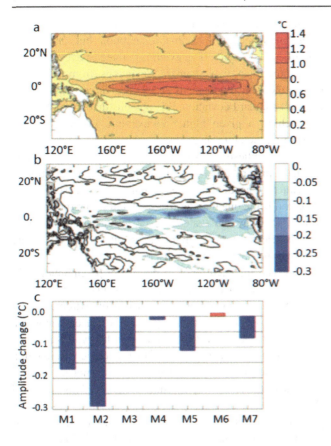

Fig. 30.11 a Interannual variability of simulated sea surface temperatures for the current period. **b** Changes in sea surface temperature variability for the Middle Holocene. **c** Changes in simulated variability by eight climate models for the Eastern Pacific. Figures (**a**) and (**b**) represent the median of the results of the various models shown in figure (**c**)

Holocene, they have come to slightly different conclusions about the origin of this evolution. All of the studies, however, attribute the main source of variation to solar radiation and a change in the seasonal cycle.

The first simulations were carried out with a model of intermediate complexity dedicated to the study of this phenomenon (Clement et al. 2000). The area covered by the model was limited to the tropical Pacific Ocean. The different equations are treated with regard to their difference from the mean state. These simulations do not therefore explicitly include changes in the seasonal cycle brought about by changes in insolation during the Holocene. Nevertheless, they have helped to improve our understanding of the response by the tropical ocean to changes in surface fluxes. The basic mechanism implies a behavioral difference between the western part of the basin, where warm waters accumulate, and the eastern part, where equatorial upwelling develops. During the northern polar summer, the increase in insolation warms the western part of the basin more than the east. It increases the temperature gradient across the basin

and, strengthens the easterly trade winds and hence favors the development of equatorial upwelling and the temperature gradient. This feedback is called the Bjerknes mechanism (1969).

Although simulations from general circulation models also show wind intensification and a reduction in El Niño events, they do not show a coupling between winds and the temperature gradient across the basin at an annual scale. Rather, the results of these simulations indicate a link with large-scale changes in circulation. In particular, the intensification of the Asian monsoon promotes the intensification of the winds in the Pacific and their convergence towards the southeast of the Asian continent. These winds counteract the development of El Niño events (Zheng et al. 2008). Also the seasonal changes in insolation prevents the development of the upwelling Kelvin wave associated with El Ninõ (Luan et al. 2012).

This result is very different from that obtained for other climates, such as the climate of the future or the climate of the Last Glacial Maximum. It shows the strong constraint exerted by the changes in large-scale dynamics, which are dominant over the changes in local heat fluxes in tropical regions. For the Mid Holocene, this decrease in intensity of El Niño events is also accompanied by a reduction in the El Niño impact on other regions of the world, such as on the Sahelian region (Zhao et al. 2007). However, it is not yet possible to estimate whether these changes in teleconnection have played a role in the environmental modifications in Africa, nor to test this result against available data.

Fluctuations in El Niño events have also been reported for the last millennium, particularly in corals in the Pacific (Bradley 1999). The results indicate that the ENSO had characteristics in the mid-seventeenth century similar to those of the current period, with events that may have been as intense as the El Niño of 1997–1998. In the period from the twelfth to the fourteenth centuries, however, events appear to have been less intense. However, it is not possible to make the connection between the average conditions and the characteristics of the El Niño events. Other authors have found statistical links between the intensity of El Niño events and volcanic eruptions (Adams et al. 2003). ENSO variability is reduced in the years following a volcanic eruption, and this is confirmed by simulations conducted by Mann et al. (2005), but had not been identified in previous studies. These relatively contradictory results come from the difficulty in representing all of the facets of El Niño events in current models. Certain systematic biases in the models interfere with the representation of interannual variability, for example, the dual 'intertropical convergence zone' (ITCZ) which is overly pronounced in the eastern Pacific in coupled ocean-atmosphere simulations, or the fact that the structure of the equatorial upwelling is too confined along the equator and extends too far into the western part of the basin [Chap. 8 of Solomon et al. (2007)].

Climate Extremes Relative to the Average

We have seen how it is possible to reconstruct average climate conditions from environmental indicators or simulate them with climate models of varying levels of complexity. By average we mean both in terms of space (hemispherically, globally or regionally) and time (seasonal, annual and decadal averages). Thus, most indicators show that the average temperature of the planet rose by ~ 0.74 °C in the course of the last century. It is obvious that behind this average variation there are significant regional and temporal disparities: not all the places on the planet heat up at the same rate (high latitudes seem to be more affected than low ones), and gradual warming does not eliminate the possibility of local cold spells that can last several days. It is even statistically possible for local cold records to be beaten in a climate that is warming globally.

If we examine precipitation patterns, we can observe a trend towards aridification in southern Europe over the last few decades. This is reflected in a lower average seasonal water balance than during the first part of the twentieth century. On the other hand, observations show that extreme precipitation events (such as the flooding episodes that occur during the autumn in southern France or the wet winters in the UK) may become more frequent and more intense.

Typical extreme climate events can be classified into several categories:

- Heatwaves and cold snaps, i.e. when the temperature exceeds certain thresholds for a certain duration. The definition of a heatwave differs depending on the location: a heatwave in Paris is a typical summer in Seville!
- Intense precipitation and droughts. On a one-month scale, it is important to distinguish between long episodes of 'average' precipitation and those that last only a few hours. They are responses to distinctly different meteorological conditions.
- Storms, cyclones and other hurricanes.

These examples illustrate the use of statistics and probabilities to describe climate phenomena (von Storch and Zwiers 2001). These are particularly relevant when describing the interactions between normal climatological magnitudes (mean temperatures and precipitations) and the occurrence of extreme or rare events. A typical paradigm for describing temperatures is to represent them as an average and a standard deviation. These are the statistical parameters necessary to carry out reconstructions of average temperatures over recent millennia.

Let's assume that temperature can be represented by a Gaussian shape, a debatable assumption but one which simplifies the subject. If an event is defined as extreme when it is more than two standard deviations from the mean (which theoretically happens ~ 2.5 times out of 100) then, at constant standard deviation, it is obvious that the probability of reaching a given higher temperature increases as the average temperature increases.

This hypothesis of Gaussian representation doesn't hold for daily precipitation, for which the standard deviation is ill-defined. One of the crucial parameters to describe precipitation is the one that describes the tail of the distribution, that is, the way rare values occur. A Gaussian possesses what is referred to as a 'light' tail, so that if big events are possible, they are exponentially rare.

However, the occurrence of rare major precipitations, for most locations on the planet, decreases much more slowly than in a Gaussian graph. The theory of extreme values makes it possible to statistically describe these phenomena (Coles 2001). Armed with this statistical theory, it is possible to model the different ways in which average precipitation and extreme precipitation behave which do not necessarily respond in the same way to climate change. This statistical characterization of climate variability makes it possible to predict the evolution of the probabilities of intense events, as well as the uncertainty associated with this prediction.

Outstanding Questions at the Beginning of the Twenty-First Century

Weather or Climate

Variations in the climate of the Holocene up to the present time have been punctuated by very short-lived meteorological events that have had a variety of consequences, but were often disastrous for societies. Since the end of the twentieth century, there appears to have been an upsurge in extreme events on the planet (the storms of 1999, record monsoons, deadly heat waves etc.). Assuming that this increase is not due to improved communication of information on the planet, the first question to be answered is whether this state of events has been 'observed' in the recent past (such as over the last millennium). To answer this question, it is essential to gather as much information as possible on extreme events (of all types) and put them into perspective. The gathering of this information is the subject of many national and international programs on historical research and the climate reconstruction of extreme events.

A parallel approach is the numerical simulation of climates at a resolution sufficiently high to enable the extremes to be adequately represented (over the current period). The most sophisticated models can be used to simulate the reported weather over recent centuries, and thus to examine how meteorological statistics (regimes, extremes etc.) respond to forcings over the long-term. The confrontation of models with observations, within the context of extremes, becomes a crucial issue, involving complex statistical theories. These mathematical tools allow for the regionalization of the climate, with the aim of forecasting climate variables on a very small scale (e.g. precipitation at the scale of a city or a field) based on understanding of large-scale variables (such as atmospheric circulation on the scale of the North Atlantic basin).

Detection and Attribution of Climate Change

Understanding the role of human activity in climate change has become a major issue in recent years. The identification of this role is not limited to superimposing demographic, greenhouse gas and temperature curves on each other and highlighting their simultaneous growth. The difficulty in assessing this role is that there is no 'control' version of the Earth on which the climate with no human activity can be evaluated.

To attribute climate change to any one factor, it is necessary to evaluate the role of all other possible factors and to show that the characteristics of the change are compatible with this factor alone. This endeavor is usually done by researchers through climate model simulations, taking into account all of the physical factors minus one, and then comparing the simulations obtained with the observations. The key is to then identify the crucial factor(s) that best allows simulations to resemble observations. The difficulty with this evaluation is defining the statistical criteria for the comparisons and to obtain the necessary confidence intervals.

In order to make progress in this direction, improved models able to take the many interactions in the climate system into account are essential. This requires considering not only physical interactions involving energy and water cycles, but also the interactions between climate and the biogeochemical cycles. In particular, land use, deforestation, different types of aerosols (natural and anthropic) are often excluded from climate simulations. The distinction between meteorology and changes in average climate also necessitates improvement in the resolution of the models used, which goes hand in hand with better representation of small-scale processes. All of these elements combined with high-resolution data make it possible to better understand the risks associated with the profound changes in our environment linked to the ongoing global warming.

The attribution of extreme events is a rather recent field of research, that poses many scientific challenges (NAS 2016). Since 2012, the Bulletin of the American Meteorological Society (BAMS) publishes a summary of the analyses of past extreme events and their attribution to climate change.

One of the teams has developed the weather@home system (Massey et al. 2015), which was devised to calculate tens of thousands of high resolution climate simulations. Those large ensembles are performed with present day greenhouse gas concentration and natural variability, and with natural variability only. They allow estimating the probability density function that a key climate variable (temperature, precipitation) exceeds a large threshold, with and without climate change. Other statistical methods have been devised since then to attribute extreme events to climate change (Stott et al. 2016; Jézéquel et al. 2018).

References

Adams, J., et al. (2003). Proxy Evidence for an El Niño-like Response to Volcanic Forcing. *Nature, 426*(6964), 274–278.

Bjerknes, J. (1969). Atmospheric teleconnections from equatorial pacific. *Monthly Weather Review, 97*(3), 163–172.

Braconnot, P., Harrison, S. P., Kageyama, M., Bartlein, P. J., Masson-Delmotte, V., Abe-Ouchi, A., et al. (2012). Evaluation of climate models using palaeoclimatic data. *Nature Climate Change, 2*, 417–424. https://doi.org/10.1038/nclimate1456.

Braconnot, P., Joussaume, S., Marti, O., & de Noblet, N. (1999). Synergistic feedbacks from ocean and vegetation on the African monsoon response to mid-Holocene insolation. *Geophysical Reseach Letters, 26*, 2481–2484.

Braconnot, P., et al. (2007a). Results of PMIP2 coupled simulations of the mid-holocene and last glacial maximum—Part 1: Experiments and large-scale features. *Climate of the Past, 3*(2), 261–277.

Braconnot, P., et al. (2007b). Results of PMIP2 coupled simulations of the mid-holocene and last glacial maximum—Part 2: Feedbacks with emphasis on the location of the ITCZ and mid- and high latitudes heat budget. *Climate of the Past, 3*(2), 279–296.

Bradley, R. S. (1999). *Paleoclimatology: Reconstructing climates of the quaternary*, 2nd edn. San Diego: Harcourt/Academic, 613p.

Brazdil, R., et al. (2005). Historical climatology in Europe—The state of the art. *Climatic Change, 70*(3), 363–430.

Cane, M., et al. (2006). Progress in paleoclimate modeling. *Journal of Climate, 19*(20), 5031–5057.

Carre, M., Sachs, J. P., Purca, S., Schauer, A. J., Braconnot, P., Falcon, R. A., et al. (2014). Holocene history of ENSO variance and asymmetry in the eastern tropical Pacific. *Science, 345*, 1045–1048.

Claussen, M., et al. (1999). Simulation of an Abrupt change in saharan vegetation in the mid-holocene. *Geophysical Research Letters, 26*(14), 2037–2040.

Claussen, M., & Gayler, V. (1997). The greening of the sahara during the mid-holocene: Results of an interactive atmosphere-biome model. *Global Ecology and Biogeography Letters, 6*, 369–377.

Clement, A. C., et al. (2000). Suppression of El Nino during the mid-holocene by changes in the earth's orbit. *Paleoceanography, 15*(6), 731–737.

Cobb, K. M., Westphal, N., Sayani, H. R., Watson, J. T., Di Lorenzo, E., Cheng, H., et al. (2013). Highly variable El Nino-southern oscillation throughout the holocene. *Science, 339*, 67–70.

Coles, S. (2001). *An introduction to statistical modeling of extreme values* (p. 208). London, New York: Springer.

Crucifix, M., et al. (2002). Climate evolution during the holocene: A study with an earth system model of intermediate complexity. *Climate Dynamics, 19*(1), 43–60.

de Noblet, N., et al. (1996). Possible role of atmosphere-biosphere interactions in triggering the last glaciation. *Nature, 23*(22), 3191–3194.

deMenocal, P., et al. (2000). Coherent high- and low-latitude climate variability during the holocene warm period, *Science, 288*(5474), 2198–2202.

Emile-Geay, J., Cobb, K. M., Carre, M., Braconnot, P., Leloup, J., Zhou, Y., Harrison, S. P., Correge, T., McGregor, H. V., Collins, M., Driscoll, R., Elliot, M., Schneider, B., & Tudhope, A. (2016). Links between tropical Pacific seasonal, interannual and orbital variability during the Holocene. *Nature Geoscience, 9*, 168-+.

Garnier, E. (2009). *Les Dérangements du temps: 500 ans de chaud et de froid en Europe*. Paris: Plon.

Gasse, F. (2000). Hydrological changes in the African tropics since the last glacial maximum. *Quaternary Science Reviews, 19*(1–5), 189–211.

IPCC. (2007). Climate change 2007: The physical science basis. contribution of working group I to the fourth assessment report of the intergovernmental panel on climate change. In: S. Solomon, D. Qin, M. Manning, Z. Chen, M. Marquis, K.B. Averyt, M. Tignor & H.L. Miller (Eds.), Cambridge, United Kingdom and New York, NY, USA: Cambridge University Press, 996pp.

IPCC. (2013). Climate change 2013: The physical science basis. contribution of working group I to the fifth assessment report of the intergovernmental panel on climate change. In: T. F. Stocker, D. Qin, G.-K. Plattner, M. Tignor, S. K. Allen, J. Boschung, A. Nauels, Y. Xia, V. Bex, & P. M. Midgley (Eds.), Cambridge, United Kingdom and New York, NY, USA: Cambridge University Press, 1535pp.

Jansen, E., et al. (2007). Palaeoclimate, dans Solomon, In S. et al., (Ed.), *Climate change 2007: The physical science basis. contribution of working group I to the fourth assessment report of the intergovernmental panel on climate change*, Cambridge: Cambridge University Press.

Jézéquel, A., Dépoues, V., Guillemot, H., Trolliet, M., Vanderlinden, J.-P., & Yiou, P. (2018). Behind the veil of extreme event attribution, *Climatic Change*. https://doi-org.insu.bib.cnrs.fr/10.1007/s10584-018-2252-9.

Jolly, D., et al. (1998). Biome reconstruction from pollen and plant macrofossil data for Africa and the Arabian Peninsula at 0 and 6000 Years. *Journal of Biogeography, 25*(6), 1007–1027.

Jones, P., & Mann, M. (2004). Climate over past millennia, *Reviews of Geophysics, 42*(2).

Joussaume, S. & Taylor, K. E. (1995). Status of the paleoclimate modeling intercomparison project, dans. In *Proceedings of the first international AMIP scientific conference, WCRP-92, Monterey, USA*, pp. 425–430.

Joussaume, S., et al. (1999). Monsoon changes for 6000 years ago: Results of 18 simulations from the paleoclimate modeling intercomparison project (PMIP). *Geophysical Reseach Letters, 26*(7), 859–862.

Kageyama, M., et al. (1999). Weather regimes in past climate atmospheric general circulation model simulations. *Climate Dynamics, 15*(10), 773–793.

Krinner, G., et al. (2005). A dynamic global vegetation model for studies of the coupled atmosphere-biosphere system, *Global Biogeochemical Cycles, 19*(1).

Kropelin, S., & Kuper, R. (2006). Climate-controlled holocene occupation in the Sahara: Motor of Africa's evolution. *Science, 313*(5788), 803–807.

Kutzbach, J. E. (1988). Climatic changes of the last 18,000 years—observations and model simulations, *Science, 241*(4869), 1043–1052.

Le Roy Ladurie, E. (1967). *Histoire du climat depuis l'an mil* (p. 381). Paris: Flammarion.

Liu, Z., et al. (2007). Simulating the Transient Evolution and Abrupt Change of Northern Africa Atmosphere-Ocean-Terrestrial Ecosystem in the Holocene, *Quaternary Science Reviews, 26*(13–14), 1818–1837.

Luan, Y., Braconnot, P., Yu, Y., Zheng, W., & Marti, O. (2012). Early and mid-Holocene climate in the tropical Pacific: Seasonal cycle and interannual variability induced by insolation changes. *Climate of the Past, 8*, 1093–1108.

Mann, M., Bradley, R., & Hughes, M. (1998). Global-scale temperature patterns and climate forcing over the past six centuries. *Nature, 392*, 779–787.

Mann, M., et al. (2005). Volcanic and solar forcing of the tropical pacific over the past 1000 years. *Journal of Climate, 18*(3), 447–456.

Massey, N., Jones, R., Otto, F. E. L., Aina, T., Wilson, S., Murphy, J. M., et al. (2015). Weather@home—development and validation of a very large ensemble modelling system for probabilistic event attribution. *Quarterly Journal of the Royal Meteorological Society, 141*(690), 1528–1545. https://doi.org/10.1002/qj.2455.

National Academies of Sciences Engineering and Medicine. (2016). éd. *Attribution of extreme weather events in the context of climate change*. Washington, DC: The National Academies Press. https://doi.org/10.17226/21852.

Philander, S. G. H. (Ed.). (1990). *El Niño, La Niña, and the Southern Oscillation* (p. 312). San Diego: Academic Press.

Prentice, I. C., & Webb, T. (1998). BIOME 6000: Reconstructing Global Mid-Holocene Vegetation Patterns from Palaeoecological Records, *Journal of Biogeography, 25*(6), 997–1005.

Renssen, H., et al. (2005). Simulating the holocene climate evolution at northern high latitudes using a coupled atmosphere-sea ice-ocean-vegetation model. *Climate Dynamics, 24*(1), 23–43.

Robock, A. (2000). Volcanic eruptions and climate. *Reviews of Geophysics, 38*(2), 191–219.

Ruddiman, W. (2007). The early anthropogenic hypothesis: challenges and responses, *Reviews Of Geophysics, 45*(3).

Shindell, D., et al. (2001). Solar forcing of regional climate change during the maunder minimum, *Science, 294*(5549), 2149–2152.

Solomon, S., et al. (Eds.). (2007). *Climate change 2007: The physical science basis: Contribution of working group I to the fourth assessment report of the intergovernmental panel on climate change* (p. 996). Cambridge, New York: Cambridge University Press.

Stott, P. A., Christidis, N., Otto, F. E. Sun, Y., Vanderlinden, J.-P., van Oldenborgh, G. J., et al. (2016). Attribution of extreme weather and climate-related events, *Wiley Interdisciplinary Reviews, Climate Change, 7*(1), 23–41, https://doi.org/10.1002/wcc.380.

Taylor, K. E., et al. (2007). Estimating shortwave radiative forcing and response in climate models. *Journal of Climate, 20*(11), 2530–2543.

Texier, D., et al. (2000). Sensitivity of the African and Asian monsoons to mid-holocene insolation and data-inferred surface changes. *Journal of Climate, 13*(1), 164–181.

von Storch, H., & Zwiers, F. W. (2001). *Statistical Analysis in Climate Research*. Cambridge: Cambridge University Press.

Wohlfahrt, J., et al. (2004). Synergistic feedbacks between ocean and vegetation on mid- and high-latitude climates during the mid-holocene. *Climate Dynamics, 22*(2–3), 223–238.

Zhao, Y., et al. (2005). A multi-model analysis of the role of the ocean on the African and Indian monsoon during the mid-holocene. *Climate Dynamics, 25*(7–8), 777–800.

Zhao, Y., et al. (2007). Simulated changes in the relationship between tropical ocean temperatures and the western african monsoon during the mid-holocene. *Climate Dynamics, 28*(5), 533–551.

Zheng, W., et al. (2008). ENSO at 6 Ka and 21 Ka from ocean-atmosphere coupled model simulations. *Climate Dynamics, 30*(7–8), 745–762.

From the Climates of the Past to the Climates of the Future

31

Sylvie Charbit, Nathaelle Bouttes, Aurélien Quiquet, Laurent Bopp, Gilles Ramstein, Jean-Louis Dufresne, and Julien Cattiaux

This chapter connects everything we have learned about past climates (both from the analysis of natural archives and from numerical simulations), and future climate projections. The models used to explore the future are similar to those used for past climates, except that the results are based on emission scenarios of greenhouse gas emissions whereas past climate simulations may be compared to reconstructions from the natural archives described in volume 1. The climate of the past 1000 years can be seen as the 'background noise' of the recent natural evolution of the climate system. On this basis, the current climate change can be analyzed. Over the longer term, analysis of ice cores allows us to trace back the history of atmospheric CO_2 concentration over the last 800,000 years, and shows that the anthropogenic disturbance (producing an atmospheric CO_2 concentration currently in excess of 410 ppm) is completely outside the documented glacial-interglacial variations over the last million years, ranging from 180 to 280 ppm. With this awareness you are well equipped to now explore the climate of the future.

Climate Observations in Recent Decades: The First Signs of Warming

The principles of the physical laws governing the temperature on the Earth's surface were formulated at the beginning of the nineteenth century by Joseph Fourier, who established

S. Charbit (✉) · N. Bouttes · A. Quiquet · L. Bopp · G. Ramstein
Laboratoire des Sciences du Climat et de l'Environnement, LSCE/IPSL, CEA-CNRS-UVSQ, Université Paris-Saclay, 91190 Gif-sur-Yvette, France
e-mail: Sylvie.Charbit@lsce.ipsl.fr

J.-L. Dufresne
Laboratoire de Météorologie Dynamique/IPSL, Université Pierre et Marie Curie, BP 99, 4 Place Jussieu, Cedex 05, 75252 Paris, France

J. Cattiaux
Centre National de Recherches Météorologiques, Université de Toulouse, CNRS, Météo-France, 42 Avenue Gaspard Coriolis, 31057 Toulouse Cedex, France

that the energy balance at the surface of our planet is dominated by incoming solar radiation which is the primary source of energy, and infrared emission exchanges which control energy losses. He concluded that any change in surface conditions could lead to a change in climate and argued that *the development and progress of human societies can notably change the state of the ground surface over vast regions, as well as the distribution of waters and the great movements of the air* and that *such effects have the ability to cause the mean degree of heat to vary over the course of several centuries*. Joseph Fourier also identified the trapping of infrared radiation by the gases in the atmosphere (Fourier 1824).

This was the beginning of the greenhouse effect theory. Starting from this seminal work, numerous studies were conducted throughout the nineteenth and twentieth centuries. Within a scientific context where the understanding of the glacial-interglacial cycles was the matter of a very hot debate, the Swedish chemist Svante Arrhenius was the first one to quantify the effect of the atmospheric carbon dioxide concentration on the average surface temperature of the Earth and to suggest that substantial variations of the atmospheric CO_2 levels could explain the glacial advances and retreats. However, it is only since the late 1970s that we can calculate precisely the radiation exchanges using radiative transfer codes and spectral databases to break down energy by wavelength.

Long time refuted, or at least underestimated, global warming has now become an incontrovertible reality: in 2001, the scientific review by the Intergovernmental Panel on Climate Change (IPCC) concluded that there was a growing body of evidence confirming global warming as well as other changes in the climate system. Based on the results of numerical experiments with coupled atmosphere-ocean general circulation models (AOGCMs), the Fifth IPCC Assessment Report, published in September 2013 (IPCC 2013), established with a probability of more than 95%, that the global warming observed in recent decades is due to anthropogenic activities. This situation has no direct equivalent in the past climates of the Earth. The last time we find a climate with the current level of atmospheric CO_2 is several million years ago. The climate at this period corresponds to a warm Earth, with a reduced cryosphere (no Greenland) and a smaller Antarctic ice sheet yielding a much higher sea level (15–30 m higher as reported by Haywood et al. 2011, 2016).

Evolution of Greenhouse Gases

We have seen in previous chapters that the greenhouse effect is above all a natural phenomenon, without which the surface temperature of the Earth would be about −18 °C, making life, as we know it today, impossible. The main greenhouse gases naturally present are water vapor (H_2O), carbon dioxide (CO_2) emitted by volcanic eruptions and forest fires, methane (CH_4) produced by wetlands and various fermentation processes, ozone (O_3), and the nitrous oxide (N_2O) emitted by soils. Industrialization has led societies to discharge massive amounts of these gases through the combustion of fossil fuels (oil, gas, coal), deforestation, agriculture, intensive livestock breeding and fertilizer production. An inhabitant of an industrialized country releases on average ten tons of carbon per year (or CO_2 equivalent) compared with only two tons per year for an inhabitant of most emerging countries. However, these estimates mask large disparities from one country to another, with differences ranging from 0.12 ton of CO_2 per year and per inhabitant for Ethiopia to 49.3 tons of CO_2 per year and per inhabitant for Qatar (from Global Carbon Atlas 2017). In addition, human activities produce fluorinated gases (CFCs, HFCs, PFCs, SF_6) used especially in refrigeration and air conditioning systems, as well as in aerosol cans. In total, more than forty of these gases have been identified by the IPCC. The greenhouse effect produced by human activities is called *additional greenhouse effect*. The contribution of each gas to the additional greenhouse effect can be estimated by taking into account the increase in their concentration and their 'radiative efficiency'. Between 1750 and 2011, the variation in their concentration increased the greenhouse effect by about 2.83 W/m², with main contributions of 64% for CO_2, 17% for CH_4, 12% for O_3 and 6% for N_2O (Fig. 31.1). Since 1750, in other words, since the beginning of the industrial era, this anthropogenic phenomenon has produced an energy imbalance of the Earth and has caused a warming of the lower layers of the atmosphere. The additional radiative forcing corresponds to approximately 1% of the total radiation received.

Analysis of air bubbles trapped in the ice of the Antarctic ice sheet revealed that over the last 800,000 years, CO_2 levels have changed by no more than 100 ppm, going from 180 ppm during glacial periods to 280 ppm in interglacial periods. For periods prior to 1950, analysis of air bubbles in the ice is the only reliable way to track the chemical composition of the atmosphere. In recent times, the first direct measurements (i.e. in situ) were obtained in 1958 at the Mauna Loa site in Hawaii (Keeling et al. 1995). These measurements revealed for the first time that not only was CO_2 increasing in the atmosphere, but that this increase was modulated in line with seasonal variations due to photosynthesis of the terrestrial biosphere. This first measurement campaign was then supplemented by campaigns covering other sites in the northern and southern hemispheres. Currently, a wide range of direct and indirect measurements confirm that atmospheric CO_2 levels have increased since the beginning of the preindustrial era, rising from 275–285 ppm between the years 1000 and 1750, to about 380 ppm in 2005 (Fig. 31.1) and to more than 410 ppm in 2018, that is a difference of more than 100 ppm compared to the pre-industrial period. About 30% of the current atmospheric CO_2 has been emitted by anthropogenic sources. In addition, the amount of annual anthropogenic emissions has been continuously increasing throughout the industrial era. In 1990, 2000, 2010 and 2017, global CO_2 emissions from human activities (fossil fuel combustion, cement production, land-use change) reached 22, 25, 32 and 37 Gt CO_2/yr respectively. The increase in annual CO_2 emissions accelerated from 1.1% per year in the 1990s to 3.3% per year in the 2000s. This growth rate of atmospheric CO_2 is ten times faster than the highest rates recorded in ice cores, and is mainly due to the rapid growth of developing countries and a drop in the efficiency of fossil fuel use in the global economy. Between 65% and 80% of CO_2 released in the atmosphere is trapped and/or dissolved in the ocean and the terrestrial biosphere in 20–200 years, depending on the various estimations. The rest is removed by slower processes, including chemical weathering and rock formation that take several thousands of years, indicating that the effect of anthropogenic CO_2 will persist for hundreds to thousands of years into the future (Archer et al. 2009).

Atmospheric methane is the third most important greenhouse gas after H_2O and CO_2 in terms of atmospheric concentration. Averaged over 100 years, the radiative efficiency of methane is estimated to be 28–36 times greater than that of CO_2 but its lifetime in the atmosphere (i.e. the time it takes for a CH_4 molecule to be removed from the atmosphere by chemical reaction) is much less (∼ 9–12 years) than that of CO_2. Ice core records indicate that CH_4 levels in the atmosphere also show variations from about 350 ppb (during glacial periods) to 700 ppb (during interglacial periods). In 2011, the level of methane in the atmosphere, established from a network of measurements covering both hemispheres, was at 1803 ppb, a level never attained, at least throughout the last 800,000 years. Since the pre-industrial era, CH_4 has increased by approximately 250%. Although the growth rate of methane was over 1%

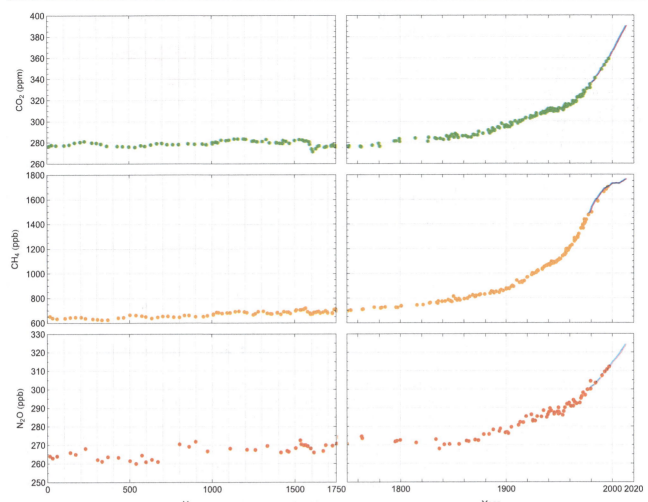

Fig. 31.1 Atmospheric CO_2 (*top*), CH_4 (*middle*), and N_2O (*bottom*) concentrations history since the beginning of the industrial era (right) and from years 0 to 1750 (left), determined from air trapped in ice cores and firn air (color symbols) and from direct atmospheric measurements (blue lines, measurements from the Cape Grim observatory) (*Source* IPCC 2013)

per year during the 1970s and early 1980s, the 1990s and early 2000s saw this rate stabilize, resulting in relatively stable concentrations. However, atmospheric CH_4 concentrations started growing again from 2007, although the true cause of this renewed increase is still unclear. CH_4 is emitted by many agricultural activities (ruminant farming, rice cultivation), by industrial activities (biomass combustion, the oil and gas industry), as well as by natural processes (wetlands, permafrost, peat bogs). There are no available data on annual CH_4 emissions from industrial activities as these are difficult to quantify. When the climate warms up, CH_4 emissions from natural processes can increase. This has been observed with permafrost thawing in Sweden, but no large-scale evidence is available to clearly relate this process to the recent increase in methane. If the observed increase is caused by the response of natural reservoirs to global warming, this could last for several decades, even centuries, and thus reinforce the enhanced greenhouse effect (positive feedback).

The greenhouse gas with the fourth contribution to radiative forcing is nitrous oxide (N_2O). Its level has steadily increased from 270 ppb in 1750 to 323 ppb in 2011. The main natural emissions of this gas come from soil microbial activity and ocean processes. As for anthropogenic emissions, they come mainly from the use of nitrogen fertilizers in agriculture, fossil fuel combustion and chemical industry.

Halocarbons (or halogenated hydrocarbons), responsible in particular for the destruction of stratospheric ozone, generate a lower radiative forcing (about 0.35 W/m² in 2011) than the three main greenhouse gases (CO_2, CH_4 and N_2O), whose total contribution is 2.30 W/m². The emissions of these gases are almost exclusively anthropogenic. To combat the destruction of stratospheric ozone, the Montreal Protocol regulated the production of halocarbons containing chlorine (chlorofluorocarbons or CFCs) and bromine. Substitute products adopted to replace CFCs, for example in refrigeration processes, do not affect the ozone layer but remain powerful greenhouse gases. Perfluorinated

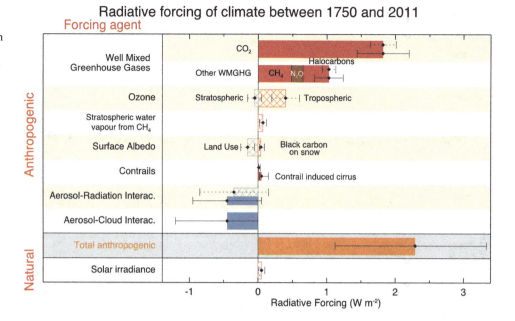

Fig. 31.2 Main components of radiative forcing involved in climate change coming from both natural processes and human activities. The values correspond to the estimated difference between 2011 and 1750. Anthropogenic contributions are responsible for most of the radiative forcing. Positive forcings cause global warming, while negative forcings lead to cooling. The black lines associated with each box show the associated uncertainties (*Source* IPCC 2013)

hydrocarbons (PFCs, such as CF_4 and C_2F_6) and sulfur hexafluoride have extremely long residence times in the atmosphere and are excellent absorbers of infrared radiation. Thus, even though these compounds are released in small quantities, their impact on the greenhouse effect and the climate is far from negligible.

Although ozone is also a greenhouse gas, it is not emitted directly, but is formed from photochemical reactions involving other precursor gases of natural and anthropogenic origin. Its impact on the radiative budget depends on the altitude at which the changes in its concentration occur, as these vary spatially. Moreover, once formed, its residence time in the atmosphere is very short, unlike the greenhouse gases mentioned previously. For this reason, it is difficult to establish precisely its role in the radiative budget.

In addition to the production of greenhouse gases, human activities also produce aerosols. These can have a direct impact on radiative forcing by absorbing or reflecting solar and infrared radiation. Some of them contribute negatively to radiative forcing, others positively. Finally, aerosols can have an indirect effect by modifying the reflective properties of clouds. Taking the totality of aerosols into account, the overall contribution is negative and therefore partially compensates the effect of greenhouse gases. In 2011, the estimated forcing of anthropogenic aerosols is about -0.9 W/m^2, albeit with large uncertainties (-1.9 to -0.1 W/m^2, IPCC 2013). Overall, taking into account both greenhouse gases and aerosols, anthropogenic activities are responsible for a positive radiative forcing since the beginning of the industrial era, estimated at 2.3 W/m^2 in 2011 (IPCC 2013). The main components involved in climate change are summarized in Fig. 31.2.

Evolution of Surface Temperatures

Instrumental observations documented for the past 150 years show an overall increase in temperature on the Earth's surface (Fig. 31.3). According to the synthesis presented in the Fifth IPCC Assessment Report (IPCC 2013), it has increased by an average of 0.89 ± 0.20 °C over the last century (1901–2012), that is an increase of about 0.08 ± 0.02 °C per decade. This multi-decadal signal (warming trend) emerges as significant from the noise of the internal climate variability – a global warming is thus detected. Furthermore, dedicated studies have shown that the observed warming cannot be explained solely by natural forcings (solar and volcanic activities), and that anthropogenic forcings necessarily contribute: global warming is thus *attributed* to both natural and anthropogenic causes. In particular, *"it is extremely likely that human activities caused more than half of the observed increase in global average surface temperature from 1951 to 2010"* (IPCC 2013). However, this increase has not been steady over time, due to decadal fluctuations in both the internal climate variability and the external natural and anthropogenic forcings. Indeed, the observations highlight two periods of accelerated warming: one from 1910 to about 1940, and the other, even more important, since 1970, while temperatures were relatively stable between 1940 and 1970 (Fig. 31.3). Trends computed over short time periods are highly uncertain. For instance, the warming rate over the 15-year period 1998–2012 is about 0.05 °C per decade, which is weaker than the 1901–2012 warming trend mentioned above (0.08 °C per decade). However, the associated confidence interval is large ($\pm$ 0.10 °C per decade), so that there is no inconsistency between both values. More

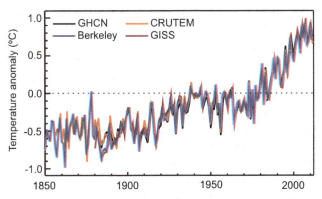

Fig. 31.3 Global annual average land-surface air temperature (LSAT) anomalies relative to the 1961–1990 climatology from the latest versions of four different data sets (Berkeley, CRUTEM, GHCN and GISS) (*Source* IPCC 2013)

generally, global warming is a long-term process that is superimposed on the internal climate variability, and it remains possible to have occasional short-term cooling trends in a warming world. Importantly, the decade of the 2000s appears to be the warmest compared to the entire period covered by instrumental data. The decade of the 2010s is likely to break this record, since the years 2014–2018 have been the five hottest years observed at global scale.

The values mentioned above are global mean annual values. There are, in fact, large disparities from one region to another, as well as large seasonal disparities. In general, warming is amplified in high latitude regions in summer, partly due to the albedo effect induced by the decrease in snow cover and/or in the sea-ice extent. In addition, warming over the continents is much faster than over the oceans because of their weaker thermal inertia but also due to changes in evaporation. Not all regions exhibit a statistically significant warming over the observational period (e.g. South Greenland and North-West Atlantic); but no region exhibits a statistically significant cooling. Climate change detection is a signal-to-noise ratio problem: the higher the temperature variability, the harder it is to detect long-term warming. In other words, internal variability can hide climate change over small spatial domains and/or short time periods; but this does not call into doubt the existence of a long-term global warming trend, nor the fact that it is attributable to the anthropogenic fingerprint.

Evolution of Temperature in the Troposphere

In response to an enhanced greenhouse effect (e.g. due to anthropogenic activities), the physical theory of radiative transfer indicates that not only the surface, but the entire lower atmospheric layer, namely the *troposphere* (altitude 0–10 km), should warm. Conversely, the atmospheric layer above, namely the *stratosphere* (altitude 10–50 km), should cool. Observations of the vertical temperature profile are more difficult to obtain than at the surface, but available data unanimously reveal a warming troposphere and a cooling stratosphere since the mid-twentieth century. This provides further evidence for the anthropogenic nature of the recent climate change. However, there has been a recent controversy about the amplitude of these changes. The IPCC's Fourth Assessment Report published in 2007 (IPCC 2007) indicated persistent uncertainty on temperature trends in the middle atmosphere since 1979. Indeed, most of the available data from radiosondes and satellite measurements indicated a lower warming in the tropical high troposphere (between 10 and 15 km altitude) than that recorded at the surface, whereas all of the climate models projected an amplified warming in this zone of the atmosphere, especially in the tropics. This apparent difference between data and numerical model outputs has been widely cited to emphasize the inconsistency between observational data and models. Some have used this example to call into question the impact of human activities on the climate, even going so far as to deny the existence of the current warming. But more in-depth analyzes have revealed that the observations on which this controversy was based were unreliable, particularly because the measurements did not take into account interannual climate variability. Since then, researchers have re-analyzed these measurements using more rigorous techniques (Allen and Sherwood 2008). New estimates of these observations show greater warming than previously reported, and this new larger set of estimates now falls within the model trends, thus removing any concerns expressed in the 2007 Fourth IPCC Assessment Report.

Precipitations and Water Balance

While the increase in mean surface temperatures is one of the most obvious manifestations of ongoing climate change, spatial and temporal changes in the hydrological cycle (e.g. precipitation, evaporation, runoff) appear to be just as important but are much more difficult to simulate. The term precipitation refers to all the meteoric waters that fall to Earth in the form of liquid (rain, mist, showers) or solid (snow, hail, sleet) water. The formation of precipitation requires the condensation of water vapor around what are called condensation nuclei, which allow the water molecules to aggregate together. This phenomenon is called coalescence. Condensation only occurs when the amount of water vapor per unit of volume exceeds a threshold value, called the saturation value. This is an increasing function of temperature. Warmer air can therefore hold more water vapor before condensation occurs: this physical law is called the Clausius-Clapeyron relationship. For the Earth's atmosphere, the additional humidity that can be held by warmer air is estimated at 7% per °C. Once the water vapor condensates, the different precipitation episodes can be categorized according to their intensity, duration, frequency, and by type (stratiform, like those caused by depressions in the mid-latitudes, or convective, like heavy rains and tropical cyclones in the intertropical convergence zone). These characteristics depend to a large extent on local temperature and weather conditions (wind speed and direction, pressure, humidity, evaporation). It is thus clear that a change in any of these parameters will affect the hydrological cycle as a whole.

Throughout the twentieth century, for example, annual precipitation increased on the eastern side of the South and North American continents. In contrast, a significant rainfall deficit has been observed in south and west Africa, as well as in the Sahel. In northwestern India, an increase of about 20% was observed for the period 1901–2005, despite a sharp decrease between 1979 and 2005. However, changes in precipitation are hard to measure using the existing records, and there is only medium confidence that observed trends are due to the anthropogenic influence.

Observed trends in relative humidity (i.e. air humidity/saturation humidity) suggest that these have remained constant through the tropospheric column down to the surface. However, if the amount of water vapor at saturation increases and the relative humidity remains constant, then this means that the absolute humidity (and thus the amount of water vapor) has increased in the atmosphere. Observations indicate that tropospheric water vapor has increased by about 3.5% over the past 40 years, which is consistent with the observed temperature change of 0.5 °C over the same time period. Climate models confirm these empirical observations: a warmer climate leads to increased moisture content in the atmosphere and more intense precipitation events (although total precipitation over a full year is reduced) and therefore flooding is more likely to occur. Thus, in winter, it is observed that for most of the extratropical land surface areas of the northern hemisphere, the greatest precipitation is linked to higher temperatures. Conversely, in areas with low rainfall, such as the Mediterranean basin, rising temperatures are associated with a higher risk of drought. This general intensification of the hydrological cycle can be summarized by: wet gets wetter, dry gets drier.

Added to these complex phenomena is the variability in atmospheric circulation. In Sect. "Climate Variability", the modes of atmospheric variability will be examined in greater detail. However, we already know that fluctuations in atmospheric circulation over the North Atlantic brought heavy rainfall in the 1990s to northern Europe, and, in contrast, led to a drying-up of the Mediterranean basin. In addition, the severe Sahelian drought over more than 20 years (1970–1990) was linked to changes in both atmospheric circulation and surface ocean temperatures in all three of the Pacific, Indian and Atlantic basins. Although this trend towards drought still persists, it has become less pronounced since the early 1990s.

There are numerous uncertainties surrounding the determination of the hydrological cycle variables. This is due to a lack of data for some regions (for example, Canada, Greenland and Antarctica, some desert regions such as the Sahara, the Tibetan plateau and over the oceans), and to the fact that accurate measurements have only been available for a very short time. In addition, it is very difficult to measure precipitation rates and to accurately quantify their changes at the global and regional scales.

In situ measurements are affected by atmospheric conditions (e.g. the effect of strong winds especially on snowfall). Spatial observations, on the other hand, provide only instantaneous measurements and are affected by the uncertainties associated with the algorithms used to convert radiometric measurements into precipitation rates. Because of these difficulties, climatologists explore the coherence between the whole set of complementary variables associated with the hydrological cycle. One way to represent precipitation changes over the past century or over decades is to calculate the Palmer Drought Severity Index. This index is a measure of drought, in other words, the accumulated surface soil moisture deficit compared to average local conditions. It is based on recent rainfall and atmospheric humidity (determined from temperatures). In general, analyzes of the last century suggest a trend towards drying for much of Africa, southern Eurasia, Canada and Alaska. Other direct or indirect measurements (e.g. from river flow estimations or ocean salinity measurements) show that during the twentieth century precipitation has generally increased on land surfaces between 30°N and 85°N, but significant reductions have been observed over the last thirty or forty years between 10°S and 30°N.

At present, the main challenge is to determine the interannual variations and trends in precipitation changes over the oceans. Global averages are often unrepresentative and mask large regional disparities. However, particularly pronounced droughts in the last 30 years, as well as heavy rainfall events in many regions clearly illustrate an intensification of the hydrological cycle.

Extreme Weather Events

Extreme weather events result from exceptional fluctuations of a climate variable, and are generally associated with significant impacts on society and the environment. As illustrated by the frequent exposure by the media of the most impressive events, the statistical evolution of climate extreme characteristics (frequency, amplitude) is of major concern for current climate change, especially in the development of adaptation strategies. The term 'weather extreme' covers a wide spectrum of events at spatio-temporal scales ranging from intense local precipitation lasting a few hours to exceptional hot years for the whole planet. In addition, because they are by definition rare, extreme events offer few case studies, and the detection of possible man-induced trends requires analysis over long periods of time and/or large spatial domains. The study of extreme events is thus often restricted by the availability of datasets of sufficient spatial and temporal resolution.

Temperature Extremes

Extreme temperature events typically affect large areas (usually several thousand kilometers), and are often accompanied by extremes in other climate variables (e.g. drought during a hot summer, snowstorms during a cold winter). Their impacts on ecosystems and human activities are thus particularly important.

One of the most striking recent examples is the European heat wave in summer 2003 which, with a temperature 2.5 °C higher than the seasonal mean, exceeded by three standard deviations the distribution of summer temperatures in Europe. It had dramatic socio-economic and environmental impacts: high mortality, loss of energy production, acute urban pollution, fires, accelerated melting of glaciers etc. The return period of this particular event was estimated at 250 years at the time it occurred (2003); it would have taken 4 times longer (1000 years) without human intervention (Stott et al. 2004).

This illustrates an expected feature of climate change: hot extremes become more frequent in a warmer world, and conversely, cold extremes become less common, but remain still possible. For instance, in the U.S., the ratio of the record of high maximum daily temperatures to the record of low minimum temperatures—which should be 1 to 1 in a stationary climate—is currently about 2 to 1 and is likely to reach about 50 to 1 by the end of the twenty-first century (Meehl et al. 2009).

At the global scale the geographical distribution of changes in the frequency and amplitude of temperature extremes is consistent with the distribution of average warming (IPCC 2013). This finding is nonetheless nuanced by the fact that changes in temperature distributions are often more complex than a uniform shift towards warmer values: spreading, tightening, and/or asymmetry of values may also show up in the statistical distribution. For instance, in Europe and Central U.S., the variability of summer temperatures is expected to increase in a warmer climate, due to the increase in evapotranspiration which leads to drier soils (Douville et al. 2016). This would result in a widening of the temperature distribution, and further amplify the increase in the frequency of hot extremes in these regions.

Precipitation Extremes, Tropical Cyclones and Extra-Tropical Storms

Changes in the frequency and intensity of droughts and floods, in response to global warming, are also of major concern, as our industrialized societies become increasingly vulnerable to rainfall extremes. Although the study of droughts is similar to that of heat waves, since they both have impacts over large areas, the analysis of extremes of intense precipitation is more difficult because it requires data with finer spatio-temporal resolution.

As illustrated by the record floods in summer 2002 in Europe, intense precipitation events have been on the increase since 1950 in the mid-latitudes of the northern hemisphere. Even in the Mediterranean Basin, where rainfall is decreasing on average, the episodes of heavy rainfall are more intense. A recent example is given by the intense precipitation that occurred in South of France in autumn 2018. This upward trend is noticeable on a global scale, although the increases are more moderate than those observed for temperature. Changes in precipitation extremes are consistent with the Clausius-Clapeyron relationship: warmer air can hold more humidity, so there is more water to be mobilized by condensation when rainfall events occur.

Paradoxically, the extent of regions affected by drought is also increasing, illustrating the fact that climate change not only affects the mean of statistical distributions, but also its variability. Africa, southern Eurasia and North America are, according to the Fifth IPCC Assessment Report, the regions most affected by these droughts in recent times, as illustrated by the dry conditions persisting from 2014 to 2018 over the west of the United States (California). The increase in the frequency and/or intensity of both intense rainfall episodes and droughts is consistent with the overall intensification of the hydrological cycle in a warmer world (wet gets wetter, dry gets drier).

Lastly, climate change may also affect meteorological systems like tropical cyclones or extra-tropical storms. In a warmer world, the former are expected to occur less frequently in general, due to less frequent atmospheric conditions favorable to the cyclogenesis (reduced temperature difference between the surface and the high-troposphere). However, once triggered, future cyclones should get more energy from a warmer ocean: the intensity of the strongest tropical cyclones is therefore expected to increase. Unfortunately, so far, trends are difficult to detect from the past due to the lack of homogeneous data. For extratropical storms, forecasting is even more difficult: projections performed with climate models do not unanimously agree on their frequency and/or intensity. The only robust future signal seems to be a poleward shift of the storm tracks, but again, available observations are insufficient to capture any past trend.

Evolution of the Cryosphere

The exact estimate of the influence of human activities on climate is still limited because it is critically dependent on our ability to distinguish the signal related to this additional radiative forcing from the natural variability of the climate. However, there is a growing number of tangible factors indicating that man has had a perceptible influence on the climate. In particular, the global warming observed over the past century has been accompanied by a rise in sea level (Clark et al. 2016), largely attributed to the thermal expansion of the oceans, but also to significant changes in the cryospheric components of the climate system over the whole planet. The cryosphere represents all the water in solid form on Earth and contains more than 70% of the Earth's freshwater reservoir. It is an excellent indicator of climate change. It includes ice sheets, floating ice-shelves, mountain glaciers, snow and sea ice, but also the water in rivers and lakes that freezes in winter, as well as the permafrost, that is to say the permanently frozen ground, covered by an 'active' soil layer which melts each summer and whose thickness is variable.

Each of these components interacts in various ways with the other components of the climate system over a wide range of time scales, from seasonal (snow, permafrost, rivers, lakes, sea ice) to a hundred thousand years for the glacial-interglacial cycles. While the cryosphere is particularly important in the polar regions, there are also many glaciers in the low and mid-latitude regions, which provide an overview of the relationship between climate change and changes in the cryosphere.

Snow Cover

Estimating the extent of snow cover and the physical properties of snow is of paramount importance for both hydrological applications, such as modeling or predicting runoff due to snowmelt, and for the understanding of local or regional weather patterns. Typical snow parameters, derived from radar data, include the extent of snow cover, the water equivalent of the snow layer, and the state of the snow (wet or dry).

The extent of snow cover has a direct influence on the energy balance on the Earth's surface, but also on the soil water content. Fresh snow reflects between 80 and 90% of incident solar radiation. The warming trend decreases the snow cover, which in turn decreases the fraction of solar energy reflected back to space, and increases the absorption of incoming radiation, thereby increasing warming, which in turn accelerates the snow melting. This amplifying mechanism is known as the 'temperature-albedo' effect. Thus, the surface temperature is strongly dependent on the presence or absence of snow. Another important aspect of snow cover is the role it plays in thermal insulation. In winter, snow covered grounds cool much less quickly than bare grounds, hence the importance of snow depth for plant and animal life. Finally, melting snow in spring and summer requires a high latent heat of fusion, so that snow cover represents a significant heat loss for the atmosphere during the melting season. As a result, seasonal snow produces thermal inertia within the climate system, as it involves significant energy exchanges, with little or no change in temperature.

In the northern hemisphere, snow cover varies seasonally with a maximum in winter and a minimum in summer, but with large inter-annual variations. Since the end of the nineteenth century, daily records of snowfall and snow depth have been kept by many countries. Nevertheless, these measurements were only fully developed after 1950, and in particular from 1966 onwards, with the arrival of satellites.

All of these data series reveal that snow cover has decreased in spring and summer since the 1920s, with an even more striking decrease since the end of the 1970s. According to the Fifth IPCC Assessment report, this decrease in March-April snow cover extent ranges from -0.8% per decade over the 1922–2012 period to -2.2% per decade between years 1979 and 2012. For the fall and winter seasons, the signal is less clear: some data sets suggest positive trends as a result of increased snowfall in a warming climate, while others suggest negative trends similar to what is observed in spring and summer. Nevertheless, there is a consensus that the mean annual snow cover has decreased with a shift from February to January of the maximum extent, an earlier onset of melting (~ 5.3 days since winter 1972–1973) and thus a reduction of snow cover duration. This drop in snow cover is mainly observed in the northern hemisphere. In the southern hemisphere, very few data exist outside of Antarctica, and these are often of much lower quality than in the northern hemisphere.

Evolution of Sea Ice

Sea ice is frozen seawater. When freezing occurs, salt is expelled from the ice crystals, thus raising the density of the surface ocean waters. The formation of sea ice can therefore have a direct impact on the intensity of the thermohaline circulation. Sea ice is a highly reflective surface with a high albedo of about 0.8. Conversely, when sea ice melts, the ice-free ocean surface absorbs about 90% of the radiation due to large albedo changes, causing the ocean to warm up, followed by a further increase in surface temperature. This phenomenon is a positive feedback between temperature and albedo. Thus, sea ice regulates heat exchanges between the atmosphere and the polar ocean. It isolates the relatively 'warm' ocean waters from the much colder atmosphere, except when there are winter 'leads' occurring as a result of sea-ice break-up. These leads allow the exchange of heat and water vapor between the atmosphere and ocean which can affect local cloud cover and precipitation rate.

The extent (i.e. area of the sea covered by sea ice), and the thickness of sea ice are the two indicators of sea ice conditions. Typically, the average sea-ice extent ranges from 14 to 16×10^6 km^2 at the end of winter (7 to 9×10^6 km^2 at the end of summer) for the Arctic and from 17 to 20×10^6 km^2 (3 to 4×10^6 km^2 at the end of summer) for Antarctica. The first systematic monitoring of sea ice conditions began in 1972 with the first satellite observations. For older periods, only a few scattered data are available. Most of the analysis of the variability and trends in sea ice cover relates to the post-1978 period. The results from different types of satellites give very consistent results, and all highlight an asymmetry between the Arctic and the Austral Ocean, with a clear decreasing trend for the Arctic and a slight growth for Antarctica, although the latter trend is not statistically significant.

Before the satellite era, sea-ice data are relatively sparse and inconsistent. Nevertheless, although these data indicate significant regional variability, all confirm an increase in sea-ice cover during the nineteenth century and the first half of the twentieth century, and a sharp decline in sea ice from 1970 onwards. According to the assessment by the National Snow and Ice Data Center, sea ice coverage in September has decreased at a rate of $\sim 12.8\%$ per decade from 1979 to 2018 relative to the 1981–2010 average (Fig. 31.4). One striking result was the record reached in 2007 when the total area of sea ice fell to 4.1×10^6 km^2 in mid-September. In 2012, this record was broken with a minimum sea ice coverage of 3.4×10^6 km^2 (i.e. $\sim 20\%$ below the previous record) making 2012 the lowest minimum since the beginning of satellite era. Although measurements did not reach new records between 2012 and 2018, the recorded September minima for sea ice extent for the 2015–2018 period are all below 5.0×10^6 km^2. The persistence of these minima shows that the decrease in Arctic sea ice is accelerating. In the context of global warming, this observation is of growing concern for scientists, especially since the phenomenon is amplified by the temperature-albedo feedback. Moreover, in the past, a year marked by a very small sea-ice extent was generally followed the year after by a return to normal conditions. Observations made in recent years suggest that a progressive disappearance of the oldest ice could occur if the decrease in sea ice continues at the same rate as in the last two decades.

The decline in sea ice observed in recent years seems to be mainly driven by rising temperatures. However, atmospheric variability, and in particular the Arctic oscillation, may also favor the decreasing trend if the Arctic oscillation is in a positive mode. This has the effect of shifting the jet stream from the mid-latitudes to higher latitudes and causes the older, thicker sea ice to be pushed out of the Arctic. More

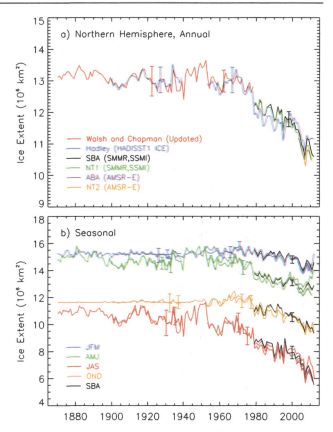

Fig. 31.4 Evolution of Arctic sea-ice extent from 1870 to 2011. **a** Annual sea-ice extent and **b** seasonal ice extent using averages of mid-month values derived from in situ measurements and remote sensing observations In (**b**), data from the different seasons are shown in different colors to illustrate variation between seasons (blue: January–February–March; green: April-May-June; red: July-August-September; orange: October–November–December). The black lines in **a** and **b** correspond to data coming from the Scanning Multichannel Microwave Radiometer and passive microwave data from the Special Sensor Microwave Imager (*Source* IPCC 2013)

recent thinner ice remains in the region and is more prone to disappear during the melt season (Rigor and Wallace 2004). Between 1989 and 1995 the Arctic Oscillation entered a very positive mode, thereby reducing the extent of sea ice. Since the mid-1990s, only a few years have seen a positive or a neutral mode. Meanwhile, the thickness of Arctic sea ice continues to drop due to increasing temperatures. In March 1985, the fraction of first-year ice was $\sim 50\%$ against 70% or more in 2015, and during the 1985–2015 period, the multi-year ice (4 years and older) dropped dramatically from 20% to only 3% (Tschudi et al. 2016).

The situation is different in Antarctica mainly because of its geographic location (surrounded by the Southern Ocean and centered on the South Pole). Atmospheric winds and the oceanic circumpolar currents act as barriers to warmer air and warmer waters coming from the north. Changes in sea-ice coverage are more tenuous, with an increase of 1.2–1.8% per decade between 1979 and 2012 (IPCC 2013). However, changes in the distribution of Antarctic sea ice show major regional differences. For example, the Weddell and Ross seas are experiencing an increase in sea ice extent due to large-scale changes in atmospheric circulation, while in West Antarctica, the surface covered by sea ice is drastically decreasing, consistent with the observed warming in this region.

Permafrost

Permafrost is defined as soil that remains permanently frozen for at least two consecutive years. It is topped by a so-called 'active layer' that thaws each summer, and whose thickness can vary from a few centimeters to a few meters, depending on altitude and latitude. In areas where it has persisted for several glacial-interglacial cycles, the permafrost can be several hundred meters thick, and even exceeds 1000 m in some parts of Siberia and Canada. During the last glacial-interglacial cycle, there have been large variations in area and depth of permafrost over North America (Tarasov and Peltier 2007) or Europe (VandenBerghe 2011). Currently, permafrost covers 22.8×10^6 km^2 of the northern hemisphere, or about 24% of the continental areas. Permafrost occurs mainly in polar and circumpolar areas and in mountain regions at lower latitudes (e.g. Chile, the Alps, the Himalayas). It can also be found in the seabed of the Arctic Ocean in the continental shelf areas. When surface conditions are not spatially homogeneous (e.g. snow cover, vegetation) permafrost can occur in patches. Such permafrost areas are called 'discontinuous permafrost zones'. When temperatures drop at higher latitudes, gaps in permafrost are less frequent. When surface conditions become homogeneous, permafrost is referred to as 'continuous permafrost'.

The presence of permafrost is critically dependent on the soil temperature, which is itself controlled by the surface energy balance, and thus, by several climatic factors such as incoming solar radiation, cloudiness, snow cover, vegetation cover, surface and subsurface hydrology, and carbon exchanges between the soil and the atmosphere. One of the key factors affecting permafrost distribution is the insulating effect of snow. When snow is present, ground temperatures are generally warmer than those which would occur under smaller snow cover or snow-free conditions. In continuous permafrost areas, snow cover exerts a direct influence on the active layer thickness. For example, it has been shown that a doubling of snow cover from 25 to 50 cm may increase the mean annual surface soil temperature by several degrees (Fig. 31.5). On the other hand, if seasonal snow melting occurs in late spring or early summer, ground warming is delayed.

The permafrost thermal state is also influenced by rainfall. Firstly, ground temperatures can be increased through the energy flux released by liquid precipitation penetrating into the soil. Secondly, rain falling on a snow covered surface may alter the snow insulating effect causing

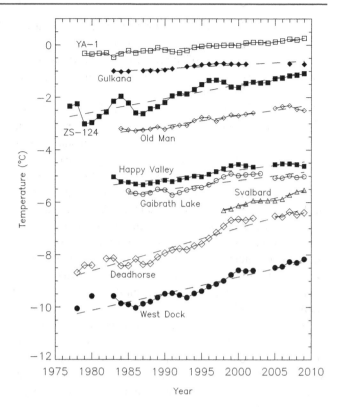

Fig. 31.5 Time series of mean annual ground temperatures at depths between 10 and 20 m for boreholes located throughout the circumpolar northern permafrost regions (Romanovsky et al. 2010a). Data sources are from Romanovsky et al. (2010b) and Christiansen et al. (2010). Measurement depth is 10 m for Russian boreholes, 15 m for Gulkana and Oldman, and 20 m for all other boreholes. Borehole locations are: ZS-124, 67.48°N 063.48°E; 85-8A, 61.68°N 121.18°W; Gulkana, 62.28°N 145.58°W; YA-1, 67.58°N 648°E; Oldman, 66.48°N 150.68° W; Happy Valley, 69.18°N 148.88°W; Svalbard, 78.28°N 016.58°E; Deadhorse, 70.28°N 148.58°W and West Dock, 70.48°N 148.58°W. The rate of change (degrees Celsius per decade) in permafrost temperature over the period of each site record is: ZS-124: 0.53 ± 0.07; YA-1: 0.21 ± 0.02; West Dock: 0.64 ± 0.08; Deadhorse: 0.82 ± 0.07; Happy Valley: 0.34 ± 0.05; Gaibrath Lake: 0.35 ± 0.07; Gulkana: 0.15 ± 0.03; Old Man: 0.40 ± 0.04 and Svalvard: 0.63 ± 0.09. The trends indicate the very likely range, 90% (*Source* IPCC 2013)

temperatures to lower. Interception of precipitation by vegetation also has an impact on ground temperatures through evaporation and transpiration and the associated turbulent heat exchanges between atmosphere and surface layers. However, the direct effect of vegetation on ground temperatures is less important than its role on snow cover. Interception of snow in boreal forests reduces snow cover on soils and acts to reduce ground temperatures. These examples show that the formation or degradation of permafrost is strongly influenced by climate. It is thus studied as an indicator of climate change by a global network of researchers (Romanovsky et al. 2010a, b) who rely on temperature measurements taken from boreholes and from satellite tracking. In the framework of the International Polar Year (2007–2009), a large permafrost- monitoring network was developed and ground temperatures (which control the thermal state of the permafrost) have been measured in 575 sites located in arctic regions (North America, Nordic Eurasian regions and Russia). In most sites belonging to the network, permafrost temperatures have increased in recent decades. The observed rate of change of mean annual ground temperatures from mid-1970s to 2010 ranges from 0.15 ± 0.03 to 0.82 ± 0.07 °C per decade, depending on the site (Fig. 31.5). In cold permafrost regions (mostly in continuous permafrost zones), the mean annual ground temperatures have increased by up to 2 °C, compared to less than 1 °C in warm, forested permafrost areas (discontinuous permafrost). This can be explained by the fact that the amount of ice in warm permafrost is usually larger than for cold permafrost. Indeed, in case of ice melting, the overall warming trend is counteracted by the latent heat effect. Moreover, in forested areas, the snow insulating effect is limited due to the interception of snow by vegetation.

One of the main consequences of permafrost warming is increased thickness of the active layer, although some permafrost areas exhibit only modest thickening or even a thinning. Indeed, a study based on the analysis of 169 circumpolar and mid-latitude sites revealed that only 43.2% of them have experienced an increase of the active layer thickness since the 1990s (Luo et al. 2016). However, there is great spatio-temporal variability from one site to the other ranging from a few tenths of cm/yr to more than 10 cm/yr. Thickening of the active layer is a matter of great concern since it may have large consequences on the stability of the surface due to the melting of shallow ground ice. Potential impacts include thaw settlement, soil creeps, slope failures and ponding of surface water. All these features can cause severe damages to infrastructures, such as roads, dams or structural building foundations but also to vegetation. In forested areas, thaw modifies the hydrological conditions and can lead, for example, to the destruction of tree roots, causing drastic changes in the ecosystems. When permafrost thaws and the active layer thickens, more organic matter is likely to be decomposed by bacteria that produce either methane or carbon. In both cases, the bacterial action enhances greenhouse gas emissions and thus promotes global warming. However, the magnitude of this thaw-related feedback is a great unknown. The total amount of carbon stored in the permafrost has been estimated at 1672 Gt, of which 277 Gt is found in peat bogs. This is about twice the amount of carbon in the atmosphere.

We still do not know for sure if the increase in atmospheric methane concentration observed in recent years, after about ten years of relative stability, is due to the warming of the high northern latitudes. Another amplification reaction observed at high northern latitudes involves the microbial transformation of nitrogen trapped in soils into nitrous oxide, which could increase with increasing temperatures and thus, in turn, amplify global warming.

In any case, even if there are still many uncertainties, the Fifth IPCC assessment report emphasized on the positive feedback of permafrost melting on climate warming: *"the release of CO_2 or CH_4 to the atmosphere from thawing permafrost carbon stocks over the twenty-first century is assessed to be in the range of 50 to 250 GtC for RCP8.5"* (IPCC 2013). The RCP scenarios are defined in the section "Projecting the Future of the Climate System" of this chapter.

Glaciers

A glacier is a mass of ice formed by the successive accumulation of layers of snow year after year. Over the years, under the pressure of its own weight, the snow hardens and becomes granular (firn), then turns into ice and expels the air it contained. Under the action of gravity, the ice flows along the slope, thus supplying the lower parts of the glacier. A glacier is in constant movement and carries mass from high altitudes to lower altitudes. In winter, the glacier grows due to snow accumulation on its surface. During the following summer, the glacier loses all or part of the mass it gained during winter. The disappearance of ice through surface processes is called ablation. The difference between accumulation and ablation determines the surface mass balance of the glacier. This brings about a change in ice volume. A positive balance overall causes the glacier to grow while a negative balance leads to a loss of ice volume which can be accompanied by a retreat of the glacial front.

The mass balance of temperate glaciers in the mid-latitudes is mainly dependent on winter precipitation, summer temperature and summer snowfalls (temporally reducing the melt due to the increased albedo). In contrast, the glaciers in low latitudes, where ablation occurs throughout the year and multiple accumulation seasons exist, are strongly influenced by variations in the atmospheric moisture content which affects incoming solar radiation, atmospheric long-wave emission, albedo, precipitation and sublimation. In monsoon areas, such as the Himalayas, accumulation and ablation occur mainly in summer. Glaciers at high altitudes and in polar regions can experience accumulation in any season.

The retreat of mountain glaciers is one of the most visible examples of climate change. A compilation of observations made by the World Glacier Monitoring Service on 19 regions around the world, gathering more than 40,000 observations, shows that the retreat is an almost global phenomenon, despite some intermittent readvances related to dynamical instabilities (e.g. Island, Svalbard) or to specific climatic conditions (i.e. increased winter snowfall) observed on a few individual glaciers in Scandinavia or New Zealand in the 1990s. However, the periods of glacier front advances are short compared to the overall ice retreat. By compiling the data obtained on 169 glaciers since 1700, Oerlemans (2005) shows that the retreat of glacier fronts began in the nineteenth century, and accelerated strongly from 1850 onwards, with a continuation throughout the twentieth century and early decades of the twenty-first century.

Regional analyses have shown that, until around 2000, the average mass balance cumulated over all European glaciers was close to zero, with significant mass losses for Alpine glaciers being compensated for by advances of glaciers in western Norway stemming from a sharp increase in precipitation in response to a positive phase of the North Atlantic Oscillation. From the year 2000 onwards, the Norwegian glaciers began to retreat in response to a decrease in precipitation. Over the period 2003–2009, the most negative mass balances occurred for glaciers in the northwestern United States and southwestern Canada, Central Europe, Southern Andes and low latitude areas. In the Alps, glaciers have been retreating since the mid-nineteenth century. In Switzerland, they currently cover only 60% of the area they occupied in 1850. Due to the heat wave, an exceptionally high loss of mass occurred in 2003, corresponding to a reduction of 2500 kg m^{-2} yr^{-1} for the nine glaciers studied. This value exceeds the previous record 1600 kg m^{-2} yr^{-1} in 1996 and is four times higher than the average measured between 1980 and 2001 (600 kg m^{-2} yr^{-1}). In Africa, the glacier area at the top of Kilimanjaro is now only 20% of what it was at the beginning of the twentieth century. In Patagonia, the 'icefields' have lost between 3 and 13 km^3/yr of ice since the 1970s. The retreat of the glaciers in Nepal and Himalayas seems to have accelerated over the past twenty years, and in Tibet, the number of glaciers in retreat has recently multiplied.

The Fifth IPCC Assessment Report estimated that the contribution of glaciers and small ice caps to sea level rise over the 1993–2010 period was about 0.76 mm/yr (IPCC 2013). Since then, new estimates based on new data indicate a higher contribution, demonstrating that mass loss from glaciers and small ice caps has accelerated significantly since the early 1990s, and currently contributes between 1.05 and 1.12 mm/yr to the rise of the global sea level.

Polar Ice Sheets

Polar ice sheets are huge masses of ice formed, like glaciers, by continual accumulation of snow in excess of ablation, which gradually turns into ice under the effect of compaction. As for glaciers, this transformation occurs in a transition zone about one hundred meters thick called firn. Currently, the ice sheets are located at high latitudes, one near the North Pole, Greenland, the other centered on the South Pole, Antarctica. The area of the Greenland ice sheet is about 1.8×10^6 km^2. In the center, the ice thickness is greater than 3000 m. The Greenland ice volume ($\sim 3.0 \times 10^6$ km^2) represents about 10% of the worldwide freshwater supplies. Antarctica is composed of two effectively distinct ice sheets in the east and west, separated by the Transantarctic Mountains. Its ice volume is close to 27×10^6 km^2 and its surface, almost 98% covered by ice, is about 14×10^6 km^2. A large part of the western ice sheet lies below sea level. The West Antarctic ice sheet extends locally over the sea to form floating ice shelves, mainly in the embayments of the coast, as in the Weddell and Ross Seas. In contrast, East Antarctica, which is larger, rests largely on bedrock. It forms a plateau with an area exceeding 10×10^6 km^2 covered by a large ice layer of more than 4000 m thick in the center.

The evolution of the part of an ice sheet grounded on the bedrock depends on its surface mass balance and its flow due to the deformation of the ice itself. When the temperature is high enough, the ice melts at the surface. As for glaciers, the surface mass balance is determined by the difference between accumulation and ablation. In addition, under the effect of its own weight, the ice flows by plastic deformation along the line of steeper slope, as well as by sliding on the bedrock when the local temperature is close to the melting point: this is called basal sliding. As ice is an insulating material, a temperature gradient is established between the colder surface and the warmer base. Furthermore, by changing the ice viscosity, the temperature also affects the flow velocities from the surface to the base of the ice sheet. Thus, the processes involved in the ice deformation are not the same at the surface and at the base of the ice sheet.

In the case of the Antarctic ice sheet, the surface temperature generally remains low enough so that ablation is negligible. The ice then drains into the ocean or feeds the floating ice shelves through ice streams which are characterized by a rapid outflow (i.e. low basal friction). The sources of these ice streams are found far upstream, and their contribution to the evacuation of grounded ice from the center of the ice sheet towards the edges is estimated at nearly 90%. The flow regime through the ice shelves is very different from that of grounded ice. In fact, whereas grounded ice is characterized by a shear regime in the vertical plane, the predominant constraints on the ice shelves are the horizontal shear and the pressure forces exerted by the sea. The destabilization of these glacial platforms is due to the increase in sea level, but also to the basal melting beneath the platforms. This melting is therefore related to the ocean temperatures under the ice shelves and to the energy released by ocean currents. The dislocation of these glacial plateaus leads to the formation of icebergs (i.e. ice calving). This is exactly what happened with the dislocation of the Larsen B ice shelf in 2002 which resulted in a surface loss of about 3250 km^2. This could also occur in the coming years with the break-up of the Larsen C ice shelf in July 2017 (6000 km^2 of surface loss) and the continuous acceleration of the Twaithes glacier reported by satellite observations since 1992. Moreover, as the ice shelves exert a buttressing effect for the upstream grounded ice, their disintegration can cause the destabilization of a large part of the ice sheet.

There are different methods to measure the amount of ice stored in an ice sheet:

(a) **The measurement of inflow** (snow accumulation) **and outflow** (ablation, discharges to the ocean). Snow accumulation is often estimated from annual layers in ice cores and interpolated between different drilling sites. The use of high resolution atmospheric models is also becoming more common. Discharges of ice to the ocean are estimated from seismic or radar measurements of the ice thickness and from measurements of ice flow velocity. Ablation is usually determined from ice models forced with atmospheric reanalyses, climatology or outputs from global climate models calibrated with surface observations. The loss of mass beneath the ice shelves remains very difficult to quantify. In general, inflows and outflows cannot be estimated with a margin error of less than 5%, which implies uncertainties of 40 and 140 Gt/yr on the estimate of the mass balance of Greenland and Antarctica respectively.

(b) **Remote-sensing techniques**. These include altimetry measurements from radio-echo sounding, interferometry and gravimetry. Interferometry provides information on ice flow velocities. The altimetric measurements provide information on the spatial and temporal variations of the topography, which makes it possible to trace the ice volume variations, after correcting for the altitude of the bedrock and for variations in the thickness and density of the firn. These measurements are strongly dependent on the nature of the terrain (flat, sloping or hilly surfaces) and the snow surface conditions (density, viscosity, etc.) and this may, in some cases, make interpretation difficult. Finally, satellite measurements of the gravity field provide, for the first time, estimates of the ice mass changes. However, there are still significant uncertainties, especially regarding the cause of the change in mass, as this could come from the ice sheet itself, the air column, the evolution of the subglacial bedrock or even from masses near the ice sheet (i.e. ocean mass, masses of water or snow contained on nearby continents). These different effects are evaluated and then corrected, but a significant uncertainty remains, mainly on how to correct for the altitude of the underlying Antarctic bedrock. Other sources of uncertainty relating to satellite measurements come from the fact that they do not provide complete coverage of the ice sheets. Moreover, in Greenland, for example, remote sensing techniques may underestimate the extent of the ablation area and the increased rainfall over some parts of the ice sheet during rainfall events.

Past published estimates of Greenland and Antarctic ice sheets obtained with these methods have often diverged. This was due to the way in which the different sources of uncertainty were estimated and to the fact that the different measurements did not cover the same regions and the same time periods. Within the IMBIE framework (Ice-sheet Mass Balance Intercomparison Exercise), scientists made a huge effort to combine, over common survey periods and common regions, various observations from satellite geodetic techniques (altimetry, interferometry, radio-echo sounding and gravimetry) with simulated surface mass balance estimates inferred from regional atmospheric models. This allowed for a reconciliation of the apparent disparities between the different methods and for a consistent picture of ice-sheet mass balance to be created (Shepherd et al. 2012).

For Greenland, this compilation effort confirmed with a high confidence level that the ice sheet is continuously losing mass and that this process now affects all sectors of the ice sheet (Fig. 31.6a–c). However, after a record mass loss in summer 2012, Greenland has seen a slight decrease in the short-term mass loss trend. The mass loss is partitioned between surface melting and dynamic ice discharges. Shepherd et al. (2012) estimate that the ice mass loss is about -142 ± 49 Gt/yr over the IMBIE time period (1992–2011) with an acceleration of the mass loss rate as illustrated by the comparison between the estimations made for 1992–2000 (-51 ± 65 Gt/yr) and 2005–2010 (-263 ± 30 Gt/yr). Using gravimetry observations, Velicogna et al. (2014) provide a more recent estimate of the Greenland ice sheet mass loss for the 2003–2013 decade of 280 ± 58 Gt/yr with an acceleration of the loss rate bringing it to 25.4 ± 58 Gt/yr^2.

The case of the Antarctic ice sheet (Fig. 31.6b) is a bit different since recent observations have shown that mass loss was mainly driven by dynamic ice discharge resulting from enhanced ice flow of marine-terminating glaciers. The main region experiencing mass loss is the West Antarctic ice sheet (WAIS), especially in the Amundsen/Bellingshausen Sea sectors (e.g. Thwaites and Pine Island glaciers) and, to a lesser extent, in the Antarctic Peninsula. According to a recent update of the IMBIE estimates (IMBIE team, 2018), the mass loss from the Amundsen and Bellingshausen Sea sectors increased from 53 ± 29 Gt/yr to 159 ± 26 Gt/yr over the 1992–2017 period, and from 7 ± 13 Gt/yr to 33 ± 16 Gt/yr in the Antarctic Peninsula. It has long been considered that the East Antarctic ice sheet (EAIS) was gaining mass due to enhanced precipitation, despite no firm consensus being established (Velicogna and Wahr 2006; Ramillien et al. 2006). However, recent estimates suggest that some sectors, such as the Wilkes Land region, are losing mass. As a result, the rate of change in ice-sheet mass is estimated to be +11 ± 58 Gt/yr in 1992 (mass gain) and −28 ± 30 Gt/yr (mass loss) in 2017 (IMBIE team, 2018). Using a different technique, Rignot et al. (2019) estimate an even larger mass loss from EAIS with a strongly reduced uncertainty.

Overall, taking the Antarctic and Greenland ice sheets together, it appears that mass losses have accelerated in recent years. This trend is correlated with an increase in surface ablation due to increasing temperatures, but also with an acceleration of ice flow and subsequent dynamic ice discharges.

Several processes are at the origin of ice mass loss. The increasing surface ablation, mainly observed in Greenland, is a direct response to increased atmospheric temperatures. However, the ocean warming also plays a key role. As oceanic temperatures rise, basal melting under the ice shelves is enhanced. Eventually, this may lead to the dislocation of ice shelves and to the removal of the buttressing effect mentioned above. This causes an inland retreat of the grounding line (i.e. the limit beyond which ice starts to float), and subsequently, an acceleration of the upstream grounded ice. Theoretically, this process is only valid for ice-shelves confined within their embayment. It is responsible for more than half of the ice mass loss at the margins of the Antarctic ice sheet. For unconfined ice shelves, another process, known as the Marine Ice Sheet Instability, may also apply. For a marine-based ice sheet, such as the WAIS, bedrock is often more depressed in the center of the ice sheet than it is at the margins. As ice flux increases with ice thickness, the position of the grounding line becomes highly unstable in areas of reverse bed slopes, and any change in ice thickness in the vicinity of the grounding line creates an irreversible retreat of the grounding line position. A second hypothesis is the lubrication of the subglacial substratum caused by meltwater produced at the surface that percolates to the base of the ice sheet. A third process that may be

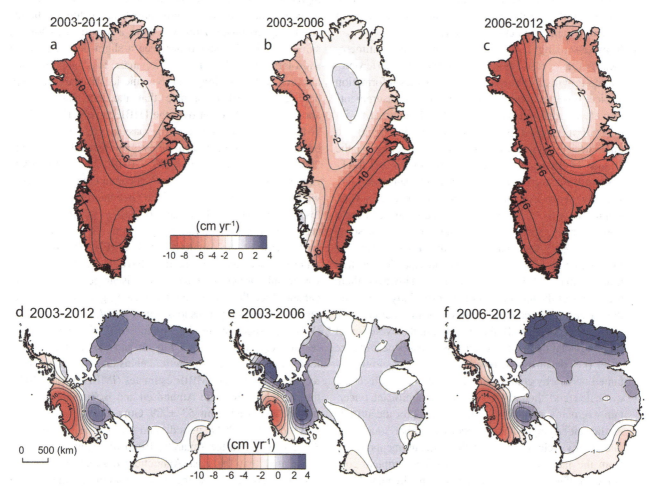

Fig. 31.6 Temporal evolution of ice loss in Greenland (top) and Antarctica (bottom) determined from time variable gravimetry observations from the GRACE satellite, shown in centimeters of water per year for the periods 2003–2012, 2003–2006 and 2006–2012, color coded red (loss) to blue (gain) (*Source* IPCC 2013)

responsible for ice flow acceleration is hydro-fracturing (DeConto and Pollard 2016). This mechanism is related to water coming from ice melting at the surface of the ice shelf that may percolate inside the ice-shelves. Crevasses can form and widen when the water pressure is high enough, thereby favoring iceberg calving. Ultimately, this process may lead to ice-shelf collapse. It may also favor the marine ice sheet instability. Indeed, once the ice shelves have collapsed, ice cliffs become unstable and fall down if their height is greater than $\sim$ 90 m. However, this process remains poorly constrained and is still a matter of debate.

A special attention is given to the evolution of ice sheets. Indeed, ice sheets strongly interact with the other components of the climate system. These interactions can lead to a highly non-linear climate response. This means that the effects of the radiative disturbance (caused by variations in insolation or in the amount of greenhouse gas in the atmosphere) can be amplified or mitigated by these feedback processes, as illustrated in Chap. 28 for abrupt events. Many studies on these interactions have been published over the past two decades. They would deserve a full chapter. Here, we only give some quick examples for time scales ranging from a few decades to a few centuries. Ice sheet melting is first of all accompanied by possible changes in albedo and therefore in the surface energy balance. This effect is almost instantaneous. In turn, a change in the energy balance can lead to changes in the mass balance of the ice sheets. Another consequence of the melting and/or mechanical destabilization of the ice sheets, widely discussed in the literature, concerns the freshwater flux released in the ocean. Locally, this release leads to a decrease in ocean surface temperatures, a change in sea ice cover and a reduction of ocean density in the vicinity of the ice sheets. Density changes also cause a disruption of large-scale ocean circulation by altering deep-water convection. For example, meltwater from Greenland has the potential to weaken the Atlantic Meridional Overturning Circulation. These changes can have effects in regions far from the polar zones. For example, the recent study by Defrance (2017) showed that for a substantial melting of Greenland under a climate change driven by the RCP8.5 scenario, Greenland meltwater could cause a massive alteration of the monsoon regime with a drastic decrease in West African rainfall, and subsequently with a significant reduction in cultivable areas. Therefore, in addition to the direct and obvious consequences on sea level, the future of polar ice sheets is of primary importance for the future of human societies, all the more so if they are already economically fragile.

Sea Level Changes

Sea level variations are the result of changes in both the volume of the oceans and ocean basins, as well as changes in the mass of water contained in the oceans. Depending on the time scale being considered, these types of variations have different origins. Over geological time scales, changes in the shape of ocean basins and in the continent/ocean distribution are the main factors affecting sea level. Over glacial-interglacial cycles, sea-level variations are mainly related to changes in continental ice volume and to the isostatic adjustment due to the vertical movements of the Earth's crust in response to changes in the mass of land ice. Following the last deglaciation initiated about 21,000 years ago, sea level rose by $\sim$ 120 m and then stabilized around 6000 years ago. Geological data indicate that sea level has not changed by more than 30 cm from that time until the end of the nineteenth century.

On time scales ranging from a few years to a few decades, variations in the mean sea level are the result of two factors, mainly related to climate change:

(a) Variation of the ocean volume caused by changes in sea temperature. As temperature increases, the volume of water expands. This process is called thermal expansion;
(b) Variations in the bodies of oceanic water resulting mainly from exchanges with continental reservoirs, such as rivers, lakes and inland seas, snowpack, ground water, but also mountain glaciers and polar ice sheets.

Other factors such as ocean circulation or atmospheric pressure can bring about local variations, without altering the mean global sea level. In addition, some human interventions have the effect of modifying regional hydrology by modifying the runoff of freshwater released to the ocean, and thus the sea level. This occurs in the case of dams, irrigation, urbanization, water extraction from aquifers and deforestation. Some of these processes have the effect of increasing runoff (urbanization, deforestation); others, such as dams and irrigation, contribute to the sequestration of freshwater on the continents. Current estimates of the net land water storage are based on observations and models, and vary between −0.33 and 0.23 mm/yr over the period 2002–2014/15 (WCRP Global Sea Level Budget Group 2018).

Measurements of current sea-level variations are based on two different techniques: tide gauges, which began to be installed in the nineteenth century, and altimetry data from satellite observations since 1992. The two main limitations of tide gauges are their inhomogeneous spatial and temporal coverage, and the inclusion of vertical land motion in their record, which needs to be corrected for to obtain relative sea-level changes. To limit the uncertainties related to these movements, which are difficult to quantify in current models, only a few tens of geologically stable sites, mainly located along the coasts of North America and Europe, are taken into account to inform us of the sea level evolution in the course of the twentieth century. Based on a compilation of the most recent estimates, the sea-level rise as indicated by tide gauge data is estimated at 1.7 ± 0.2 mm/yr between 1901 and 2010 for a total sea level rise of 0.19 ± 0.02 m (IPCC 2013).

Since 1992 and the launch of the TOPEX/Poseidon satellite, altimetric data have provided a new way of estimating sea-level variations, based on the time required for the round trip of the radar wave and on the satellite altitude defined above a standard reference surface. These measurements are thus independent on the vertical land motion, in spite of a small correction (around 0.3 mm/year) to account for the change in the reference level at the ocean bottom due to the post-glacial rebound (Peltier et al., 2015). This results in an increase of about 3.2 ± 0.4 mm/yr between 1993 and 2012 (IPCC 2013, Fig. 31.7), now reevaluated at 3.1 ± 0.4 mm/yr between 1993 and 2017 (WCRP Global Sea Level Budget Group 2018). Satellite altimetry data now cover a time period of 25 years, long enough to identify an acceleration of the sea level rise estimated at around 0.084 ± 0.025 mm/yr^2 (Nerem et al. 2018).

These altimetry data show significant regional disparity, with some regions showing a sea-level rise well above the global mean and other regions showing a decrease in sea level. This regional variability partially explains the differences between tide gauge data and altimetry data, and is due to several factors including density variations, ocean circulation, atmospheric pressure, and variations within the solid Earth or the geoid. For example, the Scandinavian shield continues to rise at a faster rate than the mean and, so paradoxically, a decrease in local sea level is recorded there.

Tide gauges and altimetry data provide ways of quantifying the global mean sea level rise and its different components. Data on ocean temperature changes allow us to quantify the contribution from thermal expansion. Recent advances in temperature measurements have greatly improved our knowledge. Since 2000, the Argo project has deployed thousands of free-drifting profiling floats measuring temperature (and salinity) in the ocean at depths between 0 and 2000 m, providing a continuous record of heat penetrating in the ocean. In addition, ship-based data were collected during the World Ocean Circulation Experiment, providing the means to estimate the deeper ocean temperature change (Johnson et al. 2007; Purkey and Johnson 2010; Kouketsu et al. 2011). These data show that, globally, the ocean has warmed significantly over the last 50 years, and in particular, over the past two decades. By vertically integrating the temperature data along the water column at each oceanic point, sea-level changes due to oceanic thermal changes can be estimated over the past 50 years. Based on data gathered over a depth of 700 m, the contribution of thermal expansion to sea level rise was estimated at 0.60 ± 0.2 mm/yr for the period 1971–2010 (0.8 ± 0.3 mm/yr if the deep ocean contribution is included). For the more recent period 1993-2017, this contribution is 1.3 ± 0.4 mm/yr (Table 10.1). In recent decades, this has been the dominant contribution to global sea level rise.

The other factors contributing to the rise in sea level come from changes in the oceanic water mass (Table 31.1). One of the most important contributions is from mountain glaciers (excluding Greenland and Antarctica) which cause a global sea level increase of 0.62 ± 0.37 mm/year for the period 1971–2010 and 0.65 ± 0.15 mm/year for the period 1993–2017. Both polar ice sheets also contribute to the recent rise in global sea level with 0.48 ± 0.10 mm/yr for Greenland

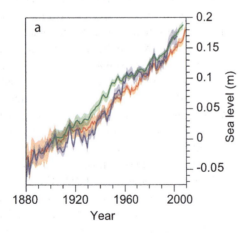
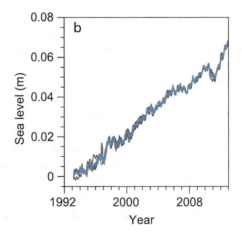
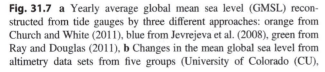

Fig. 31.7 a Yearly average global mean sea level (GMSL) reconstructed from tide gauges by three different approaches: orange from Church and White (2011), blue from Jevrejeva et al. (2008), green from Ray and Douglas (2011), **b** Changes in the mean global sea level from altimetry data sets from five groups (University of Colorado (CU), National Oceanic and Atmospheric Administration (NOAA), Goddard Space Flight Centre (GSFC), Archiving, Validation and Interpretation of Satellite Oceanographic (AVISO), Commonwealth Scientific and Industrial Research Organisation (CSIRO) with the mean of the five shown as a bright blue line (*Source* IPCC 2013)

Table 31.1 Sea level rise from different sources, adapted from IPCC (2013), Chapter 13, with additional data from WCRP Global Sea Level Budget Group (2018). The percentages are relative to the sum of contributions

Components	Sea level rise (mm/yr) 1993–2010 (IPCC, 2013)	Sea level rise (mm/yr) 1993–2017 (WCRP Global Sea Level Budget Group, 2018)	Sea level rise (mm/yr) 2005–2017 (WCRP Global Sea Level Budget Group, 2018)
1. Thermal expansion	1.1 ± 0.2	1.3 ± 0.4 (48%)	1.3 ± 0.4 (44%)
2. Glaciers (excluding Greenland and Antarctica)	0.76 ± 0.37	0.65 ± 0.15 (24%)	0.74 ± 0.1 (25%)
3. Greenland ice sheet	0.33 ± 0.08	0.48 ± 0.10 (18%)	0.76 ± 0.1 (26%)
4. Antarctic ice-sheet	0.27 ± 0.11	0.25 ± 0.10 (9%)	0.42 ± 0.1 (14%)
5. Land water storage	0.38 ± 0.11	/	−0.27 ± 0.15 (−9%)
Total of contributions (1 + 2 + 3 + 4 + 5)	2.8 ± 0.5	2.7 ± 0.23	2.95 ± 0.21
Observed global mean sea level rise	3.2 ± 0.4	3.07 ± 0.37	3.5 ± 0.2
Thermal expansion + GRACE-based ocean mass			3.6 ± 0.4

and 0.25 ± 0.10 mm/yr for Antarctica for 1993–2017. The ice-sheet contribution to sea-level rise has been increasing from 27% of the total contributions for 1993–2017 to 40% for 2005–2017.

The sum of all contributions, including land water storage, amounts to 2.8 ± 0.5 mm/yr for the period 1993–2010, which is lower than the observed global mean sea level rise of 3.2 ± 0.4 mm/yr (Table 31.1). Similarly, for the most recent period 2005-2017, the sum of contributions is 2.95 ± 0.21 mm/yr, lower than 3.5 ± 0.2 mm/yr which is the observed global mean sea level rise. This discrepancy is due to uncertainties in the estimation of the different components of ocean mass contributions (glaciers, ice sheets and land water storage). Instead, if the ocean mass contribution is taken from gravity measurements using GRACE (Gravity Recovery And Climate Experiment), the sum of thermal expansion and GRACE-based ocean mass contributions is 3.6 ± 0.4 mm/yr, which is in the error bar of the observed global mean sea level rise for this period (3.5 ± 0.2 mm/yr).

Climate Modeling and Recent Changes

Simple Radiative Climate Models and Their Limitations

To estimate the changes in the mean Earth's temperature in response to different radiative forcings (solar irradiance, greenhouse gases, energy re-emitted by the surface etc.), a first approach is to use purely radiative models. With these models, one can easily and accurately calculate the temperature changes with some simplifications: it is assumed that only the temperatures change, and that this affects only the radiation emission law, without modifying any radiative property of the atmosphere or the surface. For example, with a doubling of atmospheric CO_2 concentration, a temperature increase of 1.2 ± 0.1 °C is obtained. However, these simplified assumptions are not reliable, because when the temperature changes, all the other climate variables (e.g. humidity, wind, clouds, rain, snow cover) also change. These changes can in turn modify the energy balance of the surface and the atmosphere and thus have an additional effect on temperatures. These are called *feedback processes*. They are said to be positive when they amplify the initial disturbances, and are said to be negative when the opposite is true, when they work towards the stabilization of the system.

The first studies which took these feedbacks into account were carried out using radiative-convective models, with only one vertical dimension. For example, Manabe and Wetherald (1967) showed with their model, that the surface warming due to a doubling of the atmospheric CO_2 concentration was 1.3 °C when the absolute humidity of the atmosphere remained constant, but reached 2.4 °C in case of constant relative humidity. Numerous other studies have confirmed the crucial importance of feedback mechanisms on the magnitude of global warming: they can amplify twice to four times the temperature variation compared to the situation where no feedback is taken into account. These studies have also shown that the magnitude of these feedbacks is strongly dependent on complex physical processes (less understood than radiative transfer), such as turbulence, convection, cloud formation and precipitation (Ramanathan and Coakley 1978). These processes, and in particular, the atmospheric circulation which determines how energy and water vapor are redistributed within the atmosphere, cannot be represented in a useful way in the radiative-convective models. Thus, even rough estimates of the changes in the global mean temperature require the atmospheric dynamics to be taken into consideration and, for more precise calculations, three-dimensional models representing the general circulation of the global atmosphere of the Earth are needed.

General Circulation Models: Progress and Limitations

The Evolution of Climate Models

A general circulation climate model is a simplified representation of the climate system, but including as best as possible most processes that influence the climate. It is based on a preliminary physical analysis, in order to reduce the number of processes to be incorporated, as well as on appropriate mathematical and numerical formulations. Numerical modeling of the climate began in the 1970s and has since greatly progressed, thanks to the steady increase in the processing power of computers. The general philosophy behind this development, established by Charney and his collaborators in the 1950s, was to understand the problem on a global scale, even if this meant making very general approximations initially, with the aim of gradually improving the model by identifying its drawbacks and its limitations. The first models only described the atmosphere and the continental surfaces. In order to reduce model complexity, the oceanic surface temperature was imposed: even if the energy balance of the ocean surface is very different from the observed measurement, the surface temperature is maintained at its prescribed value. However, to investigate the variations in climate accounting for oceanic temperatures is required.

The first studies of the impact of a CO_2 doubling using this type of model were carried out in the 1970s at the GFDL (Geophysical Fluid Dynamics Laboratory, Princeton, USA) with a representation of the ocean with no circulation and zero heat capacity, so that equilibrium with the atmosphere could be achieved quickly. In this model, there was no diurnal or annual variation in insolation, and ad hoc corrections were applied to heat fluxes at the air-sea interface to keep the ocean surface temperature close to observations. The use of this type of model became widespread during the 1980s, with a gradual increase in sophistication and realism. For example, both annual and diurnal insolation variations were incorporated, modeling of cloud formation processes began, etc. At the same time, new satellites were being used to estimate global cloud cover and radiative fluxes at the top of the atmosphere, which contributed to the improvement and evaluation of atmospheric models.

In parallel, ocean general circulation models were developed to simulate heat transport and to study the role of the ocean in the Earth's energy balance. Progressively, they included sea-ice models and, from the 1990s onwards, they were coupled with atmospheric models to simulate the atmosphere-ocean-sea ice interactions. These early models did not simulate the heat and water fluxes at the air-sea interface, resulting in strong biases in the simulated oceanic surface temperatures. To fix these shortcomings, the fluxes at the air-sea interface were corrected in an ad hoc way, before these corrections were gradually removed after the end of the 1990s, thanks to continuous model improvement.

These coupled atmosphere-ocean models gradually became the basic tool for studying both past and future climate variations. For example, in preparation for the Fourth IPCC Assessment Report (IPCC 2007), some twenty of these coupled models performed a whole series of climate change simulations, and only six of them were based on flux correction at the air-sea interface. These models can simulate a natural climate variability that can then be compared to observations at different time scales: a few days, a few years (interannual variability, the best known of which is El Niño), or a few tens, or even several hundred years.

In a schematic way, climate models simulate the energy and water cycles. Progressively, representations of chemical reactions in the atmosphere, biogeochemical cycles and the transport of species were introduced into modeling so as to study new aspects of climate variations: the effect of aerosols, coupling between climate change and the chemical composition of the atmosphere, and between climate change and the carbon and methane cycles. This required advances in our understanding of each of the components of the system: atmosphere, ocean, vegetation, continental surface, sea ice. Numerical climate models progressively incorporate, in a coherent way, a wide range of physical processes governing the climate variations and the interactions between the different climate components. On the other hand, the inclusion of a growing number of physical processes has made the models more complex and more difficult to develop and evaluate.

As ice sheets interact with atmosphere, ocean and vegetation, the next key challenge for the climate modeling community is to incorporate ice-sheet models in general circulation climate models. This is a necessary step to obtain a comprehensive representation of the climate system for past, present and future time periods.

What Are the Uncertainties Inherent in Climate Models?

The climate is characterized by a very wide range of both spatial (from the micrometer to several thousand kilometers) and temporal (from the second to several thousand years or more) scales. The processes at these different scales interact with one another, and in principle, it is never possible to know which scales should be considered and how to represent the neglected scales in a simplified way. A typical example is the formation of clouds and precipitation. Let's take the example of convective clouds (of the cumulonimbus type), whose core is a rising column of moist air in which the water vapor condenses as it rises. This ascending column mixes with the surrounding drier air, and this mixing depends on many factors (for example, the intensity of the upward thermal current and wind shear). In order to take these mixtures into account, it is first necessary to know precisely the vertical profile of the atmospheric variables in the vicinity of the column. The turbulent exchanges between the column and its environment must also be calculated, as well as the coupling between these turbulent exchanges and the formation or dissipation of rain drops, hail or snowflakes. This requires a modeling on a very small scale (a few hundred nanometers to a few meters), which is not possible with global models. Therefore, a simplified model must be developed based only on large scale variables which reproduce the effect of unresolved small scale processes. This type of modeling, called parameterization, is based on important simplifications that nevertheless require a thorough physical analysis and an in-depth understanding of the phenomena. The aim is to obtain a simplified model that is not only as accurate as possible, but also justified and well understood.

There are many parameterizations in a climate model which can be directly related with the atmospheric circulation (for example gravity waves and orographic effects), the calculation of radiative exchanges, deep convection, or boundary layer phenomena. Many of these parameterizations play a key role in the water cycle, the formation of clouds and their radiative properties, precipitation, heat and water fluxes on the surface of continents or oceans, among others. All these phenomena have an impact on the simulation of the current climate and as there are strong interactions between them, it is often very difficult to identify the precise role of each of the parameterizations on the simulation of these phenomena, and in particular to understand why some of them are poorly simulated.

Parameterizations also play a very important role in the climate response to different forcings, and in the simulation of past and future climate changes. One example is the simulation of clouds. Clouds exert two opposing effects on the terrestrial radiative balance: on the one hand, they reflect part of the solar radiation, and on the other, they absorb infrared radiation and thus contribute to the greenhouse effect. The relative importance of these two effects depends on many factors, in particular, cloud altitude. Over the past twenty years, we have learned that, on average, the first effect outweighs the second, and therefore that clouds have a cooling effect on the climate, especially low-lying clouds, because they have little impact on infrared radiation.

However, this does not explain the role that clouds could play in global warming. Depending on how their properties change, clouds may attenuate or, on the contrary, amplify, global warming. The physical formation mechanisms involve so many processes and spatial scales (from a micrometer to a thousand kilometers), and their radiative properties depend on so many factors that it is impossible to conclude on the basis of theory, simple reasoning or analysis of available observations, how they will evolve in the future.

Simulation of the Current Climate and Recent Changes

Analysis of simulated climate combined to the comparison of model results with observations is an important step to establish the reliability of climate models. The aim is to evaluate not only the mean climate, but also the climate variability at different time scales (from a few days to a few decades) as well as recent climate changes. Unless otherwise indicated, this section presents the simulated climate characteristics from the twenty AOGCMs that contributed to the preparation of the Fifth IPCC Assessment Report (IPCC 2013).

Mean Climate

The difference in mean insolation is what causes the temperature difference between the equator and the poles. This difference is the main driver of atmospheric and oceanic circulations, which act to reduce the equator-pole temperature gradient. It is also influenced by the presence of clouds, reflective surfaces (snow, glaciers, sea ice), large mountain ranges and the topography of the ocean. Latitudinal temperature variations are thus a key criterion for evaluation of climate models. All the models simulate this strong equator-pole gradient: the simulated temperature is 25 °C at the equator, −20 °C at the North Pole and −40 °C at the South Pole, which agrees with observations. However, the models also show significant biases over Antarctica, Greenland, and over large mountain ranges in general, such as the Himalayas. These are due to an approximate representation of the topography (due to the limited spatial resolution of the models) and a poor representation of turbulent exchanges under conditions of strong thermal stability. Over the oceans, there is a warm bias on the eastern coasts caused by a poor representation of stratus clouds observed in these regions.

The annual variation in solar radiation is the strongest energy 'disruption', apart from the diurnal cycle, to which the surface of the Earth is subjected. The observed seasonal temperature cycle is generally well reproduced by the models: it is higher at high latitudes than at low ones (30 °C vs. 5 °C, essentially reflecting the seasonal amplitude of incoming solar radiation), and is higher over continental areas than above the oceans, mainly because of the lower thermal inertia of the continents.

The formation of precipitation involves numerous processes, some of them on a small scale, and remains one of the phenomena that models have the largest difficulties in correctly simulating. Around the equator, the area of maximum precipitation corresponding to the intertropical convergence zone, is well simulated by the models (Fig. 31.8). In the Pacific Ocean, observations show that the maximum is at 10°N, indicating that this convergence zone is located mainly in the northern hemisphere. The models, on the contrary, generally have two maxima located on either side of the equator, the southern maximum extending almost to the coasts of Peru, although observations indicate that precipitation is almost absent in this region. In observations, the maximum rainfall ranges from northern New Guinea to southern South America. On the continents, one of the biggest flaws of the models is that the intensity of rainfall over the Amazon is too low.

The extent and characteristics of sea ice are relatively well simulated by current models, both in terms of mean value and seasonal cycle. However, in the Arctic basin, few models are able to successfully replicate the distribution of sea ice thickness, which should be less than 1 m north of Siberia, to more than 5 m north of Greenland and of the Canadian archipelago. This difference in ice thickness is mainly due to winds, which move ice from the Siberian coasts to Canadian shores and Greenland. These winds allowed FRAM, the F. Nansen's boat, to cross the Arctic Ocean at the end of the nineteenth century although it was trapped in the ice. The spatial distribution of ice thickness is generally poorly simulated by the models due mainly to poorly simulated surface winds.

Climate Variability

The mean climate state gives a very incomplete picture of the climate. Indeed, climate varies continuously over a very wide range of space and time scales with fluctuations having hourly, daily, interannual, decadal or even longer time scales.

One way to characterize these fluctuations is to select those having a large-scale spatial structure (typically, the scale of an ocean basin or a continent), whether fixed or spreading, and to characterize its temporal evolution (amplitude, phase). To describe these fluctuations, we refer to what we call *modes of variability*. Some of which are well known, such as the El Niño-Southern Oscillation (ENSO) in the tropical Pacific or the North Atlantic Oscillation (NAO), which dominates the weather and climate fluctuations over Europe.

ENSO is the dominant mode of tropical variability at the interannual to decadal time scale. The warm phase, or the El Niño event, is characterized by a warming of the eastern tropical Pacific (along the equatorial cold water tongue) and an easterly displacement of the zone of maximum precipitation, usually located above Indonesia. The cold phase, or La Niña, is characterized by negative sea surface temperature anomalies (i.e. temperatures below the climatological mean) and can be interpreted as an enhancement of the seasonal climatological cycle. This mode of variability is a coupled ocean-atmosphere oscillation that affects atmospheric circulation throughout the tropical belt and ocean circulation throughout the Indo-Pacific basin. Its periodicity is between three and seven years, and the characteristics of ENSO events observed in the twentieth century can vary considerably from one event to another. All current climate models simulate a mode of tropical variability with general

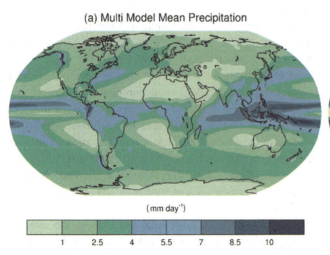
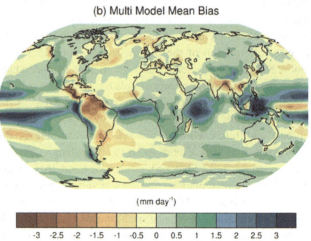

Fig. 31.8 Annual-mean precipitation rate (mm/day) for the 1980–2005 period. **a** Multi-model-mean constructed with one realization of all available AOGCMs used in the CMIP5 (Coupled Models Intercomparison Project, Phase 5) historical experiment. **b** Difference between multi-model mean and precipitation analyses from the Global Precipitation Climatology Project (Adler et al. 2003) (*Source* IPCC 2013)

characteristics resembling those of ENSO. This was not the case with the previous generation of climate models. Nevertheless, the spatial structure of these events is not generally well simulated, such as the asymmetry between the El Niño and La Niña episodes. In terms of the recurrence of these events, the periodicity simulated by the models is usually too short and too regular. In general, the strong diversity of observed spatial and temporal characteristics of ENSO is often poorly simulated. Research works are being done to identify the impact of the different atmospheric and oceanic processes on the characteristics of ENSO, and on the reasons for model errors.

The Madden and Julian—MJO—oscillation (Madden and Julian 1994) is the main mode of intra-seasonal variability in the tropical region, with a periodicity of between 30 and 90 days. Unlike ENSO, whose spatial structure is stationary, this oscillation is characterized by a wave that propagates from west to east, the intensity of the convection alternating between reinforced and reduced. An MJO- type signal is present in the results of most models, but several essential characteristics of this mode of variability (amplitude, phase, propagation) are not realistic. The roles of the different processes and their interactions in the characteristics of this mode of coupled atmosphere-ocean variability have not been well identified and several hypotheses have been proposed.

In the extratropical regions, an important mode of variability is the North Atlantic Oscillation (NAO). It is a pressure oscillation between the temperate and the subpolar latitudes, often defined as the difference in normalized pressure between the Icelandic low and the Azores anticyclone. It is associated with changes in prevailing westerly winds throughout the North Atlantic Basin, and affects the climate of Europe and its surroundings. For example, the positive phases of NAO (NAO+) are associated with a northward shift of low pressure systems with mild and rainy winters and droughts in southern Europe. Current models correctly simulate the spatial properties of the NAO, but are not as good at simulating its temporal properties. In particular, the current trend of the NAO (an increase in positive phases) is underestimated by the models. The reasons for this underestimation are diverse: poor representations of i/the interactions between the stratosphere and the troposphere, ii/the exchanges between stationary waves and transient activity (storms) and iii/the exchanges with the surface of the ocean.

Recent Evolution of the Climate

Over the last 150 years, the evolution of the global surface temperature is documented with a large set of observations. Simulating this evolution is therefore one way to test climate models. Between 1850 and 2000 the steady increase in the concentration of greenhouse gases has led to an increase in radiative forcing of about 2.5 W/m^2. The uncertainty in this forcing is estimated to be quite low at less than ±10%. Since fossil fuels contain sulfur, CO_2 emissions are accompanied by SO_2 emissions which lead to the formation of sulfate aerosols. In 2000, these aerosols produced a radiative forcing of about − 1 W/m^2, but, depending on how it is estimated, this value varies from − 0.5 to −2 W/m^2. In addition, other aerosols, such as soot or aerosols from biomass fires, may also play an important role, but their effects are even less well known. Thus, about a third of the positive radiative forcing (albeit with a high degree of uncertainty) from the increase in greenhouse gases is masked by the negative radiative forcing from aerosols. In addition to anthropogenic forcings, there are natural forcings. At the century time scale, aerosols are mainly injected into the stratosphere by variations in the intensity of incoming solar radiation, and by strong volcanic eruptions where they can remain for several months or even years. These aerosols reflect solar radiation, creating a negative forcing. At the end of the twentieth century, strong volcanic eruptions were more frequent than at the beginning, resulting in an enhanced radiative forcing. Therefore, regardless of the climate response, there is already an inherent uncertainty in the radiative forcing of about ± 50% (IPCC 2013).

When only natural forcings are considered, simulated warming is not in line with observations, especially since the 1980s (Fig. 31.9). However when all forcings are taken into account (natural + anthropogenic), all models are able to correctly simulate the increase in the mean Earth temperature over the past 150 years (Fig. 31.9). Models simulate a greater increase in temperature over the continents than over the ocean and a geographic distribution of warming in agreement with observations (IPCC 2013). Thus, temperature changes over the past 150 years make it possible to verify that the climate response simulated by the models is coherent with the observed temperature variations, although there is too much uncertainty in the forcings to be able to constrain precisely the climate sensitivity of the models.

The aerosols have two opposing effects: one warming the surface, the other cooling it. The first one, called the direct effect, is the diffusion of incoming solar radiation, where part of the solar radiation is returned to space. The second, the indirect effect, is the modification of the optical properties of clouds: the presence of a large number of aerosols increases the number of condensation nuclei. For the same amount of liquid water, more numerous drops forming the clouds tend to have a smaller radius and thus to diffuse more solar radiation. Aerosols are also likely to modify the formation of rain, and therefore the liquid water content of clouds. The complexity of the radiative properties of aerosols in the atmosphere make it more difficult to model the impact of cloud physics.

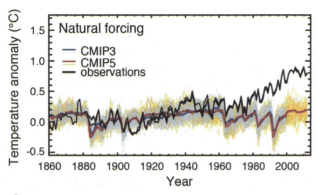

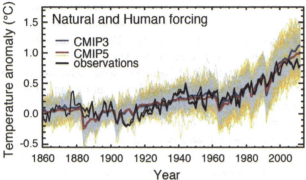

Fig. 31.9 Temporal evolution of the global surface air temperature of the Earth: observations (black line) model simulations taking only natural forcings into account (top), and taking both natural and anthropogenic forcings into account (bottom). The red and blue lines represent the multi-model average of the CMIP3 and CMIP5 (Coupled Models Intercomparison Project, Phases 3 and 5) models respectively (*Source* IPCC 2013)

Projecting the Future of the Climate System

Climate Response to a Doubling of CO_2: Forcings and Feedbacks

Perturbations that modify the energy balance of the climate system are quantified in terms of energy flux at the top of the atmosphere. Quantifying the forcing due to a variation in average insolation is immediate. For a change in the greenhouse gas concentration, a radiative model is used to calculate how these changes affect fluxes at the top of the atmosphere. Since the late 1980s, it is now possible to perform these calculations using radiative transfer codes and spectral databases, provided that all other atmospheric (e.g. clouds, aerosols) and surface (e.g. snow cover) characteristics are assumed to be fixed. For a doubling of the atmospheric CO_2 concentration, the global mean annual radiative forcing at the top of the atmosphere is 3.7 ± 0.2 W/m^2. As we focus on slow climate variations, the calculation of the radiative forcing takes into account the rapid adjustment of the stratospheric temperature.

In response to the radiative forcing ΔQ, the different climate models simulate a temperature change at equilibrium ΔT which is different from one model to another. These differences are difficult to analyze and interpret directly because of the high level of model complexity. A standard method to identify the origin of these differences is the feedback analysis (see, for example, Dufresne and Saint-Lu 2016). From the temperature change at equilibrium ΔT, a 'feedback parameter' λ is defined:

$$\lambda = -\Delta Q / \Delta T.$$

By writing this equation in the form $\lambda \times \Delta T + \Delta Q = 0$, we see that $\lambda \times \Delta T$ represents the variation in the average flux at the top of the atmosphere necessary to compensate for the radiative forcing ΔQ. We can therefore write $\lambda = -dF/dT$, with F being the net radiative flux at the top of the atmosphere, counted as positive when it is descending. This derivative can be decomposed, in the first order, as a sum of partial derivatives:

$$\lambda = -\Sigma \partial F / \partial X \times \partial X / \partial T$$

The sum over X is the sum of all the X variables affecting the radiative balance at the top of the atmosphere and that are modified when the surface temperature changes. These are mainly the three-dimensional fields of temperature, water vapor and clouds, and the two-dimensional fields of surface albedo. The change in the temperature field is generally broken down into two terms, one corresponding to a uniform temperature change, the other to the non-uniform part of the temperature change. Finally, the parameter λ can be decomposed as follows:

$$\lambda = \lambda_P + \lambda_L + \lambda_c + \lambda_w + \lambda_a$$

The terms of the right-hand side are the respective feedback parameters: Planck λ_P (uniform temperature change), the temperature gradient λ_L (non-uniform part of the temperature change), clouds λ_c, water vapor λ_w and surface albedo λ_a.

These parameters are often calculated as follows using the partial radiative perturbation method. For a given climate model, two simulations are carried out, a reference one and a perturbed one. The fluxes at the top of the atmosphere are calculated off-line from the outputs of the reference simulation using a radiative code. They are then recalculated by replacing the values of some variables (temperature, humidity, clouds, surface albedo) from the reference simulation with the corresponding values from the perturbed simulation. The difference between these two fluxes gives the sensitivity of the fluxes at the top of the atmosphere to a perturbation of each of the variables.

The particular Planck parameter λ_P corresponds to the response of the flux at the top of the atmosphere to a uniform temperature change at the surface and within the whole atmosphere. Its value is approximately − 3.2 W/m² K. There is very little change in this value from one model to another, and the sign convention used corresponds to a decrease in the Earth's energy balance when the surface temperature increases. In response to a radiative forcing, ΔQ, the surface temperature response can be calculated if the Planck parameter λ_P is the only non-zero feedback parameter:

$$T_P = -\Delta Q/\lambda_P.$$

This so-called Planck response causes an increase of 1.2 °C for a forcing of 3.7 W/m², resulting from a doubling of the CO_2 concentration. It is the response of an idealized system in which only the atmospheric and surface temperatures can change uniformly and where only the radiation emission law is affected (see Sect. "Simple Radiative Climate Models and Their Limitations" of this chapter). It can be said that this is the response with no feedback from the climate system. By combining the above equations, the increase in temperature at equilibrium ΔT can be written as a function of the Planck response:

$$\Delta T = \Delta T_P/(1-g)$$

where g is the feedback gain of the system:

$$g = -(\lambda_L + \lambda_c + \lambda_w + \lambda_a)/\lambda_P.$$

If the gain g is positive and less than 1, the feedbacks will amplify the temperature increase ΔT, relative to ΔT_P. Conversely, if the gain is negative, the ΔT increase will be attenuated. In the preparation of the Fifth IPCC Assessment Report (IPCC 2013), climate change simulations were conducted within the CMIP5 project by around forty climate models. In particular, for a doubling of the CO_2 concentration, the models simulate a global warming of 3 °C on average (between 2.0 and 4.6 °C depending on the model), until a new energy equilibrium is found. We have seen that in the absence of feedbacks, this warming would be 1.2 °C. This means that climate feedbacks amplify this warming by a factor of 2 to 4 depending on the model.

Other developments make it possible to use the feedback parameters to estimate the temperature increase due to the Planck response and to the various feedbacks (Dufresne and Bony 2008). They were applied to twelve CMIP3 (Coupled Models Intercomparison Project) models, and the results are shown in Fig. 31.10 illustrating both the average contribution of the models and the inter-model dispersion. These results were confirmed for the CMIP5 models (Vial et al. 2013).

Several of the mechanisms governing the feedback parameter values, and hence the gain, are now well identified (Bony 2006). For example, an increase in the temperature of the atmosphere increases the saturation vapor pressure of the water vapor. If there is little variation in the relative humidity, this results in an increase in the concentration of water vapor in the atmosphere, and therefore in the greenhouse effect, thus constituting a very powerful amplification mechanism of the warming: 1.7 °C on average for the models considered here.

For thermodynamic reasons (variation of the adiabatic temperature gradient as a function of humidity), it is also expected that, in the case of a humid tropical atmosphere, variations in water vapor will be accompanied by greater warming at high altitudes than close to the surface (except at high latitudes). This increases the emission of infrared radiation from the upper atmosphere and is the only negative feedback: it decreases warming (− 0.8 °C on average). As these two feedbacks are very strongly correlated for physical reasons, their combined effect is generally considered to contribute about 1 °C to the increase in mean temperature (Fig. 31.10a).

The mechanisms behind the surface albedo feedback are also well identified: an increase in temperature increases the melting of snow and ice. This decreases the area covered by surfaces that reflect solar radiation, and therefore increases the amount of energy absorbed by the Earth. This effect contributes about 0.2 °C to the temperature increase (Fig. 31.10a).

Finally, the increase in temperature is likely to impact cloud cover. As seen before, clouds exert two opposing effects on the terrestrial radiative balance: on the one hand, by reflecting solar radiation, they contribute to a cooling of the Earth, and on the other, by absorbing infrared radiation, they contribute to increase the greenhouse effect. The relative importance of these two effects depends on multiple factors. The average contribution of clouds estimated by the models considered here is equivalent to a warming of 0.7 °C (Fig. 31.10a). However, the dispersion between models is high (Fig. 31.10b): while some models predict a relatively neutral cloud response, most predict a decrease in cloud cover as temperature increases and an increase in global warming of up to 2 °C.

Recent studies indicate that this uncertainty stems mainly from the way different climate models predict the response of low clouds (stratus, stratocumulus or small cumulus clouds) to global warming. The way other clouds (including large cumulonimbus storm clouds) respond to climate change is also uncertain. This response contributes only poorly to the uncertainty on the magnitude of global warming, but contributes strongly to uncertainties in regional precipitation changes associated with global warming.

Other feedbacks exist in the climate system, for example, those potentially causing changes in the atmospheric carbon storage capabilities by the ocean and the biosphere. They will be discussed below (Sect. "The CO_2 Cycle").

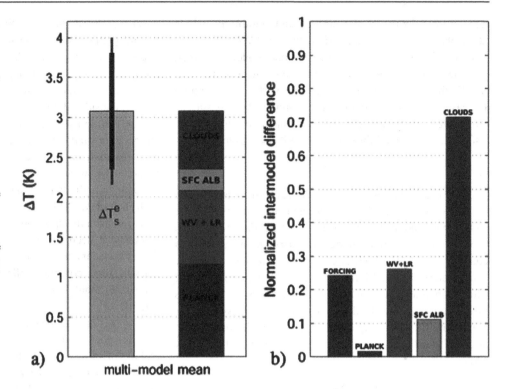

Fig. 31.10 For a doubling of atmospheric CO_2: **a** bar on the left, multi-model mean ± 1 standard deviation of increase in global temperature (ΔT_e, °C) and, bar on the right, part of this increase due to the Planck response and to the different feedbacks: combined effects of water vapor and temperature gradient (WV + LR), surface albedo and clouds. **b** Standard deviation of the inter-model difference in temperature increase attributable to radiative forcing, the Planck response, and various feedbacks, normalized by the standard deviation of the increase in global temperature. According to Dufresne and Bony (2008)

Future Scenarios

It is important to keep in mind that the future climate cannot be accurately predicted. This can be explained by several reasons including model uncertainties, uncertainties in scenarios of future greenhouse gas emissions and uncertainties in natural disturbances, such as volcanism, having a strong impact on radiative forcing. However, we can try to answer specific questions: independently of the natural forcings, how would the climate evolve if greenhouse gas emissions followed this or that emission pathway? To this end, various socio-economic scenarios for the evolution of human activities have been established. Four representative concentration pathways (RCP) scenarios have been selected, labeled by a value that corresponds approximatively to the radiative forcing in 2100: RCP2.6, RCP4.5, RCP6.0, RCP8.5 (top of Fig. 31.11). On one extreme, CO_2 emissions rapidly stabilize and then decrease down to zero before 2100 for scenario RCP2.6. On the other extreme, CO_2 emissions continue to grow until 2100 for scenario RCP8.5, which is for now the most realistic one. CO_2 emissions are mainly due to the use of fossil fuels (oil, coal, gas) and to the SO_2 emissions from the sulfur contained in these fuels. Mainly for health reasons, fuels are increasingly purified of sulfur before use, resulting in a slower increase in CO_2 emissions (or a faster reduction) than for CO_2 in almost all scenarios. For the last IPCC exercise (IPCC 2013), the concentrations of the different gases were calculated from their emissions by biogeochemical cycle models and by chemical-transport models for sulfate aerosols. The concentrations of each of these constituents can then be used to calculate the corresponding radiative forcing. For example, the evolution of the different forcings from 1860 to 2100 for the RCP8.5 scenarios is shown at the bottom of Fig. 31.11.

The CO_2 Cycle

As described in Chap. 23, the carbon cycle is strongly linked to the living world. Understanding the roles of the two main reservoirs, the ocean and the terrestrial biosphere, is therefore crucial to understanding how atmospheric CO_2 will evolve in the coming decades. The ocean is the largest carbon reservoir (holding 40 times the atmospheric content) and can regulate the atmospheric CO_2 concentration by exchanging CO_2 with the atmosphere. These exchanges depend largely on the vertical carbon gradient in the ocean: surface waters are depleted of inorganic carbon and mineral salts; conversely, the deep ocean is enriched in carbon and mineral salts. The existence of this vertical gradient is driven mainly by ocean biology, which establishes a biological pump and transfers carbon from the surface to the bottom. Any modification of this pump may ultimately influence the atmospheric reservoir.

What about anthropogenic disturbances and their impacts on the carbon cycle? These impacts are manifold: a direct chemical consequence of the increase in atmospheric CO_2 is the acidification of the oceans; a thermal consequence of the

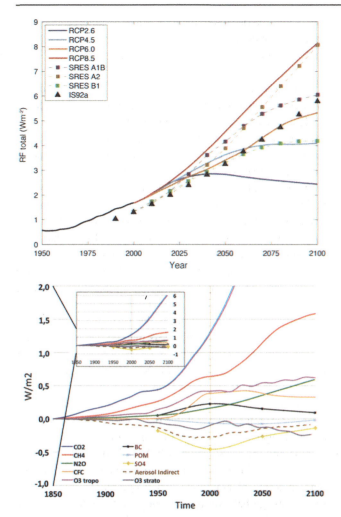

Fig. 31.11 Top: Time evolution of the total radiative forcing (in W/m^2) due to human activities for the three SRES emission scenarios used in IPCC-AR4 (IPCC 2007) and the four RCP scenarios used in IPCC-AR5 (IPCC 2013). Bottom: Temporal evolution of the radiative forcing (in W/m^2) due to the different chemical agents for the scenario RCP8.5 (Szopa et al. 2013)

warming of the water is a decrease in the capacity of the ocean to absorb CO_2, and a biological effect is a modification in the distribution of species due to environmental changes (salinity, temperature) which can cause transformations in the trophic chain and therefore in the biological pump. In terms of CO_2 sinks, it is now known that for two CO_2 molecules emitted, only one remains in the atmospheric reservoir while the other is stored either in the terrestrial biosphere or in the ocean. This ratio is the result of many processes involved in the regulation of the carbon cycle; therefore, due to the anthropogenic perturbation it is expected to vary. The capacity of continents to store carbon can change either gradually or abruptly. For example, permafrost thawing in Siberia would release the equivalent of an additional 100 ppm of CO_2 at least into the atmosphere. As for the biological pump, this will depend on how life functions and adapts in a warming world, which is not easy to predict.

Climate Projections for 2100: What the Models Say: Main Climate Characteristics

In this section, the results are derived, unless otherwise indicated, from simulations performed by the forty or so coupled atmosphere-ocean general circulation models (or climate models) that contributed to the CMIP5 exercise. The values given below are the average of the models ± 1 standard deviation.

The Amplitude of the Warming

In response to both anthropogenic (greenhouse gases and aerosols) and natural (volcanoes and solar intensity) forcings, the models simulate an average global increase in air temperature near the surface of about 0.8 °C between the beginning and the end of the twentieth century, consistent with observed measurements. Sensitivity studies, with different forcings, have shown that this warming is mainly due to anthropogenic forcings. Today, the climate system is out of balance; if the concentrations of greenhouse gases and aerosols were maintained at their 2000 values, the climate would continue to warm by around 0.4 °C through the twenty-first century. However, the simulated increase in temperature depends above all on the emission scenarios. Indeed over twenty models, when the concentration of greenhouse gases is prescribed, there is on average a variation of a little less than 1 °C for the RCP2.6 scenario and of 4 °C for the RCP8.5 scenario (Fig. 31.12). The dispersion of the simulated warming by the different models at the end of the century is ± 0.5 °C, the main reasons for which are outlined in Sect. "Climate Response to a Doubling of CO_2: Forcings and Feedbacks" of this chapter.

Geographical Distribution of Temperature Changes

The geographical distribution of the temperature increase is roughly similar for the different scenarios (although the amplitudes are very different). Figure 31.13 illustrates the RCP2.6 and RCP8.5 scenarios. It shows results that are now well established: the temperature increase is higher on land than on the oceans, and is particularly strong in the high latitudes of the northern hemisphere.

In the tropics, the greater increase in temperature over land than over ocean is partly explained by changes in

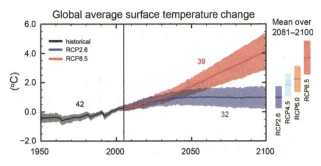

Fig. 31.12 Multi-model time series of air surface temperature anomaly (°C) (compared to the 1986–2005 average), for the twentieth and twenty-first centuries under two different scenarios with very contrasting emission scenarios: RCP2.6 (blue) and RCP8.5 (red). The mean and associated uncertainties averaged over 2081–2100 are given for all RCP scenarios as colored vertical bars. The line represents the average of the models and the light-colored regions, the spread

and where the ocean temperature remains homogeneous to a considerable depth. In order for the temperature of the ocean surface to increase, a large amount of water must be heated up.

The larger increase in temperature in the high latitudes of the northern hemisphere is partly due to the surface albedo feedback. The increase in temperature is accompanied by a decrease in snow cover and sea ice extent in spring and summer. This reduces the reflection of solar radiation at the surface, increases the amount of radiation absorbed and tends to amplify the initial increase in temperature. In regions where the sea ice thins, or even disappears, air temperature rises sharply as the temperature of the ocean surface is higher than that of sea ice. Finally, a change in atmospheric circulation (and in particular the increase in water vapor transported to the high latitudes) is another reason for the strong increase in temperature in these regions. In the south and east of Greenland, it is noted that the temperature of the air near the surface increases only slightly. This trend is more or less marked depending on the model, with some even simulating a slight local cooling. The reason for this is a change in the ocean circulation, and especially in the thermohaline circulation. In these areas, the density of sea water decreases at the surface because of increasing temperatures and/or precipitation. Surface waters are no longer dense enough to sink, reducing oceanic convection and the associated North Atlantic drift. The effect of this density reduction on temperature depends on the model, both in terms of amplitude and geographical extension. It modulates global warming locally and influences global warming slightly, but nevertheless this warming remains significant on all continents of the northern hemisphere, especially in Europe.

evaporation. Over the oceans, the amount of water available for evaporation is not limited, while on the continents it depends on the amount of water available in the soil, and thus, on the amount of precipitation. Due to latent heat exchanges, evaporation cools the surface and this cooling effect is larger over the oceans than over continental areas. For example, with the IPSL model, evaporative cooling is found to reduce the surface radiative balance by 9.8 W/m^2 over the ocean against only 0.2 W/m^2 over the continents. Other processes, such as changing cloud cover or changing atmospheric circulation, also play a role in the ocean-continent warming differential.

In the mid and high latitude regions, the smaller increase in ocean surface temperature is partly due to its thermal inertia. This is particularly true in the southern hemisphere, where very strong winds cause great mixing of the ocean,

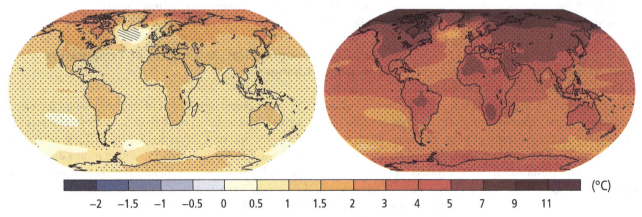

Fig. 31.13 Multi-model average of the difference in air surface temperature (°C), between the end of the twenty-first century (2081–2100) and the end of the twentieth century (1986–2005), for the RCP2.6 (left) and RCP8.5 (right) scenarios (*Source* IPCC 2013)

Changes in Precipitation

Although there is a large spread between climate models, on average, they project an increase in precipitation over the twenty-first century alongside increased temperatures (IPCC 2013). Projected precipitation changes exceed 0.05 mm/day and 0.15 mm/day with the RCP2.6 and RCP8.5 scenarios respectively. These changes are far from being spatially homogeneous and exhibit a strong seasonal variability (Fig. 31.14). However, if we consider the zonal means, there is a tendency towards increased precipitation everywhere, except in the subtropical regions (around 30°N and 30°S) and in the Mediterranean basin where they decrease. The general increase in precipitation is due to the increase in atmospheric water vapor content, while the decrease simulated in the subtropical regions is related to a change in atmospheric circulation.

It can be seen that the models consistently simulate a year-round increase in precipitation at high latitudes, and a winter increase in the mid-latitudes (Fig. 31.14). Similarly, they consistently simulate a decrease in precipitation in subtropical regions. In Europe, models simulate an increase in precipitation in the north and, on the contrary, a drying up around the Mediterranean Basin. On the other hand, changes in precipitation in the equatorial and tropical regions are not consistent from one model to the other, especially on the continents such as South America, West and Central Africa, India and South-East Asia. In these areas, some models simulate a decrease in precipitation, while others simulate an increase. These differences between models are particularly pronounced in the monsoon regions. In general, changes in precipitation on the continents remain very uncertain, even in terms of annual mean. This is due to major uncertainties in the representation of the different processes. Currently, there

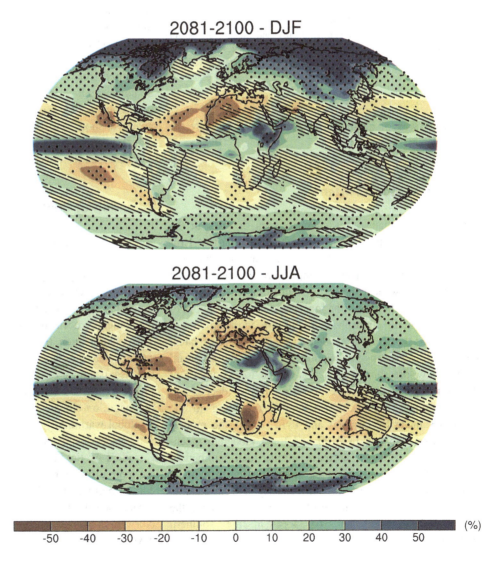

Fig. 31.14 Multi-model average of the precipitation difference (%) between the end of the twenty-first century (average between 2081 and 2100) and the end of the twentieth century (average between 1980 and 1999), for the RCP8.5 scenario for December–January–February (*top*) and (*bottom*) June–July–August. Hatching indicates regions where the multi-model mean change is less than one standard deviation of internal variability. Stippling indicates regions where the multi-model mean change is greater than two standard deviations of internal variability and where at least 90% of models agree on the sign of change

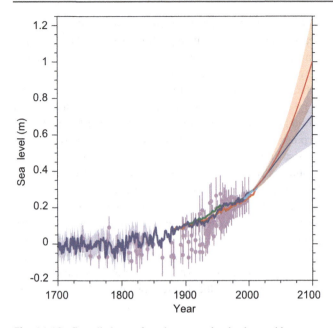

Fig. 31.15 Compilation of paleo sea level data, tide gauge data, altimeter data and central estimates and likely ranges for projections of global mean sea level rise for RCP2.6 (blue) and RCP8.5 (red) scenarios, all relative to pre-industrial values (*Source* IPCC 2013)

is no clear way of identifying which of the results are the most reliable (Fig. 31.15).

Changes in Storm Patterns

Under climate change, the characteristics of lows in the mid-latitudes (particularly those reaching the coast of Brittany) are likely to change for two reasons: the first is a change in the equator-pole temperature gradient, which tends to decrease near the surface but to increase with altitude. The second reason is an increase in the total amount of water vapor in the atmosphere, and in turn, in the amount of water vapor that can be condensed and thus release latent heat. In climate change simulations, we observe a poleward shift of low-pressure, a reduction in the total number of depressions, but an increase in the number of deeper and hence stronger troughs. For example, according to the diagnoses made by Lambert and Fyfe (2006), the models simulate, on average, in 2100 and for a moderate greenhouse gas scenario, a decrease in the total number of depressions of around 10% in the southern hemisphere and a little less in the northern hemisphere. They also simulate an increase of 20% of the number of intense depressions in the northern hemisphere and of 40% in the southern hemisphere.

Evolution of Sea Ice

The analysis of CMIP3 and CMIP5 models clearly highlights a sea-ice decline in the course of the twenty-first century, the dominant factor being the rising summer temperatures rather than the winter ones which are projected to remain negative. Though most of the CMIP5 models project a nearly ice-free Arctic Ocean by 2100 with a sea ice extent of less than 1×10^6 km^2 (for at least 5 consecutive years) at the end of summer in the RCP8.5 scenario, some show large changes occurring before 2050. However, these results are strongly dependent on the ability of models to reproduce the climatological mean state of the Arctic sea ice cover over 1979–2012. Based on the CMIP5 multi-model ensemble, projections of average reductions in Arctic sea ice extent for 2081–2100 compared to 1986–2005 range from 8% (RCP2.6) to 34% (RCP8.5) in February and from 43% (RCP2.6) to 94% (RCP8.5) in September. The evolution of sea ice around the Antarctic is more uncertain. The CMIP5 multi-model mean projects a decrease in sea ice extent ranging from 16% for RCP2.6 to 67% for RCP8.5 in February and from 8% for RCP2.6 to 30% for RCP8.5 in September for 2081–2100 compared to 1986–2005. There is, however, low confidence in those values because of the wide inter-model spread and the inability of almost all of the available models to reproduce the mean annual cycle, the interannual variability and the overall increase of the Antarctic sea ice coverage observed during the satellite era.

Evolution of Continental Ice

Retreat of the glaciers

Glaciers react quickly to climatic effects and are thus good indicators of climate change. For more than a century, and especially since the 1980s, the retreat of glaciers has been a phenomenon occurring almost everywhere on the globe. This phenomenon is likely to increase in the twenty-first century with the rising temperatures, and could even lead to the disappearance of certain glaciers in the coming decades. Projections of the future evolution of glaciers depend, on the one hand, on climate scenarios but also, on their sensitivity to global warming.

Glaciers located in dry and cold regions (such as the Canadian Arctic) have low sensitivity to global warming and are more prone to resist. On the contrary, glaciers located in coastal regions are subject to oceanic influence and have a greater sensitivity to global warming. This is the case of Norwegian glaciers, for example. In recent decades, these have tended to grow due to increased precipitation. But in recent years, this trend seems to have been reversed. Alpine glaciers are in an intermediate situation. We will not review exhaustively the melting of glaciers, but we chose to illustrate this behavior through the study of two alpine glaciers.

At the Environment Geosciences Institute of Grenoble, two studies were carried out on alpine glaciers, one on the Saint-Sorlin, the other at the dome of Goûter, in the Mont-Blanc massif.

The first study simulated the future evolution of the mass balance of the Saint-Sorlin, located in the Grandes Rousses massif whose highest point is at 3400 m altitude. The results show that even with an optimistic greenhouse gas emission scenario (+1.8 °C in 2100), the equilibrium line (i.e. the lowest limit of eternal snow) is at a higher altitude than the highest point of the glacier. This means that over a full year, the glacier does not accumulate snow (or ice) and is therefore in a state of chronic ablation. In order to simulate the dynamic response of the glacier, these results were then used as inputs for a two-dimensional ice flow model. Despite a moderate global warming, the model suggests a complete disappearance of the glacier around 2070. The dynamic response of the glaciers is complex because it depends not only on their specific morphology, which is different from one glacier to another, but also on many physical processes, which are sometimes difficult to simulate, and which determine their temporal response, i.e. on what timescale the glacier will grow or, on the contrary, disappear. Nevertheless, this study suggests that alpine glaciers similar in size and located at an altitude close to that of the Saint-Sorlin, could undergo the same type of evolution in the twenty-first century under the influence of moderate global warming.

The second study is based on borehole temperature measurements (140 m deep) located at the Dome du Goûter at 4250 m altitude. These measurements showed a warming of 1 to 1.5 °C over the first 60 m of ice, between 1994 and 2005. A physical modeling of the process of heat diffusion in the ice made it possible to show that the observed warming results not only from atmospheric warming, but also from the heat produced by the refreezing of surface melt water that percolates at depth.

Moreover, simulations carried out for different global warming scenarios confirm the twenty-first century trend and show that, regardless of the scenario, the Alpine glaciers, currently classified as 'cold' and located between 3500 and 4250 m altitude with an internal temperature ranging from 0 to −11 °C, could become 'temperate', with an internal temperature close to the melting point.

Other modeling studies using different greenhouse gas scenarios lead to similar and equally disturbing conclusions. A study conducted in the Montana region in the Glacier National Park shows that, with a doubling of CO_2, some glaciers would disappear by 2030, despite an increase of 5–10% in average precipitation in the mid and high latitudes.

The future of ice sheets

There are different approaches to determine how polar ice sheets will evolve in the future. The first one is to determine their surface mass balance which is directly dependent on the climate. To simulate as precisely as possible the mass balance of the ice sheets, high-resolution climate models are necessary to properly represent the topography of the ice sheets, especially the steep slopes at the margins. These areas correspond to the ablation zones and are also the locations where most of the precipitation falls. However, AOGCMs usually tend to overestimate precipitation and underestimate ablation. In addition, snowpack models implemented in general circulation models are often too simplistic and do not account for key processes taking place in the snowpack. Among the missing processes are the refreezing of surface melt water that percolates at depth, the transport of snow by winds, the snow metamorphism and the transformation of snow into ice. These processes are crucial for the estimates of surface mass balance. For example, water refreezing modulates surface runoff and snow metamorphism strongly modifies the albedo and thus the surface energy balance.

Regional atmospheric models generally offer a better representation of the surface energy balance than AOGCMs because of their finer resolution and the inclusion of more sophisticated snow models. However, as they are forced at their lateral boundaries by outputs from global climate models, they may suffer from the global model deficiencies. As a result, the uncertainties associated with the GCM-based forcing represent about half of the uncertainty associated with changes in surface mass balance inferred from regional climate models (RCM). All RCM-based studies project an increase in precipitation over large parts of Greenland and Antarctic ice sheets in response to global warming, but there is great uncertainty as to the magnitude of this increase. Projections suggest that over Greenland, ablation will largely exceed the increase in snowfall. Conversely, in East Antarctica, precipitation is expected to exceed ablation throughout the twenty-first century, but the key question is whether the mass gain will be offset by dynamical ice losses from the West Antarctic ice sheet.

The surface mass balance is not the only important parameter in the determination of the evolution of polar ice sheets. As explained earlier, their dynamic response must also be considered. This can be achieved through the use of three-dimensional ice-sheet models which include most of the processes responsible for the ice-sheet evolution as a function of climate forcing (temperatures and precipitation or surface mass balance). For a more thorough description of ice-sheet models, we ask the reader to refer to Chap. 24 of this volume. However, several problems may arise when using such models. As previously mentioned, the surface mass balance can be computed by a climate model (RCM or AOGCM), but frequently, ablation is still determined from an empirical formulation that relates the number of positive-degree-days (integral of positive temperatures) to the snow and ice melting rates. The main drawbacks of this approach is that it does not account for albedo changes and it has been validated on only a few Greenland sites for the present-day period. The mass balance derived from such methods is generally more uncertain than that obtained using regional models. Another difficulty is linked to the difference of resolution between climate and ice-sheet models. As temperature, precipitation and surface mass balance are highly dependent on topography, the development of appropriate downscaling techniques is required to account for the effect of altitude.

In addition, researchers still face many challenges in representing some processes related to rapid dynamics, such as those occurring at the base of the ice sheet or at the ice-ocean interface, where small disturbances seem capable of triggering strong instabilities, which can propagate up to a thousand kilometers upstream, and lead to a destabilization of the entire ice sheet. However, understanding the impact of small-scale processes on the large scale is still in its infancy. Over the last two decades, observations have shown an increase in the flow of outlet glaciers, suggesting that these could respond much more quickly than previously expected to variations in atmospheric and oceanic conditions, making possible a significant retreat of the ice sheet in the more or less distant future. The physical laws governing ice flow in these glaciers still remain poorly known. For example, although the marine ice-sheet instability is increasingly well represented in new generation ice-sheet models, there is no consensus on hydro-fracturing yet. Large uncertainties also exist on the physical laws governing iceberg calving and basal melting under the ice shelves. A last, but not least, problem is related to the initial state of the ice sheets. Indeed, starting future short-term (a few centuries) simulations from a realistic present-day state of the ice sheet is of primary importance to avoid as much as possible spurious biases in the results. Getting an accurate initial state is challenging however because of the scarcity of observations on vertical velocity and temperature profiles, and also on basal conditions (e.g. frozen or thawed bed areas, bedrock topography). Several approaches have been developed ranging from long free-evolving simulations over one or several glacial-interglacial cycles to more formal optimization approaches often based on inversion techniques, with various target criteria (either surface velocities or topography in agreement with observations, or internal properties accounting for the past history of ice sheets). Each of these techniques presents some advantages and drawbacks, but their efficiency should improve in the future as the number of observations grows. Recent remote sensing observations made over Greenland and Antarctica are expected to refine numerical simulations and thus to improve the relevance of future forecasting models. Finally, part of the uncertainty in the future ice-sheet responses comes from uncertainty in the climate forcing itself. This is due to our lack of knowledge in the future socio-economic pathways and to the biases of the climate models. Moreover, climate-ice sheet interactions are still very rarely taken into account in the models. Accounting for these interactions may have the potential to strongly modify not only the simulated climate and ice-sheet responses but also sea-level projections (Bronselaer et al. 2018; Golledge et al. 2019).

Sea Level Change

We have seen, in the first part of this chapter, that at the scale of a few decades, the main causes of sea-level variations are due, on the one hand, to the thermal expansion of the ocean (in the case of a warming climate), and on the other hand, to changes in the mass of water in the ocean due to exchanges with continental reservoirs. Based on the results from 21 AOGCMs, the sea-level projections reported in the Fifth IPCC report (IPCC 2013), all show an increase in sea-level by the end of the twenty-first century (2081-2100) compared to the 1986–2005 period, with an average global mean sea-level rise ranging from 0.28 to 0.61 m for RCP2.6 and from 0.52 to 0.98 m for RCP8.5. For the RCP8.5 scenario, the rate of sea-level rise increases throughout the twenty-first century, going from 3.2 ± 0.4 mm/yr (for the 1993–2010 period) to 11.2 mm/yr (2081–2100), whereas, in the RCP2.6 scenario, the rate increases up to 4.4 mm/yr around 2050 and slightly declines in the second half of the century.

According to the Fifth IPCC report (IPCC 2013), thermal expansion remains the main contributor (30 to 55%) to global mean sea-level rise over the twenty-first century with median rates of 0.14 ± 0.04 m (RCP2.6) and 0.27 ± 0.04 m (RCP8.5) in 2081–2100. These contributions can be estimated from changes in ocean heat uptake increasing roughly linearly with global mean surface air temperatures simulated by the AOGCMs. It should be noted that AOGCM simulations do not include volcanic forcing which may reduce the projected contribution of thermal expansion.

The AOGCM simulations forced by the different RCP scenarios highlight a spatial variability of sea level (Fig. 31.16). As they do not take into account the freshwater fluxes coming from ice sheets, this regional variability, previously observed in satellite measurements over the 1993–2010 period, is linked to a variability in the atmosphere-ocean-sea-ice system. More specifically, regional variability can be due not only to variations in temperature and salinity (i.e. precipitation/evaporation ratio) and therefore density, but also to variations in ocean and atmospheric circulations

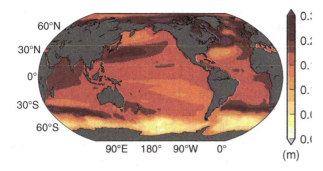

Fig. 31.16 Ensemble mean projection of the time-averaged dynamic and steric sea-level changes for the period 2081–2100 relative to the reference period 1986–2005, computed from 21 CMIP5 climate models (in meters), using the RCP4.5 experiment. The figure includes the globally-averaged steric sea level increase of 0.18 ± 0.05 m (*Source* IPCC 2013)

(variations in wind direction and intensity). From one model to another, the spatial distributions of sea level variations are not identical. However, some common features emerge in the projections for the end of the twenty-first century that are identified in Fig. 31.16. For example, the Southern Ocean is characterized by a lower-than-average sea-level rise, likely related to changes in wind stress and heat fluxes. These changes are also responsible for the greater sea-level rise in the Pacific (Bouttes and Gregory 2014). A dipole pattern with greater sea-level rise in the north and smaller sea-level rise in the south can also be seen in the North Atlantic and is attributed to changes in heat and freshwater fluxes.

While changes in sea level due to thermal expansion can easily be determined from climate models, we have seen in the previous section that there are significant uncertainties in the simulation of future continental ice. These uncertainties dominate the projections of sea-level rise resulting from ice-sheet surface melting, dynamic ice discharges or iceberg calving. As a result, these projections are not very different from one climate scenario to another. For example, in the last IPCC report (IPCC 2013), the contribution from the rapid dynamics of the ice sheets is estimated to be between 0.03 and 0.19 m for the RCP2.6 scenario and between 0.03 and 0.20 m for the RCP8.5 scenario.

To better represent the contribution of polar ice sheets to sea-level rise in the future, an accurate representation of rapid ice dynamics in ice-sheet models is of primary importance. The resolution (5–10 km) of the previous generation of three-dimensional large-scale ice-sheet models is not fine enough to properly capture the rapid ice streams whose characteristic spatial scale may be smaller than 1 km in some cases. Moreover, these models are based on approximations resulting in simplified equations of ice dynamics. While these approximations are crucial to investigate the ice-sheet behavior at the multi-millennial time scale, they may cause deficiencies in the representation of rapid ice dynamics as well as in the representation of grounding line migration, which plays a crucial role in ice-sheet evolution as mentioned earlier. However, significant progress has been made in the development of ice-sheet models, such as the Full Stokes models (e.g. Gillet-Chaulet et al. 2012). These models do not rely on simplifying assumptions and have an adaptive mesh that may be less than 1 km at the ice-sheet margins, thereby providing a more accurate representation of ice-sheet dynamics. However, these models are highly expensive in terms of computational time and have not yet been applied to the whole Antarctic ice sheet. A second limiting factor is related to the fact that some processes remain poorly constrained due to the scarcity of observations (e.g. basal conditions or hydro-fracturing), or poorly understood (e.g. basal melting under the ice shelves or iceberg calving). Finally, as mentioned in the previous section, there is a crucial need to develop Earth System Models which account for the feedbacks between climate and ice sheets so as to refine sea-level projections.

It is important to note that since the publication of the last IPCC report (IPCC 2013), many projections coming from individual studies have provided higher estimates of the projected sea-level rise compared to those reported by the IPCC (0.52 to 0.98 m for the highest emission scenario). The main reasons for this disagreement are related to the large uncertainties associated with the future ice-sheet evolution and to the fact that the IPCC only selected the most likely range of sea-level rise estimates and excluded the most extreme (considered as less likely) outcomes. All the projections together give a range of 0.16–2.54 m in 2100 (Garner et al. 2018) which reflects the large uncertainties in the maximum contribution of Greenland and Antarctica (DeConto and Pollard 2016) and suggests the possibility for these ice sheets to become the main contributor to global mean sea-level rise in the course of the twenty-first century.

The Climate of the Next Millennium: Towards Integrated Modeling of the Earth System

Climate Change: Anthropogenic Disturbance Versus Variations in Insolation

In Chap. 28, we analyzed in detail the climate variations induced by the orbital parameter variations over the scale of tens of thousands of years. We are interested here in much longer time scales ranging from 100,000 to millions of years. The amplitude of warming projected by climate models over the next decades and even the next century raises questions about the impact of this profound change in the climate system over several millennia, especially as the acceleration of CO_2 emissions suggests that the magnitude of the perturbation may be even greater than that predicted by the most pessimistic scenario.

Data from Vostok and EPICA ice cores in Antarctica, showed that atmospheric CO_2 varied by about 100 ppm during the transition from an ice age to an interglacial period, and that these variations were correlated with changes in air temperatures, suggesting a link between CO_2 and climate. Today, this link is commonly accepted and attention is focused on global warming due to the increase in greenhouse gases in the atmosphere. However almost forty years ago, in 1972, the climatological community met in the United States to discuss the imminence of the next ice age. This question arose because geological data showed that for about a million years, the Earth had alternated between cold episodes corresponding to the glaciation phases, and the warmer interglacial periods, with a pseudo-periodicity of 100,000 years.

Data available in 1972 showed that the previous two interglacial periods had lasted about 10,000 years. Yet, the Holocene, the interglacial period we are currently experiencing, has been going on for 10,000 years. It seemed therefore reasonable to think that this warm phase would soon end and give way to a new ice age. This reasoning was based on the assumption that all interglacial periods are of equal duration. However, although the first numerical experiments carried out with statistical models indicated that the cooling initiated 6000 years ago (after the Holocene climate optimum) would continue in the future, Oerlemans and Van der Veen (1984) of the University of Utrecht showed, with an ice-sheet model, that the transition to a new glacial phase would not occur before 50,000 years.

Based on data from marine sediments and ice cores, it is now well established that the duration of an interglacial period varies considerably from one climatic cycle to the next. The same models used to predict the future were used to simulate the last interglacial-glacial transition. Indeed, taking into account the atmosphere-ocean feedbacks, these models produce perennial snow in the Canadian archipelago (Khodri et al. 2001). It is also known that the Earth's orbital parameters (see Chap. 28) that govern the seasonal and latitudinal distribution of solar energy can vary considerably from one cycle to another. For example, geological data tell us that the interglacial period occurring 400,000 years ago (marine isotopic stage 11) was exceptionally long. This situation corresponds to a weak eccentricity where the seasonal and latitudinal distribution of solar energy varied very little. Celestial mechanics tell us that a similar situation should recur within 20,000 years, and for this reason the marine isotopic stage 11 is often considered to be one of the best analogues for the future climate. Many internal feedbacks generated by the different components of the climate system amplify or reduce the effect of the latitudinal and seasonal distribution of insolation. Thus, climate projections on the scale of a few hundreds of thousands of years require the variations of both insolation and atmospheric CO_2 (and other greenhouse gases) to be taken into account in models including representations of the atmosphere, ocean, cryosphere, lithosphere and vegetation.

One of the first models of this type was developed for the northern hemisphere at the University of Louvain-la-Neuve, Belgium (Gallée 1992). It successfully reproduced the main characteristics of the current climate, as well as the variations in ice volume during glacial-interglacial cycles. Since the model has been shown to provide reasonable simulations of the 100,000-year cycle, it has been applied as a second step to the simulation of future climates. Several tests have thus been carried out on the climate of the next 130,000 years, either by keeping the CO_2 constant at different levels (290 ppm, 200 ppm and 250 ppm), or by using the CO_2 variations of the last glacial-interglacial cycle (Loutre and Berger 2000). The results of these simulations suggest, on the one hand, that the climate of the next 50,000 years is particularly sensitive to the level of atmospheric CO_2 concentration and that our current interglacial period will be much longer than any other one in the past. It could last more than 55,000 years with CO_2 levels between 230 and 290 ppm: the first glacial stage would appear around 60,000 AD, and the next glacial maximum (in terms of ice volume) would be around 100,000 AD, followed by a deglaciation phase that would end around 120,000 AD. These results suggest that the marked differences between our interglacial period (present and future) and the previous Quaternary warm periods are due to the small insolation variations that characterize the former. Based on another scenario designed to reproduce the natural variations of CO_2 and not the anthropogenic contribution, other simulations have been carried out, either with a more elaborate version of the Loutre and Berger's model, or with a climate model of intermediate complexity coupled with a more sophisticated model of the evolution of polar ice sheets (Ganopolski et al. 2016). The results of these simulations are in agreement with those presented previously, and show that the next glacial inception will be postponed by at least 100,000 years.

Although, the variations in summer insolation at 65°N have long been considered as the pacemaker of glacial-interglacial transitions, the above results show that the level of atmospheric CO_2 will be a key parameter in the future. While the natural variations of CO_2 along with the insolation forcing exclude the possibility of a glacial episode in the future for at least 50,000 years, it is possible that the impact of the anthropogenic contribution will lead to a complete disappearance of polar ice sheets, and that a return to a glacial phase could occur only in 100,000 years. Coupled climate-ice sheet models, validated on the last glacial-interglacial cycle, will make it possible to explore the threshold values of CO_2 which could have long-term effects on the fate of ice sheets.

The Long-Term Future of the Polar Ice Sheets: Impact and Irreversibility

As seen earlier, the response times of the different components of the climate system are extremely variable, ranging from a few minutes to a few days for the atmosphere and from a few months to several hundreds of years for the ocean. While the ice sheets have long been considered as a slow component of the Earth system (with characteristic time scales ranging from thousands to hundreds of thousands of years), recent observations provide evidences that they can react to climate change far more quickly than previously thought. The last glacial-interglacial cycle demonstrates how deeply the climate is influenced by the slow development and rapid collapse of the ice sheets.

The ice core records retrieved from the Vostok and Dome C sites in Antarctica show that, for 800,000 years, the world has alternated between four ice sheets (during glacial periods) and only two ice sheets (during interglacial periods). In other words, Greenland and Antarctica withstood the warming that led to the disappearance of the Fennoscandian and North American ice sheets. There is no doubt that Greenland and West Antarctica have not always emerged unscathed from glacial-interglacial cycles. Indeed, during the last interglacial period (i.e. 130–115 ka ago), the sea level was 6 to 9 m higher (Kopp et al. 2009). It is therefore possible that anthropogenic activity could lead to the partial or total melting of Greenland (Charbit et al. 2008).

Recent observations of Greenland, and, more surprisingly, of Antarctica mass balances (Velicogna 2009) show that these ice sheets have become one of the main contributors to the increase in sea level (Cazenave et al. 2009). Their complete disappearance would lead to a sea-level rise of nearly 60 m: 6.6 m for Greenland and 52.8 m for Antarctica, of which 3.3 m would come from West Antarctica. These figures should be compared to the 120 m sea-level rise corresponding to the disappearance of the North American and Fennoscandian ice sheets in response to very small changes in insolation compared to the additional radiative forcing from anthropogenic activities. Also, the disappearance of past ice sheets was spread over 14,000 years (from the Last Glacial Maximum to the Mid-Holocene). The long-term effects (several hundred years) of anthropogenic forcing on ice sheets are not easy to model, partly because of uncertainties in the future socio-economic pathways over the twenty-first century and therefore in the evolution of the greenhouse gas concentrations in the atmosphere. It is therefore even more difficult to establish scenarios over several centuries (Charbit et al. 2008). However, several assumptions can be made about the long-term evolution of greenhouse gas emissions to explore the sensitivity of the present-day ice sheets with numerous scenarios. Performing simulations over several centuries or several millennia cannot be achieved through the use of general circulation models, which are too expensive in terms of computational time, and simplified climate models coupled with efficient ice-sheet models are required.

In fact, there are three different time frames relevant in the study of the evolution of ice sheets. The evolution over the twenty-first century, the basis of all IPCC analyzes, will very likely bring about a rise in sea level of several tens of centimeters (or even more), although there is a great deal of uncertainty linked to the emission scenarios, to climate model biases and to difficulties of ice-sheet models to capture the processes causing rapid dynamical change. The second horizon, beyond the twenty-first and the following centuries, is one where the CO_2 level may stabilize at three or four times that of the pre-industrial level, likely resulting in a massive retreat of Greenland and West Antarctica. These changes are also likely to modify the ocean circulation and the global climate. If this high level of CO_2 persists for a long time in the atmosphere, the ice-sheet melting could be irreversible, in the sense that there would no longer be any perennial snow in Greenland (Charbit et al. 2008). Finally, there is the much more distant third horizon, which raises the following question: for the last million years, our climate has oscillated between ice ages (long periods of about 100,000 years) and interglacial periods (short periods of about 10,000 years). Is it possible that the anthropogenic perturbation might cause a switch to another climate mode with strongly reduced ice sheets or even no ice sheet at all, similar to the hot climate mode of the pre-Quaternary era? In other words, is it possible that anthropogenic disturbance could induce modifications such that the next marked decline in summer insolation due in about 100,000 years (Loutre and Berger 2000) might not bring about an ice age?

Even if this question seems 'futuristic', we have good reasons to believe that the glacial-interglacial cycles of the last million years will no longer occur because the expected decline of insolation will not be large enough to compensate for the radiative forcing due to the high atmospheric CO_2 levels. This is what is suggested by the results of simulations carried out for periods such as the Pliocene around 3 Ma where the CO_2 level is estimated to have been around 405 ± 50 ppm.

In these scenarios, a key factor is the long time required to reach atmospheric CO_2 equilibrium which can be several tens to hundreds of thousands years (Archer et al. 1997). The anthropogenic disturbance is almost instantaneous, but its consequences will last for a very long time, and the

Milankovitch cycles that have so far allowed glacial-interglacial transitions through profoundly non-linear processes, can become inoperative in the future. Is it a bit presumptuous though to think that a couple of hundred years of 'energy profligacy' leading to a very large increase in atmospheric CO_2 could still be felt tens of thousands of years later? Maybe, but still, it is important to remember that studies of the Earth past climates show that the presence of ice sheets is associated with low CO_2 levels. It is therefore logical to think that, in a world that sustains high CO_2 concentrations, the behavior of present-day ice sheets will change in the short and medium term.

The world at the beginning of the Cenozoic (with 1120 ppm of CO_2 in the atmosphere) was much hotter than today. Antarctica was already in its polar position since the end of the Cretaceous (70 Ma ago), but far from forming an ice sheet, it was covered with forests. There was no sign of an ice sheet, until the CO_2 level dropped sufficiently to trigger the progressive freeze-up of Antarctica (DeConto and Pollard 2003). A planet with no ice sheet is not a figment of the imagination, rather their presence being the exception when viewed over geological times.

This book shows how the face of our planet changed over these time scales: from plate tectonics (millions of years) to the development of huge ice sheets at temperate latitudes (hundreds of thousands of years), with very strong variability in glacial climates at the scale of a few thousand years. Our own interglacial period, the Holocene, has been much more stable in terms of climate and has contributed to the extraordinary expansion of the human population. With an Earth inhabited by more than 9 billion men and women, it will be essential to manage climate change in order to protect societies and their environment. The past teaches us that our little blue planet has undergone many changes. It is not the planet that is in danger, it is rather the populations, especially as they do not have an equal standing in the face of climate change. And so, paradoxically, if man has become a major actor in climate change through industrial development and massive use of fossil fuels, he could also suffer at his own expense from upheavals and difficult situations that he has created himself.

References

Adler, R. F., et al. (2003). The Version 2 Global Precipitation Climatology Project (GPCP) monthly precipitation analysis (1979–Present). *Journal of Hydrometeorology, 4*, 1147–1167.

Allen, R. J., & Sherwood, S. C. (2008). Warming maximum in the tropical upper troposphere deduced from thermal winds. *Nature Geoscience, 1*, 399–403.

Archer, D., et al. (1997). Multiple timescales for neutralization of fossil fuel CO_2. *Geophysical Research Letters, 24*, 405–408.

Archer, D., et al. (2009). Atmospheric lifetime of fossil fuel carbon dioxide. *Annual Review of Earth and Planetary Sciences, 37*, 117–134. https://doi.org/10.1146/annurevearth031208.100296.

Bony, S., et al. (2006). How well do we understand and evaluate climate change feedback processes? *Journal of Climate, 19*(15), 3445–3482.

Bouttes, N., & Gregory, J. M. (2014). Attribution of the spatial pattern of CO_2-forced sea level change to ocean surface flux changes. *Environmental Research Letters, 9*, 034004. https://doi.org/10.1088/1748-9326/9/3/034004.

Bronselaer, B., et al. (2018). Change in future climate due to Antarctic meltwater. *Nature, 564*, 53–58. https://doi.org/10.1038/s41586-018-0712-z.

Cazenave, A., et al. (2009). Sea level budget over 2003-2008: A reevaluation from GRACE space gravimetry; Satellite altimetry and argo. *Global and Planetary Change, 65*, 83–88.

Charbit, S., et al. (2008). Amount of CO_2 emissions irreversibly leading to the total melting of Greenland. *Geophysical Research Letters, 35*, L12503. https://doi.org/10.1029/2008GLO33472.

Church, J. A., & White, N. J. (2011). Sea-level rise from the late 19th to the early 21st century. *Survey of Geophysics, 32*, 585–602.

Christiansen, H. H., et al. (2010). The thermal state of permafrost in the Nordic area during the International Polar Year 2007–2009. *Permafrost Periglacial Processes, 21*, 156–181.

Clark, P. U., et al. (2016). Consequences of twenty-first-century policy for multi-millennial climate and sea-level change. *Nature Climate Change, 6*, 360–369.

DeConto, R. M., & Pollard, D. (2003). Rapid cenozoic glaciation of Antarctica induced by declining atmospheric CO_2. *Nature, 421*, 245–249. https://doi.org/10.1038/nature01290.

DeConto, R. M., & Pollard, D. (2016). Contribution of Antarctica to past and future sea-level rise. *Nature, 531*, 591–597. https://doi.org/10.1038/nature17145.

Defrance, D., et al. (2017). Consequences of rapid ice-sheet melting on the Sahelian population vulnerability. In *Proceedings of the National Academy of Sciences* (Vol. 114, Issue 25). https://doi.org/10.1073/pnas.1619358114.

Dufresne, J.-L., & Bony, S. (2008). An assessment of the primary sources of spread of global warming estimates from coupled atmosphere-ocean models. *Journal of Climate, 21*(19), 5135–5144. https://doi.org/10.1175/2008jcli2239.1.

Dufresne, J.-L., & Saint-Lu, M. (2016). Positive feedback in climate: Stabilization or runaway, illustrated by a simple experiment. *Bulletin of American Meteorological. Society, 97*(5), 755–765. https://doi.org/10.1175/bams-d-14-00022.1.

Fourier, J.-B. (1824). Remarques générales sur les températures du globe terrestre et des espaces planétaires. *Annales de Chimie et de Physique*, 2e série, XXVII, 136–167.

Gallée, H., et al. (1992). Simulation of the last glacial cycle by a coupled, sectorially averaged climate-ice-sheet model 2. Response to insolation and CO_2 variations. *Journal of Geophysical Research, 97*(D14), 15713–15740.

Ganopolski, A., et al. (2016). Critical insolation-CO_2 relation for diagnosing past and future glacial inception. *Nature, 529*, 200–203. https://doi.org/10.1038/nature16494.

Garner, A. J., et al. (2018). Evolution of 21st century sea level rise projections. *Earth's Future, 6*, 1603–1615. https://doi.org/10.1029/2018EF000991.

Gillet-Chaulet, F., et al. (2012). Greenland ice sheet contribution to sea-level rise from a new-generation ice-sheet model. *The Cryosphere, 6*, 1561–1576. https://doi.org/10.5194/tc-6-1561-2012.

Golledge, N. R., et al. (2019). Global environmental consequences of twenty-first-century ice-sheet melt. *Nature, 566*, 65–72. https://doi.org/10.1038/s41586-019-0889-9.

Haywood, A. M., et al. (2011). Pliocene Model Intercomparison Project (PlioMIP): Experimental design and boundary conditions (experiment 2. *Geoscientific Model Development, 4*, 571–577. https://doi.org/10.5194/gmd-4-571-2011.

Haywood, A. M. et al. (2016). The Pliocene Model Intercomparison Project (PlioMIP) Phase 2: Scientific objectives and experimental design. *Climate of the Past, 12, 663–675.* https://doi.org/10.5194/cp-12-663-2016IMBIE team (2018). Mass balance of the Antarctic ice sheet from 1992–2017. *Nature, 558*, 219–222. https://doi.org/10.1038/s41586-018-0179.

IPCC. (Ed.). *Climate Change 2007: The Physical Science Basis; Contribution of Working Group I to the Fourth Assessment Report of the Intergovernmental Panel on Climate Change*, Cambridge, United Kingdom, and New-York, USA, Cambridge University Press.

IPCC. (Ed.). *Climate Change (2013): The Physical Science Basis; Contribution of Working Group I to the Fifth Assessment Report of the Intergovernmental Panel on Climate Change*, Cambridge, United Kingdom, and New-York, USA, Cambridge University Press.

Jevrejeva, S., Moore, J. C., Grinsted, A., & Woodworth, P. L. (2008). Recent global sea level acceleration started over 200 years ago? *Geophysical Research Letters, 35*, L08715.

Johnson, G. C., et al. (2007). Recent bottom water warming in the Pacific Ocean. *Journal of Climate, 20*, 5365–5375.

Keeling, C. D., et al. (1995). Interannual extremes in the rate of rise of atmospheric carbon dioxide since 1980. *Nature, 375*, 666–670.

Khodri, M., et al. (2001). Simulating the amplification of orbital forcing by ocean feedbacks in the last glaciation. *Nature, 410*, 570–574.

Kopp, R. E., et al. (2009). Probabilistic assessment of sea level during the last interglacial stage. *Nature, 462*, 863–867. https://doi.org/10.1038/nature08686.

Kouketsu, S., et al. (2011). Deep ocean heat content changes estimated from observation and reanalysis product and their influence on sea level change. *Journal of Geophysical Research (Ocean), 116*, C03012.

Lambert, S. J., & Fyfe, J. C. (2006). Changes in winter cyclone frequencies and strengths simulated in enhanced greenhouse warming experiments: results from the models participating in the IPCC diagnostic exercise. *Climate Dynamics, 26*, 713–728. https://doi.org/10.1007/s00382-006-0110-3.

Loutre, M.-F., & Berger, A. (2000). Are we entering an exceptionally long interglacial? *Climatic Change, 46*, 61–90.

Luo, D., et al. (2016). *Environmental Earth Science, 75*, 555. https://doi.org/10.1007/s12665-015-5229-2.

Madden, R. A., & Julian, P. R. (1994). Observations of the 40-50 day tropical oscillation: A review. *Monthly Weather Review, 122*, 814–837.

Manabe, S., & Wetherald, R. T. (1967). Thermal equilibrium of the atmosphere with a given distribution of relative humidity. *Journal of the Atmospheric Sciences, 24*(3), 241–259. https://doi.org/10.1175/1520-0469(1967)024%3c0241:TEOTAW%3e2.0.CO;2.

Meehl, G. A., Tebaldi, C., Walton, G., Easterling, D., & McDaniel, L. (2009). Relative increase of record high maximum temperatures compared to record low minimum temperatures in the U.S. *Geophysical Research Letters, 36*, L23701. https://doi.org/10.1029/2009gl040736.

Nerem, R. S., Beckley, B. D., Fasullo, J. T., Hamlington, B. D., Masters, D., & Mitchum G. T. (2018). Climate-change-driven accelerated sea-level rise. *Proceedings of the National Academy of Sciences, 115*(9), 2022–2025. https://doi.org/10.1073/pnas.1717312115.

Oerlemans, J., & Van der Veen, C. J. (1984). *Ice Sheets and Climate*, Reidel, 217 pp.

Oerlemans, J. (2005). Extracting a climate signal from 169 glacier records. *Science, 308*(5722), 675–677.

Peltier, D. F., et al. (2015). ICE-5G and ICE-6G models of postglacial relative sea-level history a Space geodesy constrains ice age terminal deglaciation: The global ICE-6G_C (VM5a) model. *Journal of Geophysical Research (Solid Earth), 120*, 450–487.

Purkey, S. G., & Johnson, G. C. (2010). Warming of global abyssal and deep southern ocean waters between the 1990s and 2000s: Contributions to global heat and sea level rise budgets. *Journal of Climate, 23*, 6336–6351.

Ramanathan, V., & Coakley, J. A., Jr. (1978). Climate modeling through radiative-convective models. *Reviews of Geophysics and Space Physics, 16*(4), 465.

Ramillien, G., et al. (2006). Interannual variations of the mass balance of the Antarctica and Greenland ice sheets from GRACE. *Global and Planetary Change, 53*, 198–208.

Ray, R. D., & Douglas, B. C. (2011). Experiments in reconstructing twentieth-century sea levels. *Progress in Oceanography, 91*, 495–515.

Rignot, E., et al. (2019). Four decades of Antarctic ice sheet mass balance from 1979–2017. *Proceedings of the National Academy of Sciences, 116*(4), 1095–1103. https://doi.org/10.1073/pnas.18128883116.

Rigor, I. G., & Wallace, J. M. (2004). Variations in the age of sea ice and summer sea ice extent. *Geophysical Research Letters, 31*. https://doi.org/10.1029/2004GL019492.

Romanovsky, V. E., Smith, S. L., & Christiansen, H. H. (2010a). Permafrost thermal state in the polar Northern Hemisphere during the International Polar Year 2007–2009: A synthesis. *Permafrost Periglacial Processes, 21*, 106–116.

Romanovsky, V. E., et al. (2010b). Thermal state of permafrost in Russia. *Permafrost Periglacial Processes, 21*, 136–155.

Shepherd, et al. (2012). A reconciled estimate of ice-sheet mass balance. *Science, 338*, 1183. https://doi.org/10.1126/science.1228102.

Stott, P. A., et al. (2004). Human contribution to the European heatwave of 2003. *Nature, 432*, 610–614.

Szopa, S., et al. (2013). Aerosol and ozone changes as forcing for climate evolution between 1850 and 2100. *Climate Dynamics, 40*(9–10), 2223–2250. https://doi.org/10.1007/s00382-012-1408-y.

Tarasov, L., & Peltier, W. R. (2007). Coevolution of continental ice cover and permafrost extent over the last glacial-interglacial cycle in North America. *Journal of Geophysical Research, 112*, F02S08. https://doi.org/10.1029/2006jf000661.

Tschudi, M. A., Stroeve, J. C., & Stewart, J. S. (2016). Relating the age of Arctic sea ice to its thickness, as measured during NASA's ICESat and IceBridge campaigns. *Remote Sensing, 8*(6). https://doi.org/10.3390/rs8060457.

VandenBerghe, J. (2011). *Permafrost during the Pleistocene in North West and Central Europe*. Permafrost Response on Economic

Development, Environmental Security and Natural Resources (pp. 185–194).

Vial, J., et al. (2013). On the interpretation of inter-model spread in CMIP5 climate sensitivity estimates. *Climate Dynamics, 41*(11–12), 3339–3362. https://doi.org/10.1007/s00382-013-1725-9.

Velicogna, I., & Wahr, J. (2006). Measurements of time-variable gravity show mass loss in Antarctica. *Science, 311*, 1754–1756.

Velicogna, I. (2009). 'Increasing Rates of Ice Mass Loss from the Greenland and Antarctic Ice Sheets Revealed by GRACE'. *Geophysical Research Letters, 36*(L19503). https://doi.org/10.1029/2009gl04022.

Velicogna, I., et al. (2014). Regional acceleration in ice mass loss from Greenland and Antarctica using GRACE time-variable gravity data. *Geophysical Research Letters, 41*, 8130–8137. https://doi.org/10.1002/2014GL061052.

WCRP Global Sea Level Budget Group. (2018). Global sea-level budget 1993–present. *Earth Syst. Sci. Data, 10*, 1551–1590. https://doi-org.insu.bib.cnrs.fr/10.5194/essd-10-1551-2018.